The Student's Companion to Geography

The Student's Companion to Geography

SECOND EDITION

Edited by
Alisdair Rogers and *Heather A. Viles*

Blackwell
Publishing

350 Main Street, Malden, MA 02148-5018, USA
108 Cowley Road, Oxford OX4 1JF, UK
550 Swanston Street, Carlton South, Melbourne, Victoria 3053, Australia
Kurfürstendamm 57, 10707 Berlin, Germany

First published 1992 by Blackwell Publishing Ltd.
Reprinted 1993, 1994, 1995, 1996, 1997, 1998, 1999
Second edition published 2003

Library of Congress Cataloging-in-Publication Data

The Student's companion to geography / edited by Alisdair Rogers and
Heather A. Viles – 2nd ed.
    p. cm.
Includes bibliographical references and index.
  ISBN 0-631-22132-8 (alk. paper) – ISBN 0-631-22133-6 (alk. paper)
  1. Geography. I. Rogers, Alisdair. II. Viles, Heather A.
  G116 .S78 2002
  910–dc21
                                                        2002002202

A catalogue record for this title is available from the British Library.

Set in 9 on 11 pt Plantin
by Ace Filmsetting Ltd, Frome, Somerset
Printed and bound in the United Kingdom
by MPG Books Ltd, Bodmin, Cornwall

For further information on
Blackwell Publishing, visit our website:
www.blackwellpublishing.com

Contents

Part III   Studying Geography

Contributors

Michael Batty is Professor of Spatial Analysis and Planning, and Director of the Centre for Advanced Spatial Analysis (CASA) at University College, London. He holds a joint appointment between the Bartlett School of Architecture and the Department of Geography. He has made many contributions to the development of computer models of cities and regions, his recent work being focused in two areas: dynamic models of urban development and the visualization of cities using virtual reality methods. His most recent book is *Fractal Cities* (1994) and he is editor of *Environment and Planning B*.

Lawrence D. Berg is Associate Professor of Geography at Okanagan University College, British Columbia, Canada. He has written extensively on a range of critical geographical issues, especially 'race' and gender issues and colonial and post-colonial relations in Canada and Aotearoa/New Zealand. He is co-editor of *ACME: An Online Journal for Critical Geographies*, co-moderator of the Critical Geography Forum and a member of the Steering Committee of the International Critical Geography Group. He lived and worked in Aotearoa/New Zealand from 1992 to 1998.

John Boardman is Director of the M.Sc. in Environmental Change and Management at the Environmental Change Institute, University of Oxford. His research interests are land degradation, soil erosion and off-site impacts of erosion. He currently works in southern England and South Africa.

Gary Bridge is a geographer at the School for Policy Studies, University of Bristol.

He researches on urban geography, with particular reference to gentrification and public space, and maintains an interest in theories of rational action. He is the co-editor (with Sophie Watson) of *A Companion to the City* (2000).

Susan M. Brooks is Senior Lecturer in Physical Geography at Birkbeck College, London, with research interests in geomorphology, hydrology, landslides, soil erosion and climatic change.

Jacquelin Burgess is a Reader in the Department of Geography, University College, London. Her research interests are the study of environmental meanings, values and knowledges, especially in the contexts of nature conservation, landscape change and sustainable development. She has also helped to develop methodologies for decision making which deploy participatory approaches in the conduct of negotiations between experts and local people.

Sue Burkill is Principal Lecturer in Geography at the College of St Mark and St John, Plymouth, UK.

Noel Castree is a Reader in Geography at the University of Manchester. His research interests are geographical political economy (with a particular interest in Marxism); geographical scale and labour unions; social theory and the environment; and the philosophy and politics of geographical thought. His publications include *Social Nature: Theory, Practice and Politics*, edited with B. Braun (2001), and *Remaking Reality: Nature at the Millennium*, edited with B. Braun (1998).

Tim Coles is Lecturer in Human Geography at the University of Exeter, where he teaches courses on sustainable tourism and tourism research methods. He is currently working on projects that examine issues of sustainability in urban tourism in heritage cities in Britain and eastern Germany. He is the honorary secretary of the new Geography of Leisure and Tourism Research Group (GLTRG) of the Royal Geographical Society (with IBG), and a member of the editorial board of the journal *Tourism Geographies*.

Danny Dorling is Professor of Quantitative Human Geography in the School of Geography, University of Leeds. His interests include the human geography of Great Britain and social medicine.

Claire Dwyer is a Lecturer at University College, London, where she teaches courses on social and cultural geography. Her research interests are in the geographies of gender and ethnicity and she is currently working on a project on British–South Asian transnational commodity culture. She is the co-author of *Qualitative Methodologies for Geographers* (2001) and *Geographies of New Femininities* (1999).

Sally Eden is a Lecturer in Geography at the University of Hull. She has research interests in environmental policy and business activity, public perceptions of nature and environmental management and restoration, particularly in Europe. She is the author of *Environmental Issues and Business: Implications of a Changing Agenda* (1996).

Richard Field is at the School of Geography, University of Nottingham, UK.

Stuart Franklin completed a D.Phil. in Geography at Oxford University in 2001. He works at the interface between geography and photography, and his research interests lie in visual representation and political ecology.

Peter W. French is a Lecturer in Physical Geography at Royal Holloway, University of London.

Wardlow Friesen is Senior Lecturer in the Department of Geography, the University of Auckland. His research interests include the study of migration, ethnicity and population, particularly in the Asia-Pacific region.

Stephen J. Gale is in the Division of Geosciences, University of Sydney, Australia, and has research interests in Quaternary environmental history, human environmental impact during the late Holocene, long-term geomorphic evolution and sedimentary geomorphology.

Wendy Gill is Senior Lecturer in Geography at the College of St Mark and St John, Plymouth, UK.

Andrew S. Goudie is Professor of Geography in the University of Oxford and an expert on topics ranging from the impacts of climate change on geomorphology to arid zones.

Richard Harris is a Lecturer in Human Geography at Birkbeck College, London. His interests are in geodemographic and social area analysis, urban geography, spatial analysis and GIS.

Stephan Harrison is Reader in Geomorphology at Coventry University. His research interests lie in mountain geomorphology and he has worked extensively in Patagonia and the Tian Shan mountains of Kazakhstan.

Iain Hay is Professor of Geography at Flinders University, Adelaide. He is author or editor of six books on geographies of oppression and research and study skills. He is also Asia-Pacific commissioning editor for *Ethics, Place and Environment* and *Journal of Geography in Higher Education*.

David L. Higgitt is a Lecturer in Physical Geography at the University of Durham. His research interests are soil erosion and sedimentation, which he has studied in subtropical China, Jordan and the southern Mediterranean. He specializes in the application of radionuclide tracing techniques and gamma spectrometry. He is on the editorial board of the *Journal of Geography in Higher Education* and is interested in the application of computer-based learning in geography and innovation in fieldwork practice.

Brian Hoskin is Lecturer in Geography at the College of St Mark and St John, Plymouth, UK.

Peter Jackson is Professor of Human Geography at the University of Sheffield. His research and teaching interests focus on consumption cultures and the social geography of 'race' and racism. Recent publications include a co-edited book on *Commercial Cultures* (2000) and a co-authored book on *Making Sense of Men's Magazines* (2001).

Robin Kearns is Associate Professor in the Department of Geography, University of Auckland. He has published (with Wilbert Gesler) *Putting Health into Place: Landscape, Identity and Well-being* (1998) and *Culture/Place/Health* (2002). He is on the editorial boards of journals such as *Health and Place* and *Health and Social Care in the Community*.

Christopher Keylock gained an M.Sc. from the University of British Columbia, Vancouver, Canada and is now a Lecturer at the University of Leeds.

Pauline E. Kneale is a Senior Lecturer at the University of Leeds. She researches in hydrology and peat, and into how geography students learn at university and their career paths.

David B. Knight is Professor of Geography at the University of Guelph, Ontario, Canada. He is a cultural and political geographer who researches and writes on geography students' writing and presentation skills; political identities, territory, territoriality and self-determination; perceptions and meanings of place. He is also a former Editorial Geographer with the *Encyclopaedia Britannica*, Director and General Editor of the Carleton University Press, Dean of Social Sciences at the University of Guelph, and Chair/Président of the IGU Commission on the World Political Map.

Stuart Lane is Professor of Physical Geography at the University of Leeds. His research interests are in fluvial geomorphology and numerical modelling and monitoring of environmental processes.

Mark Lawless obtained an M.Sc. from York University, Canada, and is currently studying for a Ph.D. at the Department of Geography, University of Leeds.

Loretta Lees is Lecturer in Human Geography at King's College, London. She is currently Chair of the RGS-IBG Urban Geography Research Group. She has written about gentrification and other urban socio-cultural issues in London, New York City, Vancouver, Canada and Portland, Maine, USA. Her current research focuses on urban community in Brooklyn Heights, New York City.

George C.S. Lin is Associate Professor in the Department of Geography, the University of Hong Kong.

David N. Livingstone is Professor of Geography at the Queen's University of Belfast. He is the author of several books, including *The Geographical Tradition*. He has also just completed a book on the historical geography of science entitled *Spaces of Science: Chapters in the Historical Geography of Scientific Knowledge*.

Rachael A. McDonnell is Lecturer in Physical Geography at Hertford College, University of Oxford, and has wide research interests in GIS and its application to hydrological issues.

Robert J. Mayhew was trained at Oxford University. A former fellow of Corpus Christi College, Cambridge, he is currently a Lecturer at the University of Wales, Aberystwyth. His first book, *Enlightenment Geography*, was published in 2000.

John Morgan runs the PGCE Geography course at the University of Bristol. His interests are in social and cultural geography.

Catherine Nash is a Lecturer in Cultural Geography in the Department of Geography, Royal Holloway, University of London. She co-edited *Modern Historical Geographies* (2000).

David J. Nash is Reader in Physical Geography at the University of Brighton. His research focuses on the geomorphology of dryland environments and environmental change in southern Africa.

Fiona O'Carroll is Assistant Director of the International Office at the University of Durham. She is the university's institutional coordinator for the SOCRATES programme and has been involved in European exchange programmes for the last six years.

Miles Ogborn is Reader in Geography at Queen Mary, University of London. He teaches courses on cultural geography and global historical geography, and is the author of *Spaces of Modernity: London's Geographies, 1680–1780* (1998).

Scott Orford is a Lecturer in GIS and spatial analysis in the Department of City and Regional Planning, University of Cardiff. His research interests include spatial statistical modelling, urban geography and visualization in the social sciences.

Ben Page recently completed a doctorate on the commodification of water supplies in twentieth-century Cameroon. He is currently a Lecturer at St Hugh's College, Oxford University, and is working on a study of public participation in the production of European water policy.

Rachel Pain is a Lecturer in Social Geography at the University of Durham. Her interests include crime, fear of crime and community safety, geographies of health, youth, ageing and gender, and teaching.

Joe Painter is Reader in Geography at the University of Durham. He is SOCRA-TES-ERASMUS Coordinator for the Department of Geography at Durham, which has one of the longest-running ERASMUS programmes in British geography.

Allan Pentecost is Reader in Geomicrobiology at King's College, London, and has research interests in micro-organisms and their role in the environment.

Alisdair Rogers is a Fellow of Keble College and teaches at the School of Geogra-

phy and the Environment, Oxford University. His research interests are in urban and social geography, with particular reference to migration and transnational communities. He is editor of *Global Networks: A Journal of Transnational Affairs*.

Robert Schindler obtained an M.Sc. from York University, Canada, and is currently studying for a Ph.D. at the Department of Geography, University of Leeds.

Pamela Shurmer-Smith is Principal Lecturer at the Department of Geography, University of Portsmouth. She is author of *India: Globalization and Change* (2000), *Worlds of Desire, Realms of Power: A Cultural Geography* (with K. Hannam, 1964) and editor of *Doing Cultural Geography* (2001). Her current research is on social movements in India.

Ian Simmons is a Professor in the Department of Geography, University of Durham, studies human–environment relations past, present and future, and is the author of several books on the human impact on the environment.

Michael C. Slattery is at the Department of Geology, Texas Christian University, Fort Worth, Texas. He has studied at the Universities of Witwatersrand (South Africa), Toronto (Canada) and Oxford (UK). His research interests are soil erosion and sediment delivery in fluvial systems, as well as the use of space shuttle photographs in environmental analysis.

Mark Patrick Taylor is Lecturer in Physical Geography at Macquarie University, Sydney, Australia. He teaches courses on resource and environmental management, Earth surface processes, and also runs a field sedimentology course to New Zealand's southern Alps. His research interests include the evolution of floodplain systems, heavy metal pollution and the physical degradation of riverine habitats, and the deposition and weathering of calcium carbonate sediments.

Peter J. Taylor is Professor of Geography and co-director of the Globalization and World Cities (GaWC) Study Group and

Network at Loughborough University, UK.

Chris Thomas is Sustainability Policy and Projects Officer for Reading Borough Council. Before that he worked for ten years as a researcher and a lecturer in human geography, specializing in social and cultural geography. His particular interests are in sustainable development and the geographies of local community/landscape relations.

Tim Unwin is Professor of Geography at Royal Holloway, University of London. In 2001 he began a three-year secondment to the UK's Department for International Development, where he is leading Imfundo: Partnership for IT in Education, which uses ICT to enhance education in Africa.

Heather A. Viles is Reader in Geomorphology at the University of Oxford and a Fellow of Worcester College. She carries out research in geomorphology, particularly on weathering in arid, coastal, karst and urban environments. She is co-editor of *Area*.

Robert L. Wilby is a Lecturer in the Department of Geography at King's College, London, and works in the field of hydroclimatology – the interface between long-term climate variability and freshwater environments.

Stephen Williams is a Principal Lecturer in Geography at Staffordshire University, where he teaches on recreation and tourism. His recent work on exclusion forms part of a wider agenda of research on urban recreation and post-industrial change.

He is the author of *Outdoor Recreation and the Urban Environment* and *Tourism Geography*.

Katie Willis is a Lecturer in Geography at the University of Liverpool. Her research interests are gender, households and development, with regional foci in Latin America and Pacific Asia. She is also interested in wider debates in feminist geography, particularly in relation to feminist methodologies. She has done field research in Singapore and China, as well as Latin America. With Brenda Yeoh she is the editor of *Gender and Migration* (2000).

Hilary P.M. Winchester is Pro-Vice Chancellor (Academic) at Flinders University, Adelaide, Australia.

Theresa Wong recently graduated with an honours degree from the Department of Geography, National University of Singapore.

Jamie C. Woodward is at the School of Geography, the University of Leeds, UK.

Brenda S.A. Yeoh is Associate Professor in the Department of Geography, National University of Singapore.

Henry Wai-chung Yeung, Ph.D., is Associate Professor at the Department of Geography, National University of Singapore. He is the co-editor of *Environment and Planning A*, associate editor of *Economic Geography* and Asia-Pacific editor of *Global Networks*.

Chris Young is Senior Lecturer in Physical Geography, Canterbury Christ Church University College, UK.

Introduction

Heather A. Viles and Alisdair Rogers

All the people who have contributed to this book believe that geography is a fascinating, stimulating and rewarding subject. The aims of the book are to provide you with a companion to many aspects of the subject which will accompany you before, during and after studying geography at university level. Alongside this, we hope to convince you that geography is as worth studying as we all think. The book is designed to be dipped into and consulted regularly, rather than to be read in one sitting. It won't answer a whole essay question for you, nor provide all the answers to every question you'll come across about geography and its relevance. What it will do is provide you with up-to-date information on some of the most important aspects of studying and researching geography today. It will get you started with choosing options and undertaking project work, and guide you on how to write essays and give classes. Every chapter contains recommended further reading to help you go further.

Part I provides a brief answer to the question 'why study geography?', giving several reasons why geography is a worthwhile, relevant and challenging subject. Part II provides a series of reviews of recent progress in various aspects of the vast subject that is today's university-level geography. We have invited a host of leading specialists in their fields to give an insight into what are the key issues and research themes. All of the authors of these essays teach geography at university level somewhere in the world, and all of them bring their research interests into their teaching. These short surveys provide a series of examples of what geography is all about at university level, and will be useful not only for those who are still thinking about doing geography as a degree course, but also for those who are considering which options to take or which research directions to follow.

Part III provides more practical information on how to study geography. The first four chapters introduce some important techniques which are commonly introduced early on in geography courses and often provide the basis for optional courses later on. The next four chapters discuss some core study skills, such as essay writing and coping with exams, and provide practical tips on how to succeed. These skills are useful at all stages of any geographer's career. The rest of Part III deals with designing and implementing an independent research project or dissertation. Most geography courses encourage students to do some independent research, and many have a compulsory dissertation. Skills discussed range from interviewing to data analysis and doing archival investigations on historical topics, and practical advice is given on how to approach each component. The final chapter in Part III looks at the important ethical issues raised by much independent research and should provide an invaluable and thought-provoking guide for those doing research at all levels.

Part IV broadens the picture a little, by looking at geography in context in terms of its history and current relations with other subjects. Many geography courses teach students about the context of geography, and the chapters in this section will give you a brief introduction to some of the key issues and debates. Part V provides a listing of some major sources of information for geography students, focusing mainly on resources available on the Internet. The major geography gateway sites are covered, as well as a selection of sites providing data, images and facts relevant to geographers, e-journals and discussion groups. The aim of this section is to provide an entrée into the online resources available to geographers, not to be comprehensive, especially given the rapidly changing nature of online information.

The final section of the book, Part VI, looks at how you can broaden your experience of geography before, during and after doing a geography course at university. There is a chapter on opportunities for studying abroad during a degree course, mainly in Europe, and one on getting funding for independent travel and fieldwork. There are seven chapters dealing with the possibilities of doing research at Master's level and beyond in a range of English-speaking countries. All provide some different insights into the joys and challenges of further study, and point you in the right direction to find out more. Finally, the last chapter in the book gives practical advice on how to market yourself as a geographer when you are looking for jobs, by writing a good CV.

Part I
Why Study Geography?

Some of you reading this book may be in the position of deciding on your future in further education, at university or even beyond. Picking the right course, whether single or joint honours, is a difficult decision. Although we would have no hesitation in recommending geography as a degree, it is important that each person makes his or her decision for the right reasons. We hope that the chapters in this book, especially in Part II, will make that decision easier. Here we simply outline some good reasons for studying geography. We suggest that geography is diverse and relevant, and that it opens new horizons personally and intellectually, posing new challenges and providing new skills as you go along. To support our case we have commissioned a photographic essay from Stuart Franklin, who turns a 'geographer's eye' on a region of England. His images, and the accompanying text, reveal how geographical interest lies in the details of everyday life, the people and things we see around us. A geographical imagination means looking at the world anew, sometimes as a scientist and sometimes as an artist.

1

Why Study Geography?

Heather A. Viles and Alisdair Rogers

Deciding which subject(s) to study at university is always a challenge and you will receive a barrage of advice from teachers, family, friends and other advisers. Open days and prospectuses also tell you a lot, but it can be hard to make comparisons between subjects and judge what you personally will gain from each course. Everyone who gives you advice is biased to some extent, and no one can really answer the question 'what should I study?' except you. Our experiences as students and, more recently, lecturers in geography departments have convinced us that geography is an exciting and worthwhile subject to study for the reasons that we outline below. Read on and see what you think . . .

Geography as a subject is incredibly diverse. You'll get some idea of the breadth of current geographical research and teaching by looking at the chapter headings in Part II of this book. Walking through any geography department, or browsing through their web pages, you will find research specialisms ranging from the hard-core scientific (such as using pollen from cores collected from peat bogs to reconstruct past environmental conditions) to the skilful and careful understanding of different cultural contexts at home and abroad. Some geography lecturers and researchers lock themselves away in front of computer screens all day looking at output from complex computer models, whilst others trudge up and down arctic hillslopes monitoring water movements.

Some are living among and participating with the communities that intrigue them. Others spend hours in libraries doing scholarly research. This inherent diversity at the geographical research frontier feeds directly into course structures and content. There are few subjects in universities today that will provide you with the wide range of information and understanding that is essential for making sense of the world around you. Geography straddles the so-called divide between social and natural sciences, enabling you to make the kinds of informed and critical judgements demanded of citizens in the twenty-first century.

Geography as a subject is relevant. Geographers are tackling many of the big questions facing the world today in terms of how the environment works and how human societies interact with it. Geography is all around you in newspapers, on television and in your environment. This relevance means that geography is always changing and developing, as a result of new challenges coming from the real world. For example, the growing concern over global warming and climatic change has had a real impact on geographical research and teaching, as have the evolution of the new global economy and the protests against globalization. Whether the issue is global security, genetic engineering, flood risk, or the diffusion and impact of new communications technologies, geographers are contributing significant research and often original insights. Lecture courses on topics such as natural hazards or international migration, for example, have to be continually updated in the light of new events, changed policies and scientific findings.

Studying geography can change your life, by opening up new horizons, exposing you to new challenges and giving you new skills. The diversity of subject matter on offer can allow students to pursue a very wide range of topics, mixing and matching from a whole host of different areas of the subject. This can be very rewarding and lead to your making connections and finding insights that would not be possible in a narrower course of study. Alternatively, the diversity of geographical topics on offer can enable students to focus on one or two areas in ever-increasing detail, so that they become real experts in a narrow range of topics which they might never have imagined wanting to pursue at school. As well as being diverse in subject material, geography at university is also diverse in terms of study methods and skills. On most geography courses you will be exposed to some aspects of computing, field research (often in groups), social survey, interpretation of texts, laboratory work, debate and discussion, reading and essay production as well as individual project or dissertation work. Geography, from its earliest days as a subject, has always had a core focus on maps and spatial patterns and in recent years computer-based geographical information science (GIScience) has become an important component of many geography degree courses. Life as a geography student is never dull and repetitive, and many of the skills that you learn will enhance your career prospects and be of lasting value.

Geography as a subject is challenging. Never believe anyone who tells you that geography is a subject for duffers who can only colour in maps and can't cope with a 'real' subject. Albert Einstein, in his diaries, is quoted as saying that he originally wanted to study geography but found it too difficult and so decided to study physics instead! Studying geography will not only improve your knowledge of the world, but it will also make you re-evaluate your ideas and challenge your creativity. In the first term of our geography degree courses one of us fell into a muddy salt marsh creek in what seemed like a blizzard, read some fascinating and inspiring books, and was encouraged to defend our ideas in small group discussions. Many geography courses require, or encourage, students to produce a dissertation based on their own, individual research, and some provide opportunities for relevant work experience or study elsewhere. Doing a dissertation (as you can see in more detail in Part III) is both a challenge and an opportunity. To take a subject that interests you and investigate it in detail (often contributing to a lecturer's bigger project, and sometimes producing results that are publishable) can be one of the most stimulating parts of the course. One of our recent students, for example, wanted to study volcanoes and ended up going to Lanzarote (with her family as field assistants!) and doing fieldwork on the weathering of old lava flows. On her return, she spent some time in our laboratories investigating her samples in detail. Her results were of such interest that we have now written a paper for a scientific journal on them, which should be published soon. Doing the dissertation, and making such a good job of it, convinced the student that she wanted to go on and do further research – which is what she is now doing at another university.

Geography opens doors. Despite (or perhaps because of) the fact that geography is not a vocational subject, there are many career and further study options open to geographers. Employers seem to like geographers – perhaps because we are generally outgoing, enthusiastic and have basic numeracy, literacy and often project design skills. According to the RGS-IBG, geographers are among the most employable of all graduates. GIS techniques are used widely by many businesses and government agencies, and thus geography students with GIS interests and experience have specific employment prospects. However, most geography students enter careers in administration and management, marketing or finance. Many other students start legal training after a geography first degree. Teaching geography is another career option, and there are many jobs available with an environmental dimension which geographers commonly apply for. Many students are pursuing further studies after their undergraduate degree, with Master's programmes (as described in Part VI of this book) attracting many who want to develop their interests and skills in particular areas, such as GIS and remote sensing, area studies or environment and development. Not all students, by any means, end up in careers which directly use their geography degree, but most find the skills they have developed of help in the future. Organizations such as the RGS-IBG (http://www.rgs.org.uk) and the Geographical Association (http://www.geography.org.uk), which run popular lecture programmes on geographical subjects, help

those interested keep up with geography, and provide a positive forum for promoting geography at all levels to all people.

Like any degree course, geography is what you make it and how much you put in determines how much you will get out. You've heard our opinions, and no doubt other people will highlight other aspects of geography which make it particularly good in their eyes. The proof of what geography is all about is in the subject itself, not in what we say about it, and there's a lot on offer in today's geography, as the rest of the chapters in this book demonstrate.

Further Reading

Craig, L.E. and Best, J. (eds) 2000: *Directory of University Geography Courses 2001*. London: Royal Geographical Society with Institute of British Geographers. A comprehensive guide to geography courses at undergraduate and graduate level in the UK, with information on course topics, structures and staff.

[The following chapter, pp. 6–10, is a photographic essay, entitled 'A Geographer's Eye . . . (Four Days in Newcastle)'. See p.10 for the commentary on the photographs.]

2

A Geographer's Eye . . . (Four Days in Newcastle)

Stuart Franklin

The photo-essay presented on the preceding pages hints at the diversity of geography. The photographs were taken in and around Newcastle, all (except one) over a four-day period – with geographers, without geographers, but always with geography in mind, with a *geographer's eye*. The essay begins (appropriately) in the map room in Newcastle University's Department of Geography. We pass urban graffiti, then GIS mapping. On to look at changing gender roles, at Tyneside gentrification where a Balkan Flour Mill (why Balkan?) metamorphoses into art-space; downriver to a heritage museum excavated from back-to-back houses. To Wallsend: cruisers that fought at Jutland once docked here. Now it hosts a large job centre. On to the productive Nissan car plant in Sunderland; visit a Saturday class studying Punjabi, the laundrette and the shops. A huge wealth gap exists between those queuing for buses, sifting for bargains in the market, and those sipping coffee at Starbucks, browsing in Waterstones. At the Victoria Royal Infirmary a baby is born. On Diana Street a large Sikh congregation remembers a community leader. The cultural, social and economic struggle of everyday life (contextualized in time and space) constitutes the heartland of human geography.

Geographers attempt to unravel the confusions of place. 'Mediterranean Village', glowing and warm, stands amid a grey, edge-city shopping centre. Africa and the neotropics (in wildfowl) appear in a wetlands reserve on the Wear. Go to Florida or Halkidiki? Who takes up these offers? Who grows those kiwi fruit in the Grainger market, and where? And what does the fishing fleet in North Shields catch? Who's driving the short distance to school in Jesmond? In Whitley Bay geography students learn to describe glacial deposits. The sun shines (after rain) on hard-hats, on the marram-grass protecting the Northumberland dunes, on a public footpath in the Cheviots, on an 8,000-year-old hazelnut unearthed from a raised mire near Berwick, on the geographers tasting the peat, and on *geography* that is, of course, everywhere.

The author would like to acknowledge the cooperation and support of staff and students of the Department of Geography, Newcastle University.

Part II
What are Geographers Doing?

It has become a cliché that geography is a diverse or wide-ranging subject. Indeed, it is usually approached through the study of special topics; it is only at the end of your course that the interrelationships between them become fully apparent. Here we present 24 short chapters, each one addressing a particular topic. The choice of topics is made on the assumption that, wherever the department and whatever the nature of the individual course, there are certain common themes. This selection is by no means comprehensive, and doubtless much has been left out in the interests of economy. We have asked experts in their fields to concentrate in some depth on one or more burning issues in their field rather than cover everything. These may be pressing scientific or social problems, new techniques and methods, or new theoretical insights. Their aim is to identify the important findings and unanswered questions that define the 'research frontier' in their specialism. The authors have all aimed to make their chapters accessible to beginning students in geography. After reading these short accounts, therefore, you should have a reasonable idea of what geographers have been doing and where they think their research is going.

The chapters are organized into two main blocks, physical and human geography, although many – such as geocomputation and environmental knowledge – straddle the boundary between them. Each chapter includes ideas for further reading and/or web searches on the topic.

3

Long-term Environmental Change: Quaternary Climate Oscillations and their Impacts on the Environment

Andrew S. Goudie

Today's geography has been revolutionized by recent improvements in our understanding of long-term environmental change (over the Quaternary period and even longer). New dating technologies (including isotopic and luminescence methods), combined with long stratigraphic sequences (from the oceans, lakes, peat bogs, loess deposits and ice caps), have enabled the construction of lengthy, high-resolution histories of environmental change. What has become apparent is that these changes have been frequent, abrupt, of great magnitude and of great significance for the environment and for humans.

The Frequency of Change

In the Quaternary (which comprises the last 2 million or so years of the Pleistocene and the Holocene) the gradual and uneven progression towards cooler conditions, which had characterized the Earth during the Tertiary (the so-called Cenozoic Climatic Decline), gave way to extraordinary climatic instability. Temperatures oscillated wildly

from values similar to, or slightly higher than, today in interglacials to levels that were cold enough in glacials to treble the volume of ice sheets on land. Not only was the degree of change remarkable, but so also, according to evidence from the deep-sea sedimentary record, was its frequency. In all, there have been about 17 glacial/interglacial cycles in the last 1.6 million years. The cycles tend to be characterized by a gradual build-up of ice volume (over a period of $c.90,000$ years), followed by a dramatic glacial 'termination' in only about 8,000 years.

The last glacial cycle reached its peak about 18,000 years ago, with ice sheets extending over Scandinavia to the north German plain, over all but the southernmost parts of Britain, and over North America to 39°N. To the south of the Scandinavian ice sheet was a tundra steppe underlain by permanently frozen subsoil (permafrost), and forest was relatively sparse to the north of the Mediterranean. Studies of the oxygen isotope composition of ocean cores suggest that deglaciation started at around 15,000 to 14,500 years BP in the North Atlantic and at 16,500 to 13,000 years BP in the Southern Ocean (Bard et al. 1990).[1] The years between the glacial maximum and the beginning of the Holocene are usually termed the Late Glacial, and they were marked by various minor stadials and interstadials, but their character, identification and correlation is a matter which is still in need of clarification (as reviewed by Anderson 1997).

The ending of the Last Glacial period was, however, not the end of substantial environmental change. Indeed, as the Holocene progressed the impact of climatic change was augmented as a cause of environmental fluctuation by the increasing role of the human impact (see chapter 4 by Ian Simmons, this volume). Some portions of the Holocene may have been slightly warmer than now, and terms like 'climatic optimum' have been used to denote the possible existence of a phase of mid-Holocene warmth, when conditions may have been 1–2°C warmer than now. There may also have been a Medieval Climatic Optimum (between AD 750 and AD 1300). However, there have also been times which have been rather colder than today, as is made evident by phases of glacial readvance (neoglaciations) in alpine valleys. The latest of these so-called neoglaciations was the 'Little Ice Age', which peaked around AD 1700 and ended towards the end of the nineteenth century (Grove 1988).

Holocene fluctuations of climate also occurred in lower latitudes, and of particular importance was the early to mid-Holocene pluvial, which 'greened' the Sahara (as described by Ritchie et al. 1985). This belt of great aridity more or less disappeared for one or two millennia before 7,000 BP. The northern limit of the Sahel shifted about 1,000 km to the north at 18,000 BP and about 600 km to the south at 6,000 BP compared to the present.

The Abruptness of Change

One of the features of palaeoclimatic research in the past decade has been the realization of just how abruptly climatic change

can occur (Adams et al. 1999). Crucial here have been ice cores, such as that from Vostok in Antarctica (Petit et al. 1999), which provide a particularly fine temporal resolution and contain large numbers of valuable environmental indicators. High-frequency swings in indicators such as isotopic composition and dust content suggest that dramatic oscillations have taken place in environmental conditions over quite short periods of time. These are known as Dansgaard-Oeschger events. Dansgaard et al. (1993), for example, documented no fewer than 24 interstades (warmer phases) in the last glacial period from the GRIP core from Greenland. However, high-frequency shifts are also known from the last interglacial (the Eemian of Europe), but some workers fear that the ice-core records could have been corrupted by deformation within the ice.

High-frequency abrupt changes have also been identified from ocean cores (as shown by the work of Oppo et al. 1998; Maslin and Tzedakis 1996), where the observed sawtooth patterns of climatic variation have been named Bond cycles. Also within the ocean-core sediment record there are layers of sediment that are rich in dolomite and limestone detritus but poor in small marine organisms called foraminifera. Each layer is interpreted as being the result of deposition by massive armadas of icebergs released from ice caps around the North Atlantic (Bond et al. 1992). The records of these iceberg flotillas are termed Heinrich events (Andrews 1998) and it is evident that they represent cold phases (stadials) of short duration (less than 1 ka). The Younger Dryas stadial towards the end of the Last Glacial was an example of a very short-lived event that came and went with great rapidity (Anderson 1997), perhaps in response to changes in the major ocean currents.

The Magnitude of Change

During glacial phases, ice covered nearly one-third of the Earth's land area, but the additional ice-covered area was almost all in the northern hemisphere, with no more than about 3 per cent in the southern. Nonetheless, large ice sheets developed over Patagonia and New Zealand. The thickness of the ice sheets may have exceeded 4 km, with typical depths of 2 to 3 km. Very major transformations also took place in the oceans. During the present Holocene interglacial, the Atlantic Ocean is at least seasonally ice-free as far north as 78°N in the Norwegian Sea. This condition reflects the bringing of warm water into this region by the Gulf Stream. During the Last Glacial Maximum, however, the oceanic polar front probably lay at about 45°N, and north of this the Atlantic was mainly covered by sea ice during the winter. In particular, attention has been drawn to the tendency for changes in the temperature and salinity of the North Atlantic to be brought about by massive discharges from melting ice sheets in the northern hemisphere. These great inputs of cold, fresh water upset what is called the thermohaline circulation or the 'Atlantic conveyor', by altering the density of different layers in the ocean. This in turn alters the state and location of major currents, and so influences climate.

The degree of temperature change that occurred over land was particularly great in the vicinity of the great ice sheets. The presence of evidence for permafrost in southern Britain (Ballantyne and Harris 1994) suggests a temperature depression that may have been of the order of 15°C. Mid-latitude areas probably underwent a lesser decline (perhaps 5–8°C), though in areas subject to maritime air masses temperatures were more likely to have been depressed by 4–5°C.

In general terms the Quaternary interglacials, which at their peak may have been a degree or two warmer than now, were short-lived but appear to have been essentially similar in their climate, fauna, flora and landforms to the present Holocene interglacial (Birks 1986). One of their most important characteristics was that they witnessed the rapid degradation of the great ice sheets

and saw the replacement of tundra conditions by forest over the now temperate lands of the northern hemisphere.

Some Impacts of Climate Change

The events which led to these expansions and contractions of the great ice sheets were accompanied by major changes in lower latitudes. Periods of greater moisture (pluvials) alternated with periods of lesser moisture (interpluvials). The evidence for such changes is particularly apparent on the margins of great deserts, where dry phases saw the development and advance of great sand seas, whereas in moister phases the dunes were stabilized by vegetation and large lakes filled with water in areas that had previously been salty wastes.

Some of the lakes that developed in pluvial phases were enormous. One of the greatest concentrations of pluvial lakes developed in the Basin and Range Province of the American Southwest. Between 100 and 120 depressions, formed by faulting, were occupied wholly or in part by large, freshwater bodies during various phases of the Pleistocene. By contrast, in interpluvials large dune fields expanded. Relict forms occur in areas where there is now a well-developed vegetation cover and annual precipitation totals of around 800 mm. The dunes probably formed when the cover of vegetation was less capable of inhibiting sand movement under annual precipitation totals that were less than 100 to 300 mm (Goudie 1999).

Temperature depression and reduced glacial atmospheric CO_2 levels also played a role in changing environmental conditions in low latitudes (Olago et al. 1999). The substantial degree of change in the altitudinal zonation of vegetation on tropical mountains during cold phases can be demonstrated by detailed pollen analysis from lakes and swamps from tropical highlands. Vegetation zones may have moved through as much as

1,700 m of altitude. A combination of temperature and precipitation change had a dramatic impact on the nature and extent of rainforest in Africa and South America (Goudie 1999).

In addition to the climatic and vegetational changes which have been discussed so far, it is important to remember that the Quaternary has also been a time of major changes in sea level. Many factors influence sea level in any particular location, but among the most important are two: glacial eustasy and glacio-isostasy. The former results from the waxing and waning of the great ice sheets. Thus since the last glaciation, the melting of the ice caps has caused world sea levels to rise by something between 100 and 150 metres. The *Flandrian transgression*, as this event is called, flooded the continental shelves and created many of the embayments of present-day coastlines. Sea level became stabilized at around its present height (give or take a few metres) at 6,000 years before the present, and since that time some of these embayments have been infilled by terrestrial and nearshore sedimentation processes.

The other mechanism, glacio-isostasy, results from the application and release of pressure by the weight of glaciers on the Earth's crust. As already noted, many of the ice caps were very thick and were capable of causing crustal depression of the order of 200–300 m in parts of the high latitudes. Since the ice burden has been released, the land has risen (and continues to do so), but in areas more distant from the former ice sheets some compensatory depression has occurred, leading to accelerated flooding of low-lying coastal areas, as in the southern North Sea.

Conclusions

Over the three or so millions of years during which humans have inhabited the Earth, conditions such as those we experience today have been relatively short-lived and

atypical of the Quaternary as a whole. The world has seen many environmental changes and has been in a state of constant flux. To understand landforms, biogeography and human history, we need to be aware of the exciting and complex history that the Earth has undergone.

Note

1   BP is before the present.

References

Adams, J., Maslin, M. and Thomas, E. 1999: Sudden climate transitions during the Quaternary, *Progress in Physical Geography*, 23, 1–36.

Anderson, D. 1997: Younger Dryas research and its implications for understanding abrupt climatic change, *Progress in Physical Geography*, 21, 230–49.

Andrews, J.T. 1998: Abrupt changes (Heinrich events) in late Quaternary North Atlantic marine environments, *Journal of Quaternary Science*, 13, 3–16.

Ballantyne, C.K. and Harris, C. 1994: *The Periglaciation of Great Britain*. Cambridge: Cambridge University Press.

Bard, E., Labergrue, L.D., Pichon, J.J., Labracherie, M., Arnold, M., Duprat, J., Moyes, J. and Duplessy, J.C. 1990: The last deglaciation in the southern and northern hemispheres. In V. Bleil and J. Thiede (eds), *Geological History of the Polar Oceans: Arctic versus Antarctic*, Dordrecht: Kluwer, 405–15.

Birks, H.J.B. 1986: Quaternary biotic changes in terrestrial and lacustrine environments, with particular reference to north-west Europe. In B.E. Bergland (ed.), *Handbook of Holocene Palaeoecology and Palaeohydrology*, Chichester: Wiley, 3–65.

Bond, G. and 13 others 1992: Evidence for massive discharges into the North Atlantic Ocean during the last glacial period, *Nature*, 360, 245–9.

Dansgaard, W. and 8 collaborators 1993: Evidence for general instability of past climate from a 250 k yr ice-core record, *Nature*, 364, 218–20.

Goudie, A.S. 1999: The Ice Age in the tropics. In P. Slack (ed.), *Environments and Historical*

Change, Oxford: Oxford University Press, 10–32.

Grove, J.M. 1988: *The Little Ice Age*. London: Methuen.

Maslin, M.A. and Tzedakis, C. 1996: Sultry last interglacial gets sudden chill, *Eos*, 77, 353–4.

Olago, D.O., Street-Perrott, F.A., Perrott, R.A., Ivanovich, M. and Harkness, D.D. 1999: Late Quaternary glacial–interglacial cycle of climatic and environmental change on Mount Kenya, Kenya, *Journal of African Earth Sciences*, 29, 592–618.

Oppo, D.W., McManus, J.F. and Cullen, J.L. 1998: Abrupt climate events 500,000 to 340,000 years ago: evidence from sub-polar North Atlantic sediments, *Science*, 279, 1335–8.

Petit, R.J. and 18 collaborators 1999: Climate and atmospheric history of the past 420,000 years from the Vostok ice core, Antarctica, *Nature*, 399, 429–36.

Ritchie, J.C., Gyles, C.H. and Haynes, C.B. 1985: Sediment and pollen evidence for an early to mid-Holocene humid period in the eastern Sahara, *Nature*, 314, 252–5.

Further Reading

Cronin, T.M. 1999: *Principles of Palaeoclimatology*. New York: Columbia University Press.

Lowe, J.J. and Walker, M.J.C. 1997: *Reconstructing Quaternary Environments*, 2nd edition. Harlow: Longman.

Williams, M.A.J., Dunkerly, D., De Dekker, P., Kershaw, P. and Chappell, J. 1998: *Quaternary Environments*, 2nd edition. London: Arnold.

Wilson, R.C.L., Drury, S.A. and Chapman, J.L. 2000: *The Great Ice Age: Climate Change and Life*. London and New York: Routledge and the Open University.

4

Human Impacts on the Environment

Ian Simmons

Nobody doubts that the physical environment of the world changes: climates alter, and vegetation follows them; volcanoes erupt, glaciers melt and so sea level rises. Alongside such realizations, as discussed in chapter 3 by Andrew Goudie in this volume, we must set at least 10,000 years of environmental changes brought about by human societies. Should anyone have remained unconvinced that such things happen, then the maps in the American Association for the Advancement of Science's *Atlas of Population and Environment* (Harrison and Pearce 2000) bring home the present-day reality in which the expansion of the human population since AD 1700 has resulted in great transformations of the Earth's surface by the 1990s.

The Burning Issue: Does it Matter?

If it is historically normal for humans to bring about changes in their surroundings, if it benefits them, if it produces stable systems that go on yielding resources, or beautiful places or environments that can process wastes, then is there a problem? Are we also interested in keeping some of the world's natural biological diversity, some of its wild and attractive places; and do we care that some societies are much richer than others

and able to choose more easily which environmental metamorphoses are acceptable? All these questions are bound up with human impacts and they happen in a world where none of them is entirely separate from the others, and where some of them are remote from most of our daily lives, whereas others are part of that experience. Further, some of them are firmly in the realm of scientific investigation and data collection, whereas others are questions of how we ought to behave – i.e., of ethics (see chapter 45 by Tim Unwin in this volume).

At what seems to most people a local scale of interest, consider the building of houses on hillsides. In some parts of the world such sites are used because any flatter land is taken up for agriculture or possibly for highway construction; in others, the views may be paramount in the eyes of potential buyers; in yet more, the land peripheral to a city is hilly but that is where the commuters want to be and where the developers can buy land. The negative consequences are well known: if there are earthquakes then the houses slide down the slope or are engulfed by landslides; if in high mountains then landslides or even avalanches are a risk; in temperate environments, the concreting and asphalting throw off the rain and cause floods downstream in their catchments. So the criteria of stability are not met and in the last case we face the ethical consideration of whether one group should benefit at the expense of others. At a regional scale, there is a great deal of concern about the loss of forests in the tropics. Developing countries in need of money, or with greedy ruling groups, or with expanding populations that need land for crop production, all regard timber as a mineable resource and so the native forests of the tropics have shrunk rapidly since 1950. Even with the greater awareness of recent years, 56 million ha of forest were lost between 1990 and 1995: 65 million ha of deforestation in developing countries being offset a little by reafforestation in the developed world. Many questions then revolve around the stability of crop and animal agriculture on the cleared areas, since the soils tend to become depleted of available nutrients; the liability of soils on steep slopes to erode, producing landslides locally and floods regionally; and the loss of biological diversity which might become a better source of long-term income than cheap beef, for example. The transpiration and evaporation effects of forest canopies may have a role in regional climate that may also alter when clearance is effected. The ethical questions consider whether these forests are part of the common ecology of humanity (as sinks for atmospheric carbon, for example), which is a global-scale issue, or whether native populations should be pushed aside to make room for new economies. At the truly global level of change, most eyes are upon the matter of climatic change and the likely consequences, with increased carbon loading of the atmosphere as the major driver of warming tendencies that seem to outstrip the natural rate of change. Most societies would somewhere be affected by one or more of sea-level rise, faunal loss, new pests and increased UV radiation. All would probably be the losers from a climate that 'flipped' from state A to state B in decades rather than making a stately progression over centuries. International conferences raise 'ought' questions like trying to restrict the carbon production of emerging nations whose greater prosperity depends upon higher per capita energy consumption.

Impact studies must include the social dimension, and that means thinking about how we think. Much of the discussion of impact is conducted in terms of the western worldview, which holds, for example, that it is permissible to alter the natural world using technology, that humans are at the apex of an evolutionary pyramid, and that the western industrialized lifestyle is the norm to which developing countries should aspire. One role of universities is constantly to hold up such worldviews to criticism and to search for alternatives.

Major Findings

Impact studies can be conducted on a series of timescales as well as on different spatial scales. Geographers, ecologists and historians alike have been concerned to look at human alterations over very long periods so as to give perspective to our current worries. Has this happened before? Is today's activity in a totally different quantitative category from that of the seventeenth century? These are the kinds of questions that are being asked (McNeill 2001).

One finding relates to the antiquity of human alterations of their surroundings. It seems that as soon as hominids could control fire at the landscape scale (i.e., beyond the hearth), they used it for hunting and to improve the yield of edible plants. Some scholars put this at about 0.5 million years ago with the evolution of *Homo erectus*, but others suggest that the emergence of *Homo sapiens* at more like 40,000 years ago was the necessary step for this technique. In any case, large areas of the world have undergone controlled burning at human hands and we know that the practice was continued into the Holocene as long as hunter-gatherers were a major category of human society. The transition from food collecting to food production (often called 'the agricultural revolution') enabled human populations to grow rapidly, and indeed started the growth which took our species from very low levels in Africa through bursts of expansion in numbers after about AD 1650 to the 6 billion mark in 1999. Though it is unpopular to attribute all intensities and types of impact simply to population growth (Mannion 1999), there is no doubt that the surface of the Earth would be markedly different if for some reason the totals had remained at their AD 1600 level. (Equally, the surface may well come to be different if the human population levels out at the predicted 10 billion figure.) The relation between a population and its resources has probably been a concern since earliest times in Africa, though we often associate it with its most famous written expression in the works of Thomas Malthus in the eighteenth century. What 'proved' Malthus wrong at the time was the development of technology, and it is clear that machines are potent elements of the ability to produce impact of all kinds. The bulldozer might stand as symbol for most of these: what they have in common is that they are powered by fossil fuels (Smil 1993). In the nineteenth century this was mostly coal; in the twentieth century and now, it is mostly oil. Not only is the field levelled by the bulldozer, but in western countries all the steps from the field to the mouth are underpinned by energy-using technology: think of the stages which convert Canadian wheat to British toast, including all the energy consumed in making the machines used. So perhaps impact can be measured in a general way by multiplying technology by population, using commercial energy use as a surrogate measure for technology. The general concept of an 'ecological footprint' of a society is gaining ground (Chambers et al. 2000).

Yet even a superficial reading of a news bulletin brings home the notion that human impacts are often high in poor countries with low levels of access to technology. We hear of the devastation caused by floods in Bangladesh when the monsoon rains flood off the deforested Himalaya and no defences are adequate, of the soil erosion on the slopes of the Andes caused by peasant farmers, or of coastal floods where tropical mangroves

have been converted to shrimp farms. Here, of course, the poor are converting marginal systems because they have no choice: the steep slopes are all that is left for a growing population perhaps, or financial stringencies force people into producing export crops no matter what the consequences will be next year. So high levels of environmental impact are not the sole province of the technology-intensive rich; but they have the choice of whether to do it or not. In the overproducing economies of Europe and America, for example, agricultural land can be set aside for wild birds without anybody starving: not something that would appeal in the Sahel zone of Africa.

Key Questions

There are a number of these, but two will suffice here. The first comes out of the totality of change and the rapidity with which it can come about: enhanced carbon levels in the atmosphere are mostly a product of the last 100 years; chemicals that feminize fish in rivers are much newer. What emerges is the anxiety produced by unpredictable instability. If a coral atoll-based society in the Pacific is going to drown in 100 years, then there is time to adapt. If sea levels take a sudden and unforeseen surge upwards, then there is more likely to be chaos. If the Atlantic conveyor-belt gradually fades, then we in Britain have time to install proper double-glazing; if it switches off next week, then next winter there will be energy rationing. From studies of past climatic change we know that rapid transitions are quite common. If virtually uniform GM crops all fall to the same pathogen, then how much food security will the world have? The second key question then comes in two parts. Will any of this affect me, and do I affect any of it? The first part has an obvious answer, and all students can profitably sit down and draw a diagram of their linkages to the rest of the planet and its peoples (recalling John Donne's remark of 1624 that 'No man is an

Island, entire of it self'). The second part follows and is uncomfortable. The consumption patterns to which most of us are accustomed are intimately linked to many of the changes that can be chronicled, and the dominance of the West in economic and political terms ensures that our reach is long. Knowing the science, as geographers must, does not absolve us from asking the ethical questions, of ourselves first of all.

References

Chambers, N., Simmons, C. and Wackernagel, M. 2001: *Sharing Nature's Interest: Ecological Footprints as an Indicator of Sustainability.* London and Sterling, VA: Earthscan Publications.

Harrison, P. and Pearce, F. 2000: *AAAS Atlas of Population and Environment.* Berkeley, Los Angeles and London: University of California Press.

McNeill, J. 2001: *Something New Under the Sun: An Environmental History of the Twentieth-century World.* London: Penguin.

Mannion, A.M. 1999: Global change: prospects for the next 25 years, *World Futures*, 54, 211–30.

Smil, V. 1993: *Global Ecology: Environmental Change and Social Flexibility.* London and New York: Routledge.

Further Reading

Clarke, R. (ed.) 2000: *Global Environmental Outlook 2000.* London: Earthscan Publications. The official end-of-millennium overview by the United Nations Environment Programme (UNEP). Most material is on the state of the environment and the policy responses, and there is a strongly regional cast to the organization of the material. A generally politically anodyne tone is, for obvious reasons, maintained. Most of the material is also available on the Internet via UNEP.

IHDP-WCRP 2002: *Earth System Atlas*, at the Potsdam Institute for Climatic Research. An evolving online interactive atlas about global environmental change and its impacts. Testing in 2001, it was planned to be available to the public in 2002; updating will only be possible after peer review so it will possess some of the standards of book and journal material.

Simmons, I.G. 1996: *Changing the Face of the Earth*, 2nd edition. Oxford: Blackwell. A straightforward book on historical principles, for those students still able to face the printed page. It adopts a chronological narrative based on access to energy sources with special emphasis on societies that are solar-powered as distinct from those that have access to fossil fuels.

5

Growing on Trees: Evidence of Human-induced Global Warming

Robert L. Wilby

The Big Issue

Global mean temperatures near the Earth's surface increased by 0.2°C/decade over the last three decades, and by about 0.6°C during the course of the twentieth century. This rise might not seem impressive at first glance, but it is unusually rapid when compared to temperature estimates for the last 1,000 years (Jones et al. 2001). Add to this the fact that the global mean temperature record was broken four times during the 1990s (1990, 1995, 1997 and 1998), and it is not surprising that global warming has become headline news. Indeed, the 1998 temperature anomaly of +0.58°C above the 1961–90 average has probably not been matched since the tenth century AD, or even earlier.

When confronted with such statistics, opinion generally divides into one of two camps. The prevailing scientific viewpoint is that natural factors (such as increasing solar irradiance and fewer explosive volcanic eruptions) were important in early twentieth-century warming, but human factors (such as emissions of greenhouse gases) contributed to observed warming since the 1960s (Stott et al. 2000). Global warming sceptics now concede that rising concentrations of atmospheric carbon dioxide can raise aver-

age temperatures, but argue that the significance of the human contribution has been overstated. Thus, the global warming debate is not so much about the *detection* of trends but rather the *attribution* of observed warming to specific causes.

This chapter examines some of the techniques being used by climatologists to tease out and model the relative significance of human and natural factors to twentieth-century warming. The outcome of this new research has major implications for our ability to simulate credible climate change scenarios for the twenty-first century.

Climate Variability

Multi-decadal climate variability can be divided into three constituents: (natural) internal variability, externally forced variability and unexplained 'noise'. Knowledge of the relative contribution of each component is crucial for explaining past climate variability, and for predicting future climate change.

Natural internal variability results from the continuous redistribution of heat between the ocean and the atmosphere, and between different states, such as vapour to liquid, or water to ice. Such exchanges occur over timescales from seconds to millennia. The two most important modes of variability are the El Niño-Southern Oscillation (ENSO) and the North Atlantic Oscillation (NAO). Both display characteristic patterns of climate anomalies and decadal variability. For example, the NAO is an index of the north–south pressure gradient over the eastern Atlantic that correlates with many aspects of the winter climate across northwest Europe (Wilby et al. 1997). A particularly strong trend towards the positive phase of the NAO over the past few decades has coincided with marked winter warming. The strength of the NAO typically varies over timescales of six to ten years, although some researchers claim to have detected further cycles with periodicity of 60 to 70 years.

Externally forced variability may be due to either natural or human-induced changes to the atmosphere's energy balance. For example, there is growing evidence that sulphate aerosols, injected into the stratosphere by the volcanism of the seventeenth century, significantly reduced downward shortwave radiation, leading to climate cooling in the Little Ice Age (see chapter 3 by Andrew Goudie in this volume). Conversely, emissions of greenhouse gases (carbon dioxide, methane, nitrous oxides and chlorofluorocarbons) since the industrial revolution have contributed to a net radiative forcing of the atmosphere, by reducing the amount of upward emission of infrared radiation at the tropopause, and by increasing downward emissions from the stratosphere.

Unexplained noise includes the climate variability left once internal and external forcing have been accounted for. This component is generally greater for precipitation than temperature. Second-order factors such as land-surface changes (which affect the amount of energy reflected at the Earth's surface) or changes in ocean circulation (which redistribute heat from equatorial regions to high latitudes) are thought to modify hemispheric temperatures but are at present difficult to quantify.

Unfortunately, interpretations of global temperature change are hindered by sparse instrumental networks prior to 1850. However, recent extensions to climate records before the dawn of instrumentation are making important contributions to understanding how rapidly climate can change, how unusual recent climate conditions really have been, and the extent to which internal or external forcing factors are responsible.

Climate Reconstruction

Several high-resolution palaeoclimate records have now been reconstructed for the last 1,000 years (Jones et al. 1998). Two types of temperature-sensitive (or 'proxy') data are employed: documentary archives and biological/physical evidence. For example, figure 5.1 shows reconstructed northern hemisphere summer temperatures for the last millennium. This record was produced using a statistical relationship between observed climate variability and multiple proxy sources of temperature information (combining tree rings, ice cores, corals and historical documents) to extrapolate earlier temperatures. The record indicates that temperatures at the beginning of the millennium were about 0.1°C above the millennial average, but still 0.1°C cooler than the 1961–90 mean.

Long-term temperature reconstructions are even more informative when accompanied by records of potential internal and external forcing factors. Proxy records of ENSO and the NAO, however, are difficult to derive because no proxy translates directly into atmospheric circulation. Instead, the proxy is measuring the environmental *influence* of the circulation patterns, so there is a real danger of circular argument when relating the record to a reconstructed temperature series. Indirect cause–effect relationships (say between patterns of atmospheric pressure and tree-ring growth) are also known to vary in strength over long periods of time. It is not surprising, therefore, that there is little agreement between

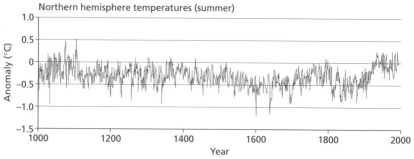

Figure 5.1   Jones et al. (1998) northern hemisphere summer temperature estimate. The temperature anomaly is calculated with respect to the 1961–90 average.
Data source: World Data Center for Palaeoclimatology (http://www.ngdc.noaa.gov/palaeo/data.html).

various indices of the NAO for periods prior to 1820.

External forcing factors have also been reconstructed with a view to interpreting long-term temperature trends. For instance, there has been some debate about the effects of solar variability and volcanic activity on centennial scale climate variability. Ice-core records of the cosmogenic isotope [10]Be imply generally increasing solar activity during the twentieth century. Similarly, records of electrical conductivity and sulphate deposition in ice cores from Greenland and Antarctica reveal a cluster of volcanic eruptions in the early twentieth century and a quiescent period between 1920 and 1960 (Crowley 2000). Taken in isolation or in combination, neither solar variability nor volcanism entirely explains twentieth-century temperature changes. Contention, therefore, surrounds the extent to which the missing climate variability is attributable to anthropogenic increases in greenhouse gases, or due to internal climate variability, or a combination of both.

Climate Model Experiments

Such hypotheses can be tested by quantifying the effects of natural variability, along with solar, volcanic and anthropogenic forcing, in climate model experiments. Two types of model are typically employed. Energy balance models (EBMs) calculate global mean temperature changes from the difference between incoming and outgoing radiation, given the planetary albedo and heat capacity, and solar influx.

More sophisticated EBMs incorporate latitudinal temperature gradients, simple representations of ocean and atmosphere layers, land-surface feedbacks and thermal inertia of deep oceans. These models have minimal computing requirements and are well suited for climate sensitivity analyses. However, they must be tuned to fit observations and can produce implausible results if 'pushed' too far beyond their calibrated range.

Energy balance models have recently been used to investigate the causes of northern hemisphere temperature changes over the past millennium. In one such experiment the EBM was run using 1,000-year reconstructions of solar and volcanic forcing, with anthropogenic greenhouse gas forcing after 1850 (Crowley 2000). Comparisons of temperature reconstructions and instrumental records for the same periods suggest that between 41 and 64 per cent of pre-1850 decadal-scale temperature variations were due to changes in solar irradiance and volcanic eruptions. The remaining variance was consistent with climate model estimates of unforced, internal climate variability. However, when the EBM was used to remove the effects of solar variability and volcanoes (but *not* greenhouse gases) after 1850, unprecedented warming was still found in the twentieth century. In this case, the temperature increase far exceeded the maximum range of variability witnessed in the previous 1,000 years. This result implies that the greenhouse effect has already established itself above the level of natural variability in the climate system.

Despite their simplicity, EBMs can provide useful insights into climate behaviour at global scales, but they do not deliver information on *spatial* patterns of climate change due to internal/external forcing. This task is better addressed by the second group of models which simulates the three-dimensional, time-varying behaviour of the climate system (using primary equations governing mass, energy, momentum and water vapour) at spatial resolutions of 200–300 km. Given observed changes in human and natural forcings, state-of-the-art coupled ocean/atmosphere general circulation models (O/AGCMs) successfully simulate the observed global mean temperature response (Stott et al. 2000). Results from O/AGCM experiments have also been an important line of evidence in support of the claim that human activities have caused global warming. They have further been used to calculate natural levels of climate variability for the pre-industrial era.

For example, the UK Meteorological Office's Hadley Centre coupled climate model (HadCM2) was used to investigate the winter NAO given present-day radiative forcing applied throughout a 1,400-year control period (Osborn et al. 1999). Whether or not a climate model can faithfully simulate the NAO (and other modes of natural variability) is a good test of its physical realism, as well as the extent to which it might be used to address other important questions – such as how unusual was the recent positive phase of the NAO, or how might the NAO manifest, or be affected by, global warming.

The study found that HadCM2 exhibits a realistic NAO in terms of spatial patterns of sea-level pressure, temperature and precipitation anomalies. However, the model did not capture the strong trend towards positive NAO between 1960 and 1990. In fact, the observed trend lies outside the range of variability produced by the entire 1,400-year run, implying that either the model is deficient or the cause of the trend is not internally generated natural variability. External forcing due to anthropogenic greenhouse gases and sulphate aerosols is thought to explain about 20 per cent of the trend, but forcing by solar or volcanic activity has yet to be fully considered.

Conclusion

The idea that changes in atmospheric composition can affect climate is nothing new. In 1896 the Swedish scientist Svante Arrhenius calculated that a doubling of carbon dioxide concentration would increase surface air temperatures by 5.7°C (about double current estimates using climate models). Whereas Arrhenius had a benign view of climate change, the modern perspective is rather different. Human-induced climate change is now generally regarded as damaging to human welfare and a serious threat to environmental systems.

However, it is clear from emerging studies of proxy data sets and climate model output that natural forcing (both internally and externally generated) made significant contributions to twentieth-century climate variability. Nonetheless, climate model results suggest that human factors will swamp natural causes of global warming in the twenty-first century. But a major uncertainty remains as to how future global warming will manifest itself: whether by amplifying natural modes of the climate system such as ENSO or the NAO, or by shifting the climate system into new states. Additional uncertainty surrounds how best to incorporate solar and volcanic forcing in these model experiments.

Even if we cannot be very sure about future levels of radiative forcing (both human and natural), successful hindcasts of climate variability over the last century have, at least, increased our confidence in the *potential* for models to predict twenty-first-century temperature changes. This research has only been possible because of complementary advances in palaeoclimate reconstruction and climate system modelling.

References

Crowley, T.J. 2000: Causes of climate change over the past 1,000 years, *Science*, 289, 270–7.

Jones, P.D., Briffa, K.R., Barnett, T.P. and Tett, S.F.B. 1998: High-resolution palaeoclimatic records for the last millennium: interpretation, integration and comparison with general circulation model control-run temperatures, *The Holocene*, 8, 455–71.

Jones, P.D., Osborn, T.J. and Briffa, K.R. 2001: The evolution of climate over the last millennium, *Science*, 292, 662–7.

Osborn, T.J., Briffa, K.R., Tett, S.F.B., Jones, P.D. and Trigo, R.M. 1999: Evaluation of the North Atlantic Oscillation as simulated by a coupled climate model, *Climate Dynamics*, 15, 685–702.

Stott, P.A., Tett, S.F.B., Jones, G.S., Allen, M.R., Mitchell, J.F.B. and Jenkins, G.J. 2000: External control of 20th century temperature by natural and anthropogenic forcings, *Science*, 290, 2133–7.

Wilby, R.L., O'Hare, G. and Barnsley, N. 1997: The North Atlantic Oscillation and British Isles climate variability 1865–1995, *Weather*, 52, 266–76.

Further Reading

Harvey, D.L.D. 2000: *Global Warming: The Hard Science*. Harlow: Prentice Hall.

Houghton, J.T., Ding, Y., Griggs, D.J., Noguer, M., van der Linden, P.J. and Xiaosu, D. (eds) 2001: *Climate Change 2001: The Scientific Basis*. Contribution of Working Group I to the Third Assessment Report of the Intergovernmental Panel on Climate Change (IPCC). Cambridge: Cambridge University Press.

6

Biodiversity: The Variety of Life

Richard Field

Why are there more types of organism at the equator than at the poles? The latitudinal diversity gradient is one of the most general patterns in biogeography: a grand cline. While the details of the pattern vary, the trend of reduction in diversity with increasing latitude is found in all of the world's major regions, at all taxonomic levels (sub-species, species, genera, families, orders, etc.), and is repeated within most major taxonomic groups (such as trees, insects, mammals, fish and plankton). The pattern has also existed throughout the history of life on Earth, as far as we can tell from the fossil record. We have known about it for more than two centuries, but there is still no consensus about its cause. Explanation of the grand cline is seen by many as the 'holy grail' of biogeography. We will return to this question later, after exploring the meaning of biodiversity and the history and breadth of its study. Biogeography's focus is on patterns of life in space and time, which is more general than biodiversity itself. How-ever, biodiversity is one of the key central themes of biogeography, and research on biodiversity is currently tackling many key questions relating both to our understanding of life on Earth and our attempts to manage plant and animal communities. It is therefore of great importance and direct relevance to geographers.

What is Biodiversity?

There is more to biodiversity than global richness patterns. Consider, for example, how biodiversity can be measured (see Magurran 1988; Gaston 1996). The sim-plest and commonest method is to count the number of species in an area: 'species rich-ness'. The area studied can vary from the microscopic to the entire globe. A variation involves counting the number of other taxo-nomic units (taxa) in a given area – these might be families, genera, sub-species or even genes. Comparison of the patterns at these taxonomic levels often reveals inter-esting differences, which can shed light on the historical biogeography of the organisms in the study area (as found, for example, by O'Brien et al. 1998). A related measure of biodiversity is turnover (commonly known as 'beta diversity'). To understand this con-cept, consider the following example. The number of tree species is counted in two plots. Both have 20. The total number of species found could be anywhere between 20 (if there is no turnover: both plots have exactly the same species) and 40 (100 per cent turnover: no species in common).

Simply counting taxa, however, ignores much important information. First, it yields no information on the relative abundance of different taxa. One area might contain one dominant and many locally rare species,

Figure 6.1   Forest on Krakatau, 111 years after sterilization by a massive volcanic eruption. Krakatau (6°06′S, 105°25′E) is a group of four small islands, totalling only about 36 km² in area and still depauperate in the diversity of life it supports because of the effects of the eruption of 1883 and subsequent smaller eruptions. Yet it is home to more than 100 tree species, which is more than the entire native British tree flora. The question of why there are so many more species in the tropics than at higher latitudes is one of the greatest unanswered questions in biogeography.
Photograph: Rob Whittaker (1994) (fig. 8.7 of Whittaker 1998).

while another might contain a much more even distribution of the same species. Second, there is the important issue of uniqueness. One area might be inhabited primarily by species found nowhere else on Earth ('endemics'), while another with the same species count might contain nothing but species that are globally common. Much of the study of biodiversity is concerned with patterns of endemism and relative abundance, rather than with species richness, measures that are not necessarily related (table 6.1). For instance, oceanic islands tend to have disproportionately high numbers of endemics. However, for most natural communities, nearly all measures of diversity tend to correlate closely with species richness, and this is one reason why usage of terminology is often sloppy. In turn, this failure to differentiate between biodiversity measures has probably caused much of the confusion that has arisen in trying to explain observed patterns.

A Brief History of Biodiversity

Interest in biodiversity goes back at least as far as Aristotle (384–322 BC), who asked where life came from and how it diversified and spread around the world. He also noticed an altitudinal gradient in biodiversity. The main problem for Aristotle was a lack of accumulated knowledge for him to draw on, so he could not hope to answer the questions he posed. This remained the case un-

Table 6.1 Estimates of the number of species of higher plants native to selected countries of the world, and the number of those species which are endemic to the country concerned. Islands (especially oceanic ones) can have relatively low species richness for their latitude and area, but tend to possess high levels of endemism.

Country	Continental or islands?	Area (000 km²)	No. of species of higher plant	No. of endemics	% endemic
Cameroon	continental	475	8,260	156	1.9
Canada	continental	9,971	3,018	147	4.9
Chile	continental	757	5,280	2,698	51.1
Colombia	continental	1,139	35,000	1,500	4.3
France	continental	552	4,630	133	2.9
Germany	continental	357	2,682	6	0.2
Honduras	continental	112	5,355	148	2.8
India	continental	3,288	16,000	5,000	31.3
Italy	continental	301	5,598	712	12.7
Malawi	continental	118	3,765	49	1.3
Mexico	continental	1,958	26,000	3,624	13.9
Pakistan	continental	796	4,938	372	7.5
Panama	continental	76	9,590	1,222	12.7
Rwanda	continental	26	2,290	26	1.1
Sudan	continental	2,506	3,137	50	1.6
Turkey	continental	775	8,579	2,651	30.9
Cuba	islands	111	6,514	3,229	49.6
Fiji	islands	18	1,628	812	49.9
Greenland	islands	2,176	529	0	0.0
Indonesia	islands	1,905	22,500	15,000	66.7
Japan	islands	378	5,372	2,000	37.2
New Caledonia	islands	19	3,094	2,480	80.2
New Zealand	islands	271	2,371	1,942	81.9
Puerto Rico	islands	9	2,493	235	9.4
Seychelles	islands	1	1,640	250	15.2
Solomon Islands	islands	28	3,172	30	0.9
United Kingdom	islands	245	1,623	16	1.0

Source: Data taken from Groombridge (1992).

til the Age of Exploration, when European naturalists travelled around the world collecting and documenting the organisms they found. One of the great cataloguers was Carolus Linnaeus (1707–78), who developed the binomial system of Latin species names that we still use today. He realized the importance of explaining patterns of diversity, as well as simply the numbers of species. Comte de Buffon (1707–88) argued that species (and climate) are mutable and not fixed by God. He also showed that en-

vironmentally similar but isolated regions have distinct biota – 'Buffon's Law', which can be regarded as the first biogeographic principle. Johann Reinhold Forster (1729–98), who circumnavigated the globe with Captain Cook (1772–5), observed that plant diversity tends to decrease from the equator to the poles – probably the first clear statement of a latitudinal diversity gradient. He attributed this to latitudinal trends in surface heat. Alexander von Humboldt (1769–1859) developed these ideas to a remarkable extent, demonstrating similar latitudinal gradients for plants and animals, and identifying the probable cause as climate (especially the importance of winter temperature and the fluidity of water). Both Charles Darwin (1809–82) and Alfred Russell Wallace (1823–1913) made important contributions, focusing on the staggering diversity of life more than the diversity patterns. They reinforced von Humboldt's observations, but asked why beneficial climates should increase diversity rather than increasing the abundance of a few dominant species – a question that still puzzles us today.

In the first half of the twentieth century, ecology started to emerge as a discipline in its own right. Much work was done on changes in biota over long time periods, reflecting the great impact of palaeontology (the study of fossil remains) on the thinking of the time. Similarly, new ideas about evolution (and particularly speciation) were hugely influential. Scientists also started to develop mathematical models to simulate ecological processes; without the aid of computers, these necessarily focused on patterns of variation within single species. Interest in biodiversity was renewed by a seminal paper by Hutchinson (1959), who argued that understanding biodiversity requires knowledge of how usable energy is acquired and partitioned among species.

Much research followed Hutchinson's paper, but there was confusion over scale and terminology, the central focus of energetics was largely ignored, and only limited progress was made. Meanwhile, four major developments profoundly changed biogeography. First, plate tectonics revolutionized how people thought about the Earth's history. Second, MacArthur and Wilson's *Theory of Island Biogeography* (1967) changed the direction of the subject, raising new types of question about patterns of evolution and species diversity. Third, the development of phylogenetic methods (reconstructing evolutionary history from genetic information) allowed people more accurately to trace the history and relationships of taxa. Fourth, advances in technology (computers and field instruments), coupled with advances in statistical methods and the increasing availability of good data with wide coverage, opened up new fields of study, in which macro-scale patterns of life could be studied with rigour. All these developments have impacted on the study of biodiversity and interest now is stronger than ever, bolstered by increasing concerns about human impact on the environment.

In the light of the long history of interest in biological diversity, it is interesting that the term 'biodiversity' itself was first used as recently as 1986, at a forum on 'BioDiversity' held in Washington. The speed with which the term spread is remarkable: by 1987 it was one of the most frequently used terms in conservation literature, and it was in mainstream use in the academic, political and public arenas soon afterwards. In part, this reflects the great importance attached to the topic today.

Biodiversity Study Today

The rapid growth of interest in biodiversity has been accompanied by a large amount of research. Part of this aims simply to document what species are in the world. Important work done over the last 20 years has shown that the 1.7 million species we have so far recorded represent only a small fraction of the world's species. When canopies of trees in tropical rainforests are fogged with insecticide, many of the insects collected tend to be new to science. When the bottoms of the deep oceans are trawled, many new species are typically found. When the soil under our feet is studied in detail, we realize how few species of bacteria, nematode worms and other micro-organisms we have described to date. Indeed, we have no idea how many types of organism there are. Estimates of the total number of species in the world vary by two orders of magnitude – typically from 3 million to 500 million.

This has important implications for the study of biodiversity: if we are interested in trying to explain the patterns, especially on the macro-scale, we have to limit our study to groups of organisms for which we have reasonably good coverage. In practice, this usually means trees, mammals, birds, other vertebrates and certain groups of invertebrates (such as butterflies and tiger beetles). We have to assume that other types of organisms follow similar 'rules'. A similar assumption is made in the use of 'indicator species' for nature conservation. These are conspicuous species, which are often appealing to the public, around which many nature reserves are designed in the hope that other species will be conserved in the process. Tests of the effectiveness of this approach produce mixed results, suggesting that any general conclusions we draw about biodiversity patterns on the basis of study of specific groups must be treated with caution.

Virtually every conceivable aspect of biodiversity is being studied today. Many focus on why there are so many species, while others are more interested in explaining diversity patterns in time and space. The former tend to focus primarily on speciation and its relation to extinctions. Those trying to explain grand clines in biodiversity tend to fall into two camps, emphasizing either the importance of climate and other environmental influences, or biogeographic history. There are many debates and topics involved in the study of biodiversity, including vicariance vs. dispersalism, refugia theory, patterns of endemism, biodiversity hotspots, island evolution, nestedness, assembly rules, succession, biological invasions and the influence of humans. Most or all of these have relevance to conservation theory and practice.

Biodiversity and Geographers

Biogeography is a typical sub-discipline of geography in that understanding emergent patterns requires knowledge of a broad range of specialist fields: biogeography lies at the interface of ecology, geography, evolutionary biology, palaeobiology, systematics, climatology, hydrology and geology. Because geographers are trained in this kind of interdisciplinarity, we are (or should be) well suited to the study of biodiversity and biogeography in general. The tendency of geographers to think holistically and to have a good grasp of scale also acts as an important complement to the highly reductionist, micro-scale approach typical of most biological research. Returning to the issue with which this chapter started, I believe that geographers can contribute much to an understanding of latitudinal diversity gradients, and we are beginning to do so (see, for example, Whittaker et al. 2001).

Latitudinal Diversity Gradients

It seems that we are close to important breakthroughs in this crucial issue. The literature of the last few decades has been characterized by a plethora of hypotheses about the causes of latitudinal diversity gradients – estimates of the number of these hypotheses vary from 30 to more than 100. Most are probably true in certain contexts and scales of analysis; few if any are generally true. Attempts to rationalize/simplify these hypotheses (notably Currie 1991; Rohde, 1992; Rosenzweig 1995) have been very widely cited, but have fallen short of providing convincing solutions. A very pertinent paper was published two decades ago (Brown 1981), which re-emphasized the role of energetics and advocated an approach that starts with the study of emergent patterns and works down, rather than starting with micro-scale study and working up. Brown outlined the need for 'capacity rules' and 'allocation rules' to be established. The former should define the availability and variability of usable energy (the capacity of an environment to support life). The latter should define how this energy is partitioned among species. While Brown's ideas have been adopted by some, many have ignored or dismissed them.

Factors influencing species richness vary in type and relative importance according to scale; untangling this complexity is one of the greatest challenges currently facing students of biodiversity. The most promising developments relate to frameworks that take Brown's ideas on board. These not only recognize the importance of scale, but also allow integration across scales. They aim to specify the scale(s) at which each factor affecting diversity patterns operates; they also aim to specify which factors are most likely to affect diversity patterns at any given scale. Further work is currently under way to address the problem that is probably most responsible for the current profusion of hypotheses: most of the phenomena proposed to explain diversity gradients are correlated with each other as well as with diversity. In other words, everything is correlated with everything else, but the patterns being studied are too big for manipulative experiments to be performed. If this problem can be solved, great advances in our understanding are likely.

Other Key Biodiversity Issues for the Noughties

There is still much to be learnt about biodiversity. Full explanation of biodiversity patterning should include gradients related to altitude (on land) and depth (under water), as well as latitude. Studies will therefore need greater integration than has been achieved so far. Recent interest in biodiversity hotspots (parts of the world thought to have particularly high diversity and/or endemism), patterns of endemism and biological invasions is likely to increase, with greater emphasis on the links to conservation. Accompanying this is the study of metapopulations (fragmented populations interconnected by gene flow), which relate to conservation in fragmented landscapes.

It is also essential to continue the work of determining what and how many types of organism there are. Finally, a topic of increasing study concerns the effects of biodiversity on other properties of communities such as productivity – reflecting increasing concern about human-induced reduction of global biodiversity. The degree to which biodiversity can be seen as a driving force of ecosystems, rather than a response to environmental conditions, is an interesting debate that could be prominent in the coming years.

References

Brown, J.H. 1981: Two decades of homage to Santa Rosalia: toward a general theory of diversity, *American Zoologist*, 21, 877–88.

Currie, D.J. 1991: Energy and large-scale patterns of animal- and plant-species richness, *American Naturalist*, 137, 27–49.

Gaston, K.J. (ed.) 1996: *Biodiversity: A Biology of Numbers and Difference*. Oxford: Blackwell.

Groombridge, B. (ed.) 1992: *Biodiversity: Status of the Earth's Living Resources*. London: Chapman and Hall.

Hutchinson, G.E. 1959: Homage to Santa Rosalia, or why are there so many kinds of animals? *American Naturalist*, 93, 145–59.

MacArthur, R.H. and Wilson, E.O. 1967: *The Theory of Island Biogeography*. Princeton, NJ: Princeton University Press.

Magurran, A.E. 1988: *Ecological Diversity and its Measurement*. Princeton, NJ: Princeton University Press.

O'Brien, E.M., Whittaker, R.J. and Field, R. 1998: Climate and woody plant diversity in southern Africa: relationships at species, genus and family levels, *Ecography*, 21, 495–509.

Rohde, K. 1992: Latitudinal gradients in species diversity: the search for the primary cause, *Oikos*, 65, 514–27.

Rosenzweig, M.C. 1995: *Species Diversity in Space and Time*. Cambridge: Cambridge University Press.

Whittaker, R.J., Willis, K.J. and Field, R. 2001: Scale and species richness: towards a general, hierarchical theory of species diversity, *Journal of Biogeography*, 28(4), 453–70.

Further Reading

Brown, J.H. and Lomolino, M.V. 1998: *Biogeography*, 2nd edition. Sunderland, MA: Sinauer Associates.

Gaston, K.J. and Spicer, J.I. 1998: *Biodiversity: An Introduction*. Oxford: Blackwell.

Tudge, C. 2000: *The Variety of Life*. Oxford: Oxford University Press.

Whittaker, R.J. 1998: *Island Biogeography: Ecology, Evolution and Conservation*. Oxford: Oxford University Press.

Wilson, E.O. 1992: *The Diversity of Life*. London: Penguin.

World Resources Institute: *Biodiversity* [online]. Available from http://www.wri.org/wri/biodiv/bri-ntro.html

7

Geoarchaeology

Jamie C. Woodward

Geoscientists, including both geographers and geologists, have worked closely with archaeologists for many years to advance our understanding of the human past. In recent decades this fusion of methods and ideas has become known as geoarchaeology. It is based on the principle that the record of human activity as documented by archaeological finds can only be understood if the finds are considered in their proper environmental context.

Collaboration between archaeologists and geoscientists can be traced back to the early Victorian era when geological expertise in stratigraphy was instrumental in the development of excavation methods, site description and the classification of the archaeological record. It is important to appreciate, however, that geoarchaeology involves much more than a systematic investigation of the layers of sediment that form an archaeological site. It is also concerned with the broader environmental context of the cultural record, because archaeological sites and past societies form part of wider landscapes and ecosystems that may have been modified by climate change or human action (Butzer 1982).

Modern studies of global environmental change have prompted new insights into the archaeological record and have stimulated new debates about early human origins as well as the demise of entire civilizations (see Weiss 2000). For example, recent data collected from ice cores, tropical lake sediments, speleothems and deltaic deposits all show that an abrupt climatic change took place around 2200 BC. This period saw the collapse of the Old Kingdom dynasties in Egypt, and it has been argued that a reduction in monsoon intensity in East Africa and a shift to a drier climate led to a dramatic decline in River Nile flows (Krom et al. 2002).

An Interdisciplinary Enterprise

Geoarchaeology is a diverse and developing area of research that straddles many subjects including geomorphology, soil science, sedimentology, mineralogy, palaeoecology, geophysics, geochemistry and, of course, archaeology and anthropology. No individual can claim expertise in all of these areas, and one of the attractions of geoarchaeology is the importance of teamwork – both in the field and in the laboratory – often with scientists from a range of backgrounds and traditions. A central issue in many geoarchaeological investigations concerns the impact of climate and landscape change on past human societies. A major challenge, for example, is to understand the strategies that were adopted to cope with repeated shifts in resources and habitats during the last glacial period. Significant climatic changes have also taken place during the past 10,000 years or so within the present (Holocene) interglacial period. During the course of the Holocene human action becomes an increasingly important factor in environmental change, and Ian Simmons explores this theme in chapter 4 in this volume. Unravelling the complex history of fluctuating Holocene climate and human impacts on the landscape is also a major research objective of the geoarchaeological community.

The development of modern geoarchaeological research gathered pace in the 1950s and 1960s. At this time a number of prehistoric archaeologists were advocating an ecological approach to archaeology as they explored concepts and research strategies related to the palaeoeconomy and palaeogeography of archaeological sites and past societies. Grahame Clark, a Cambridge archaeologist, was one of these pioneers, and in the third edition of his influential book, *Archaeology and Society* (1957), he wrote:

> What the prehistoric archaeologist has to study is the history of human settlement in relation to the history of the climate, topography, vegetation, and fauna of the territory in question. One of the greatest difficulties in such a study is to distinguish between changes in the environment brought about by purely natural processes and those produced, whether intentionally or incidentally, by the activities of human society, and this can only be resolved by intimate cooperation in the field with climatologists, geologists, pedologists, botanists, zoologists, palaeontologists in the comradeship of Quaternary research. (Clark 1957: 20)

Clark's agenda incorporates most facets of physical geography and this statement highlights the interdisciplinary nature of geoarchaeology as well as raising the thorny issue of causality in studies of landscape change. Several decades later this list could now be extended to include specialists in radiometric dating and satellite remote sensing, and perhaps also computer modelling of ecosystem dynamics and climate change.

Ice Cores, Past Climates and Human Activity

The study of the Quaternary ice age was revolutionized in the second half of the last century as new dating techniques and new methods of environmental reconstruction were developed for both the marine and terrestrial sedimentary records. Geoscientists began to apply this array of new tools and ideas to address archaeological questions – particularly in the study of prehistoric societies within the range of radiocarbon dating over the last 40,000 years or so. A good deal of research has focused on the global climate changes at the end of the last ice age and the behaviour of hunter-gatherer societies during the transition from glacial to interglacial conditions (Straus et al. 1996; Bailey 1997a,b). It is during this latter period that some of the earliest evidence has been found for the domestication of animals and for the development of early cropping activities (Bar-Yosef and Belfer-Cohen 1992; Whittle 1994).

Over the last decade or so data recovered from the Greenland and Antarctic ice cores have provided dramatic new insights into the nature of global climate change (see chapter 3 by Andrew Goudie in this volume). These records comprise annual layers of ice and contain startling evidence for frequent and abrupt shifts in global climate during the last glacial period. While this new framework of global climatic instability over decadal to centennial timescales poses problems for stratigraphic subdivision and regional correlation (see Walker 2001), it has led researchers to consider how such frequent oscillations may have impacted on human activity. Some geoarchaeologists have begun to search for evidence of these abrupt climate changes in the sedimentary sequences in Palaeolithic rockshelters and caves (Courty and Vallverdu 2001).

During the last cold stage the atmosphere and ocean surface were a closely coupled system that repeatedly underwent massive reorganizations on timescales of centuries or even less. There is a growing body of evidence to suggest that terrestrial ecosystems were sensitive to these abrupt climatic shifts (Walker 2001). Their impact on the global hydrological cycle, on plant and animal resources, and on human subsistence strategies, is currently a major research question. At present, however, the number of well-dated and high-resolution records of long-term vegetation change is limited. Furthermore, the sedimentary records in rockshelters and caves – which contain the bulk of the archaeological evidence about Palaeolithic humans – often contain large gaps, and the available temporal resolution is always much coarser than the ice-core signal and this makes direct comparisons between such records difficult (Woodward and Goldberg 2001).

The Radiocarbon Timescale

The development of robust dating frameworks is an integral part of geoarchaeology and Quaternary science more generally, and is a crucial element in any attempt to compare archaeological and climate records. A major problem that has hampered many studies of Late Pleistocene environmental change is the realization that radiocarbon years are not directly comparable to calendar years. A radiocarbon age of 20,000 years BP, for example, represents a calendar age of approximately 23,000 years BP. In the light of the frequent and abrupt climatic shifts revealed by the ice-core records, such differences are crucial when attempting to understand the wider environmental context of occupation sequences and resource exploitation strategies in a given region. The age control in most Upper Palaeolithic archaeological sites and Late-glacial pollen records is provided by the radiocarbon method, yet the high-resolution records from the ice cores are plotted in calendar years (see van Andel 1998; Walker 2001). This is an important and complex issue related to long-term variability in the global produc-

tion of radiocarbon and it means that a correction factor must be applied to all radiocarbon ages to ensure that they are compatible with timescales derived from other methods.

The Components of Geoarchaeological Investigation

Environmental reconstruction and radiometric dating are just two, albeit important, facets of geoarchaeology. A variety of field and laboratory techniques have been imported from the geosciences and applied in archaeological contexts. Geological provenance methods are a good example. These have provided valuable insights into the source of archaeological materials such as prehistoric stone tools and Greek and Roman marble and coins. This approach can provide important data on trading and seafaring activities in past societies. Obsidian, for example, is a black volcanic glass that can be worked into extremely sharp cutting tools and it was a much-prized raw material in the Eastern Mediterranean region during later prehistory. Obsidian has a distinctive chemical fingerprint that can be matched to the source volcano. Trace element analysis has established that obsidian found in the sedimentary sequence at Franchthi Cave on mainland Greece was actually imported from quarries on the island of Melos in the southern Aegean over 9,000 years ago.

Geoarchaeology often involves the use of geophysical methods – such as ground-penetrating radar – to detect archaeological features (such as hearths, buildings or bridges) that have been buried by aeolian, alluvial or other kinds of sediment. In the UK, Channel 4's Time Team have showcased many examples of geophysical prospecting on prime-time television within their investigations of Roman, Iron Age and medieval sites.

Geoarchaeological research is undertaken at several scales. These range from regional considerations of topographic features and site distribution and preservation (often using aerial photography or satellite remote sensing) (see Adams et al. 1981), to microscopic examination of the sediments, soils and organic remains associated with archaeological finds (Goldberg 2000). In his book *Archaeology as Human Ecology*, published in 1982, Karl Butzer describes the primary components of geoarchaeological research as:

1 Landscape context
2 Stratigraphic context
3 Site formation
4 Site modification
5 Landscape modification

These components can be considered for a variety of site types and environments and more detailed descriptions are given in table 7.1. A case study from the Upper Palaeolithic of northwest Greece incorporating information on landscape context, stratigraphic context, site formation and site modification is described below. Many of the components listed in table 7.1 have developed into specialist research areas in their own right, and the geoarchaeology of river environments (e.g. Brown 1997; Woodward et al. 2001) and the study of sedimentary records in rockshelters and caves (e.g. Woodward and Goldberg 2001) are good examples.

Site Catchment Analysis and Micromorphology

Many geographers have become involved in archaeological field projects and these collaborations have led to the development of new theories and methods for the field and laboratory investigation of archaeological and geographical data. In the field, one example is the method of site catchment analysis that was developed by Eric Higgs (an archaeologist) and Claudio Vita-Finzi (a geographer and geologist) in the late 1960s

(Vita-Finzi and Higgs 1970). This approach involves detailed field survey of the territory surrounding an archaeological site (or complex of sites) to record the type and extent of key resources in the landscape to provide information on the nature and function of the site. The survey data can then be compared to the plant and animal remains recovered from site excavation to establish the extent of the site exploitation territory, and the value placed on particular resources over time.

In the laboratory, a technique imported from soil science, which allows unconsolidated sediments and soils from archaeological contexts to be examined intact under the microscope, is known as micromorphology. Undisturbed blocks of sediment are impregnated and hardened with resin to allow the production of large-format (c.12 × 6 cm) thin sections that can be studied under the microscope (Goldberg 2000). This approach has allowed major advances in our understanding of the sedimentation processes in a wide range of archaeological sites. It yields information that cannot be obtained from traditional excavation and sampling methods and has often highlighted the significance of cultural debris in the sedimentary record.

A Geoarchaeological Case Study: Palaeolithic Settlement in Northwest Greece at the End of the Last Ice Age – The Klithi Project

My first geoarchaeological fieldwork took place in the summer of 1986 as I began the first year of my Ph.D. project in Cambridge shortly after completing a B.Sc. in geography from Aberystwyth. I spent three field seasons in the Pindus Mountains of northwest Greece within an international team of Palaeolithic archaeologists and geoscientists. We were investigating the archaeological sequence preserved at Klithi – a large rockshelter (at c.430 m above sea

Table 7.1   The primary components of geoarchaeological research (based on Butzer 1982)

I Landscape Context

1   Site microenvironment, defined in terms of the local environmental elements that influenced original site selection, the period of its use, and its immediate burial or subsequent preservation. Sediment analyses of site strata represent an obvious study procedure in a sealed site.
2   Site mesoenvironment, primarily the topographic setting and landforms of the area utilized directly for subsistence. This geomorphic information, combined with bioarchaeological inputs, helps define the adjacent archaeological mosaic.
3   Site macroenvironment, essentially the regional environment provided by a particular biome or ecotone. The constellation of effective geomorphic processes, together with biotic information, is indispensable in constructing a model of the regional ecosystem.

II Stratigraphic Context

1   Reconstruction of sequential natural events such as soil development, erosion and sedimentation, recorded by detailed sediment units (microstratigraphy) in the site and its environs.
2   Evaluation of the local physical sequence in terms of regional landscape history and potential matched with dated subcontinental or even global stratigraphies. External correlation can serve as a chronometric aid, can assist in palaeoenvironmental interpretation, can facilitate cross-checks between different categories of data, and can be used to test the temporal validity of archaeological horizons.
3   Direct palaeontological correlation and radiometric dating.

III Site Formation

1   People and animals, as geomorphic agents, produce archaeological sediments, with physical, biogenic and cultural components that require identification and interpretation.
2   Distinction of materials: (a) materials that were introduced to the site by people or animals, in their original form or as finished products; (b) materials that represent alteration products from on-site processing or biochemical decomposition; (c) materials that were transformed from primary on-site refuse and debris into new sediment through human and other physical agencies.
3   Evaluation of archaeosedimentary processes to help elucidate settlement and subsistence activities in space and time.

IV Site Modification

1   Preburial dispersal of archaeological residues through the action of running water, gravity, frost, deflation, animal trampling and deliberate human removal.
2   Postdepositional site disturbance through various agencies: burrowing animals and lower organisms, soil frost, expansion and contraction of clays, gravity and microfaulting, and biochemical alteration.
3   Site destruction and artefact dispersal caused by various forces: weathering, running water, deflation, slumping and human intervention.
4   Interpretation of sealed or exposed cultural residues in terms of primary, semi-primary or secondary context.

V Landscape Modification

1   Identification of human intervention in the soil landscape, in the form of disturbed or truncated soil profiles and redeposited soils.
2   Human intervention in the hydrological cycle, as reflected in erosional gullies, alluvial fills and lake sediment records.
3   Human constructs in the landscape: filled-in ditches, pits and postholes; earthworks and spoil heaps; roadways, terraced fields and irrigation networks; middens and burials adjacent to focal settlements.
4   Assessment of the cumulative direct and indirect impacts of human land use in spatial terms and in the temporal perspective of sustained landscape productivity or degradation.

level) in the deep limestone gorge of the Voidomatis River (figure 7.1). Geoff Bailey led the project and the team included specialists who identified the bones and flint tools recovered from the excavation (on-site work) and geomorphologists and palaeoecologists who examined the evidence for landscape and vegetation change in the wider Voidomatis catchment and beyond (off-site work). Samples of charcoal were collected from the excavation trenches in the rockshelter for radiocarbon dating. These showed that the site was used for a comparatively short period of time during the Late-glacial period with the bulk of the cultural debris bracketed between 16,500 and 13,500 radiocarbon years BP (Bailey 1997a). Interestingly, other rockshelter sites at lower elevations in the same region have much longer cultural records (>100,000 years), and one of the main geoarchaeological challenges to emerge at Klithi involved reconstructing the environmental context of the occupation history.

Geomorphological fieldwork by Mark Macklin, John Lewin and myself has shown that the highest mountain peaks in the catchment were glaciated during the last ice age, forming one of the most southerly glaciated terrains in Europe (Woodward et al. 1995). Fed by melting snow and ice, the Voidomatis River flowed in a wide braided channel for much of the last cold stage. The river was prone to large floods and occupied the entire valley floor between the walls of the gorge below the rockshelter (Macklin et al. 1997). Access to Klithi was therefore rather hazardous at this time, and the cold climate that promoted frost weathering of the gorge walls and ice build-up in the mountain headwaters would have created an extremely harsh environment in the uplands. However, as the climate warmed at the end of the last cold stage, the supply of sediment to the river declined and it began to incise, creating terraces on either side of the channel zone. With the narrowing channel now confined within the valley floor, the terraced surfaces provided access up the gorge to Klithi. After

Figure 7.1 Klithi rockshelter in the gorge of the Voidomatis River. The floor of the site is 30 m above the present level of the river. A major challenge for geoarchaeological research is to develop approaches (including robust chronologies) that allow the archaeological data recovered from site-based excavations to be integrated with local, regional and global palaeoenvironmental datasets.

*c.*16,500 radiocarbon years BP, Palaeolithic hunters were able to access the entire gorge in the spring and summer to exploit the ibex and chamois on the surrounding mountain slopes.

The excavations in the rockshelter produced a huge assemblage of stone tools and flint debris (with relatively few tool types, including blades for cutting and scrapers for processing hides) as well as thousands of ibex and chamois bones and bone fragments. This assemblage is indicative of a very specialized site where butchery, carcass processing and associated tool-making activities

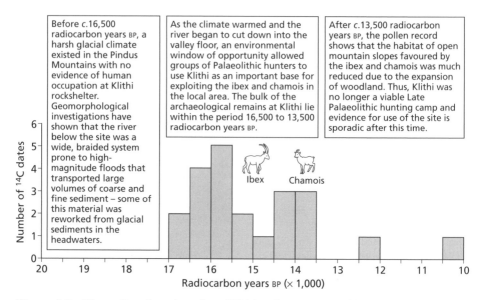

Figure 7.2   The radiocarbon dates from Klithi and a summary of the environmental conditions before, during and after the use of the shelter in the Late Upper Palaeolithic. Note that the timescale is in radiocarbon years and that ibex and chamois dominate the faunal assemblage throughout the period of occupation.

were dominant. Pollen records from lakes and peat bogs show that, by about 13,500 radiocarbon years BP, woodland expansion had replaced much of the open mountain habitat favoured by the ibex and chamois so that Klithi was no longer a viable base for hunting activities in the immediate area (figure 7.2). In short, Klithi was a seasonal hunting camp and carcass-processing station in the Pindus Mountains during a relatively brief environmental window of opportunity at the end of the last cold stage.

Crucially, various dating methods, including radiocarbon, allowed the different lines of evidence – archaeological, geomorphological, sedimentological and botanical – to be compared and evaluated to build up a picture of the local and regional ecosystems before, during and after the period of Upper Palaeolithic occupation at Klithi (figure 7.2). Data from the wider (off-site) environment were essential to make sense of the cultural and sedimentary record from the rockshelter (Woodward 1997). Much of the

project involved work on components I to IV listed in table 7.1. The results have been published in two hefty monographs (Bailey 1997a,b), which provide an excellent up-to-date example of the insights that can be gained into past societies and environmental change through multidisciplinary collaboration.

Conclusions

It is important to recognize that geoarchaeological research spans the entire archaeological record – from the origins of early humans to the industrial period – and it has only been possible to mention a few examples here. It is carried out in all regions of the world – from polar to temperate to tropical – by researchers who have developed geoarchaeological expertise in a range of geographical and sedimentary contexts, including caves and rockshelters and wetland, floodplain, coastal, hot desert and high mountain settings. Until quite recently it was common practice for a geologist or geomorphologist to be introduced to the excavation trenches in an archaeological site during the final week of the final season of a ten-year excavation campaign. Fortunately, much progress has been made over the last two decades and it is now commonplace for archaeological projects to involve geoscience expertise at all stages of project planning and field survey and excavation.

In the first edition of this book Ian Douglas stated that the relationship between people and their environment is probably the most important core theme of geography. Geoarchaeology is an exciting and rapidly developing area of research that focuses on the people of the past and their interaction with their environment. Geographers have played a key role in its development and it offers opportunities to interact with other disciplines and to consider many important geographical issues over extended timescales. There is much we can learn about the geography of the present by studying the environments and societies of the past.

References

Adams, R.E.W., Brown, W.E. and Culbert, T.P. 1981: Radar mapping, archaeology and ancient Maya land use, *Science*, 213, 1457–63.

Bailey, G.N. (ed.) 1997a: *Klithi: Palaeolithic Settlement and Quaternary Landscapes in Northwest Greece. Volume 1: Excavation and Intra-site Analysis at Klithi*. Cambridge: McDonald Institute for Archaeological Research.

Bailey, G.N. (ed.) 1997b: *Klithi: Palaeolithic Settlement and Quaternary Landscapes in Northwest Greece. Volume 2: Klithi in its Local and Regional Setting*. Cambridge: McDonald Institute for Archaeological Research.

Bar-Yosef, O. and Belfer-Cohen, A. 1992: From foraging to farming in the Mediterranean Levant. In A.B. Gebauer and T.D. Price (eds), *Transitions to Agriculture in Prehistory*, Monographs in World Archaeology 17, Madison, WI: Prehistory Press, 21–48.

Brown, A.G. 1997: *Alluvial Geoarchaeology*. Cambridge: Cambridge University Press.

Butzer, K.W. 1982: *Archaeology as Human Ecology*. Cambridge: Cambridge University Press.

Clark, J.G.D. 1957: *Archaeology and Society*. London: Methuen.

Courty, M.-A. and Vallverdu, J. 2001: The microstratigraphic record of abrupt climate changes in cave sediments of the western Mediterranean, *Geoarchaeology: An International Journal*, 16, 467–500.

Goldberg, P. 2000: Micromorphology and site formation at Die Kelders Cave 1, South Africa, *Journal of Human Evolution*, 38, 43–90.

Krom, M.D., Stanley, D.J., Cliff, R. and Woodward, J.C. 2002: River Nile sediment fluctuations over the past 7,000 years and their key role in sapropel development, *Geology*, 30, 71–4.

Macklin, M.G., Lewin, J. and Woodward, J.C. 1997: Quaternary river sedimentary sequences of the Voidomatis Basin. In G.N. Bailey (ed.), *Klithi: Palaeolithic Settlement and Quaternary Landscapes in Northwest Greece. Volume 2: Klithi in its Local and Regional Setting*, Cambridge: McDonald Institute for Archaeological Research, 347–59.

Straus, L.G., Eriksen, B.V., Erlandson, J.M. and Yesner, D.R. 1996: *Humans at the End of the Ice Age: The Archaeology of the Pleistocene–Holocene Transition*. New York and London: Plenum Press.

van Andel, T.H. 1998: Middle and Upper Palaeolithic environments and the calibration of ^{14}C dates beyond 10,000 BP, *Antiquity*, 72, 26–33.

Vita-Finzi, C. and Higgs, E.S. 1970: Prehistoric economy in the Mount Carmel area of Palestine: site catchment analysis, *Proceedings of the Prehistoric Society*, 36, 1–37.

Walker, M.J.C. 2001: Rapid climate change during the last glacial–interglacial transition: implications for stratigraphic subdivision, correlation and dating, *Global and Planetary Change*, 30, 59–72.

Weiss, H. 2000: Beyond the Younger Dryas: collapse as adaptation to abrupt climate change in ancient West Asia and the Eastern Mediterranean. In G. Bawden and R. Reycraft (eds), *Confronting Natural Disaster: Engaging the Past to Understand the Future*, Albuquerque: University of New Mexico Press, 75–98.

Whittle, A. 1994: The first farmers. In B. Cunliffe (ed.), *Prehistoric Europe: An Illustrated History*, Oxford: Oxford University Press, 136–66.

Woodward, J.C. 1997: Late Pleistocene rockshelter sedimentation at Klithi. In G.N. Bailey (ed.), *Klithi: Palaeolithic Settlement and Quaternary Landscapes in Northwest Greece. Volume 2: Klithi in its Local and Regional Setting*, Cambridge: McDonald Institute for Archaeological Research, 361–76.

Woodward, J.C. and Goldberg, P. 2001: The sedimentary records in Mediterranean rockshelters and caves: archives of environmental change, *Geoarchaeology: An International Journal*, 16, 327–54.

Woodward, J.C., Lewin, J. and Macklin, M.G. 1995: Glaciation, river behaviour and the Palaeolithic settlement of upland northwest Greece. In J. Lewin, M.G. Macklin and J.C. Woodward (eds), *Mediterranean Quaternary River Environments*, Rotterdam: A.A. Balkema, 115–29.

Woodward, J.C., Macklin, M.G. and Welsby, D.A. 2001: The Holocene fluvial sedimentary record and alluvial geoarchaeology in the Nile Valley of Northern Sudan. In D. Maddy, M.G. Macklin and J.C. Woodward (eds), *River Basin Sediment Systems: Archives of Environmental Change*, Rotterdam: A.A. Balkema, 327–56.

Further Reading

Goldberg, P., Holliday, V.T. and Reid Ferring, C. 2001: *Earth Sciences and Archaeology*. New York: Kluwer Academic/Plenum Press.

Lewin, J., Macklin, M.G. and Woodward, J.C. (eds) 1995: *Mediterranean Quaternary River Environments*. Rotterdam: A.A. Balkema.

Renfrew, C. and Bahn, P.G. 2000: *Archaeology: Theories, Methods and Practice*, 3rd edition. London: Thames and Hudson.

van Andel, T.H. and Runnels, C. 1987: *Beyond the Acropolis: A Rural Greek Past*. Stanford, CA: Stanford University Press.

Internet Resources

- *Geoarchaeology: An International Journal* was launched in 1986 and now comprises eight issues per year. The papers published in this journal represent the results of field and laboratory research from many parts of the world and provide a useful guide to the nature of modern geoarchaeology. The journal can be found on the Wiley Interscience pages (Earth Science section) at www.interscience.wiley.com/jpages/0883-6353/

- The Society for American Archaeology (SAA) has an active Geoarchaeology Interest Group (GIG). Their newsletter and details of their activities can be found at www.saa.org/membership/i-geo/

- Geoarchaeology is often referred to as environmental archaeology. In the UK the Association for Environmental Archaeology maintains a comprehensive website at www.envarch.net/ and the Channel 4 Time Team web pages can be found at www.channel4.com/timeteam

- *Current Archaeology* is one of Britain's leading archaeological magazines. It maintains a useful site with many links to ongoing archaeological and geoarchaeological projects at www.archaeology.co.uk

- Computer programmes and calibration charts have been produced for the radiocarbon timescale and the web pages maintained by the Oxford Radiocarbon Laboratory provide a valuable guide to this important area of research: www.rlaha.ox.ac.uk/orau/index.htm

- Many universities offer joint honours courses (BA or B.Sc.) in geography and archaeology and there may be opportunities to undertake geoarchaeological field research as part of the dissertation project. Modules devoted to geoarchaeology are now common within many single honours degree programmes in geography and modules on Quaternary environmental change commonly contain a significant component of geoarchaeology. The Internet provides one of the best ways to investigate the research interests of academic staff and the content of their courses.

8

Fluvial Environments

Mark Patrick Taylor

Fluvial environments describe the different sedimentary deposits that are formed by water flowing across floodplains and within river channels. While rivers and floodplains may appear superficially uniform, contemporary process and longer-term alluvial studies have shown that rivers and floodplains are complex environments. Individual fluvial environments reflect catchment morphology, valley configuration, land use and vegetation histories and the nature and availability of sediment. Fluvial environments represent an important area of geographical research because the type and range of fluvial environments are known to reflect the delicate balance between erosion and deposition within riverine systems. Indeed, they may be described as 'barometers for change' because numerous studies of floodplains have demonstrated that river channels and their associated floodplains have been constantly adjusting to the prevailing environment throughout the last 10,000 years (the Holocene period).

In addition to natural channel adjustments (intrinsic changes), some of the most significant impacts are the result of changes to major external (extrinsic) forcing factors: climate fluctuations, adjustments in vegetation cover and land-use type (e.g. grazing, forestry, mining, urban developments etc.). Numerous physical factors may combine to produce particular river morphologies and their associated fluvial environments, and hence, the range and diversity of river types

are enormous. Here we consider three principal issues that currently form the focus for fluvial studies: (1) the history of fluvial environments and the disentanglement of the forcing factors that have shaped their formation; (2) the relationship between process and channel and floodplain morphology; and (3) fluvial futures – the protection, rehabilitation and management of fluvial environments and river systems.

Fluvial Environments: The Cause and Nature of Change

Macklin and Lewin (1993) and more recently Macklin (1999) have proposed that changes in river behaviour are closely linked to climatic discontinuities during the Holocene. However, other studies have shown that local reach-scale geomorphic factors such as land-use changes, valley-floor gradient, rates of lateral reworking and sedimentation processes may result in a nonuniform, complex and sometimes unique fluvial response to Holocene environmental change (see Knighton 1998: ch. 6 for an excellent summary of this issue).

While it can be demonstrated that local geomorphic factors play a pivotal role in determining fluvial response to environmental change, it is beyond question that major climatic discontinuities such as that experienced at the end of the last glacial have had a significant impact on river systems. For example, many Late-glacial rivers in Northern and Eastern Europe were braided and as the Holocene climate warmed, vegetation recovered and sediment loads were reduced, causing channel planforms to gradually adjust, forming meandering-style rivers by the mid- to late Holocene. However, the timing of this shift was invariably non-uniform, often occurring over several millennia across adjacent regions and producing complex sedimentary sequences (cf. Kozarski and Rotnicki 1977). Variations in the timing and response of systems to morphological forming events may also occur because of the great range of channel sensitivity and their relative state of stability. For example, systems that are close to their physical threshold of change are likely to have a short lag-time between environmental impact and response (cause and effect). In contrast, other geomorphic systems may demonstrate a considerable lagged response because of their inherent stability (figure 8.1). (Readers are recommended to read Schumm 1998: 75–94 for an excellent discussion of this topic.)

Floodplain evolution can be understood through a close examination of the morphology, age and elevation of fluvial deposits, their sedimentology, rates of deposition and the environmental conditions under which the sedimentary unit was formed. The collation of such data can aid interpretation and reconstruction of climate and hydrological change. The recovery of long chronological and continuous fluvial sequences is invariably hampered by the fragmentary preservation of alluvial units and paucity of datable material, particularly from upland areas where sediment flushing and reworking is often greatest. Thus, multiple site studies encompassing the full range of catchment locations (upland, piedmont and lowland), fluvial environments, sedimentation styles (e.g. braided, anastomozing, meandering) and reach types (e.g. headwater, gorge, vertically accreted floodplain; Brierley and Fryirs 2000) are required to resolve intrabasin and regional patterns of fluvial change

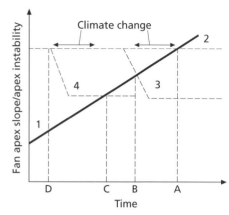

Figure 8.1   Effect of climatic change on the timing of morphologic response. Line 1 depicts the increase of fan apex slopes with time. Under unchanging conditions (line 2) fanhead slopes exceed their stability threshold at time A and incision commences. Climate change, which in this case establishes a new, lower threshold of stability (line 3), triggers incision at time B. If the same climate change was initiated much earlier, at time D, when the fan displayed greater inherent stability, then fanhead trenching at time C appears essentially unrelated to the climate change depicted by line 4 (after Schumm et al. 1987; Schumm 1998). Although this theoretical example relates specifically to alluvial fan behaviour (a subset of fluvial environments), the same principles are equally applicable to a river channel undergoing incision.

(see Macklin 1999). Such data not only provide detail on the variety of floodplain styles and types that have been present during the Holocene, but they may also reveal how floodplains have responded to cultural or climatic changes, particularly on European river systems.

Defining the relationship between external forcing factors and observed landforms is often problematic, especially when human-induced land-use changes are merged with climatic change, resulting in a 'climatically driven but culturally blurred' (Macklin and Lewin 1993) alluvial response. Thus, students of Holocene fluvial environments need carefully to consider the type of data collected, how complete and representative the alluvial record is, along with the rigour of any statistical analyses (see Tipping 2000 for a thorough discussion of these issues). Despite the inherent difficulty in deciphering the different forcing factors, fluvial environments remain an important area for research, because without such data it is impossible to gauge what the future effect of land-use change or shifts in hydro-climatic regime may have on river systems.

Floodplain Morphologies

Nanson and Croke (1992) developed a floodplain classification scheme identifying 13 different floodplain types classified under three groups: (1) high-energy non-cohesive, (2) medium-energy non-cohesive and (3) low-energy cohesive floodplains. Such a classification is particularly useful as it provides a context with which to compare individual floodplains and their environments to the wide spectrum of systems that are known to exist. Individual floodplains may be composed of single (monophase) or multiple (polyphase) units, which can be related to particular fluvial regimes (Nanson and Croke 1992). Shifts in regime can occur over a variety of timescales and may form continuous or diachronous alluvial deposits, each of which relates to a subtly different hydrological and/or climatic regime. Such changes are particularly evident in Australia, a continent characterized by droughts and floods. The low-gradient rivers and floodplains that drain the interior of Australia are distinguished by such extreme flow variability. During severe flooding, such as that which occurred on the Darling and adjacent rivers during 1990 (Gale and Bainbridge 1994), floodwater was dispersed across a 300 km area. These fluvial systems display multiple adjacent fluvial environments and, as a consequence, complex sediment sequences and landscape mor-

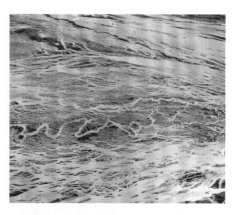

Figure 8.2 An oblique aerial photograph of the channel country, southwest Queensland, Australia taken during flooding on 6 April 1949. Multiple fluvial environments and depositional styles are visible. There are braided channels in the foreground, sinuous channel planforms in the centre, and sheet flooding in the uppermost section.

phologies (see Nanson and Croke 1992). Figure 8.2, an aerial photograph of the channel country, southwest Queensland, Australia, taken during flooding, illustrates the diversity of fluvial environments across a single floodplain zone.

Although the majority of fluvial research is focused on clastic systems (silt-sand and gravel dominated) principally because they are more common in modern and ancient sedimentary environments, freshwater carbonate (tufa) fluvial environments can also yield important environmental data. Tufa forms in spring-fed rivers and lakes where the water has been enriched with dissolved $CaCO_3$ and CO_2 relative to that found within the atmosphere. As water emerges from its spring source, CO_2 is degassed, raising the concentration of calcium carbonate to saturation point, such that precipitation occurs and tufa is formed. In contrast to clastic systems where sediments may be deposited across the whole width of a floodplain simultaneously, tufa is precipitated predominantly within the channel. The resultant depositional features include dams or barrages, cascades, fluvial crusts and cemented gravels (Pentecost and Viles 1994). Within these large-scale fluvial deposits are various smaller-scale features such as algal mats (often laminated and stromatolitic in character), phyto-clasts and biological materials introduced into the depositional system by living invertebrates (see Drysdale 1999). Because tufa is relatively resistant to erosion, especially when compared to unconsolidated fluvial sediments, its preservation allows site-specific palaeoenvironmental reconstruction (using measurement of dam orientation, geomorphology and sequence stratigraphy) and the recovery of fossiliferous materials, which can be used to decipher the prevailing biological and geomorphological environment. Not only is tufa and its preserved fossils (i.e. mollusca) datable using a variety of radiometric and stable isotopic techniques (e.g. radiocarbon, uranium-thorium ratios, $d^{13}C$ and $d^{18}O$), trace element contents (Sr, Ba and Mg) can also be used to infer past climatic conditions (e.g. Andrews et al. 2000). Thus, it could be argued that tufaceous fluvial environments are potentially more useful than purely clastic environments because of the range of palaeoenvironmental information they may yield.

Fluvial Futures

Floodplains and their associated environments are often the primary focus of human occupation because they contain some of the most fertile agricultural land. Additionally, their inherently low relief and high accessibility often result in floodplains becoming the most densely populated areas of the landscape (e.g. Calcutta on the River Ganges, London on the River Thames, New Orleans on the River Mississippi, Paris on the River Seine). Paradoxically, floodplains also represent some of the most important biophysical habitat for species diversity for they

are a primary source of food and nutrients. To achieve an understanding of the processes operating in fluvial environments, a geomorphologist must assimilate data across a range of scientific disciplines. For example, this might necessitate an understanding of the distribution of sediments and pollutants across floodplains, a quantification of sediment, nutrient and contaminant fluxes, the bio-physical impact of channelization and canalization and the role of riparian vegetation. Intertwined with such issues are natural climate variability impacts (e.g. El Niño-La Niña phenomena) along with human ones, which may alter river regime. These impacts may be manifested through changes to the flood frequency and magnitude and/or pattern and geomorphology of fluvial environments, channel shape and consequent floodplain morphology.

There is a growing research interest in deciphering the relationship between geomorphic process and habitat availability with the aim of improving channel management programmes. In particular, the setting of adequate environmental river flows in order to maintain appropriate river habitat (range, structure and size) is particularly important for environmental managers, landowners and recreational users who utilize river resources. This was central to the debate over flow allocation to the Snowy River, southeast Australia, in the wake of severe degradation following dam construction and flow regulation in 1967. The impact of flow diversion was catastrophic. Flow in the upper parts of the Snowy immediately below the Jindabyne Dam was reduced to 2 per cent of its original levels, causing massive habitat loss and a marked alteration in channel morphology (Erskine et al. 1999).

It is well established that geomorphic units (e.g. bars, pools, riffles) form relatively discrete habitat units (Brown and Brussock 1991) such that the geo-ecological functioning of a system is dependent on the range and diversity of available habitat. Disruptions to the range and diversity of fluvial environments will inevitably affect the ecological functioning of a river particularly through the loss of instream and floodplain habitat and the reduction and transfer of nutrients and organic matter. While the distribution of in-channel geomorphic units such as pools, riffles and bars records the propensity for channel sedimentation or erosion, they also provide a range of instream habitats for biota. Floodplain features such as backswamps, palaeochannels or billabongs also provide essential habitat and nutrient resources for biota. Thus, fluvial environments provide habitat for non-aquatic flora and fauna that directly affects the ecological functioning of a river through shading, litter inputs, nutrient flux (via riparian vegetation) or predation. Because the structure and dynamics of physical habitat are believed to set the template on which biological organisms evolve and communities are organized (Townsend and Hildrew 1994), it is essential that fluvial geomorphologists establish the natural range of physical habitat within and between river systems (i.e. braided, meandering, gorge) so that appropriate habitat for different river systems can be identified.

One of the most critical future issues facing fluvial geomorphologists is the establishment of proper baseline studies that link both physical and ecological structure in an integrated manner, so that river managers are able to develop sustainable practices. Recognition that catchments can be organized according to different physical scales has precipitated the development of hierarchical physical models combining the relevant components into an integrative catchment framework (e.g. Frissell et al. 1986; Brierley and Fryirs 2000). Integration of ecological and physical parameters has been achieved to some degree with the River Habitat Survey of UK rivers (Raven et al. 1998); the Index of Stream Condition in Australia (Ladson et al. 1999); and Rosgen's (1994) scheme to classify US rivers. Although the recently developed River Styles framework

(Brierley and Fryirs 2000) also integrates geo-ecological structure, it differs from the above methods because it is based on the assemblage of geomorphic units (formed under distinct fluvial environments), which have been shown to be directly related to habitat availability. The River Styles framework is a generic template that allows the relationship between geomorphic structure, functional habitat and biotic communities to be established at a scale that is ecologically meaningful. Recent research has developed the River Styles framework to incorporate hydraulic units (zones of uniform flow and substrate which equate to habitat niches) within the geomorphic template used to assess a river system (Thomson et al. 2002). The procedure now has the potential to identify appropriate reference conditions against which the quantity and quality of existing habitat can be measured, allowing management decisions to be based on data collected at a range of relevant and meaningful scales.

While it is clear that fluvial environments may themselves be complex and also form complex sequences in both space and time, they do have the potential to reveal the prevailing biophysical structure of a river system and how it has responded to environmental impacts. The role of a fluvial geomorphologist is to interpret and explain both ancient and contemporary sequences and environments such that the physical structure and the processes operating within a river system are understood. One of the major challenges facing scientists working on river systems is the protection of non-degraded rivers and the reconstruction of systems that have become damaged as a result of pollution pressure or habitat loss. Rehabilitation of river systems can only be achieved if study reaches are set within a catchment context such that the interplay of the surrounding physical and ecological components is properly understood and integrated into subsequent remedial or management plans.

References

Andrews, J.E., Pedley, H.M. and Dennis, P.F. 2000: Palaeoenvironmental records in Holocene Spanish tufas: a stable isotope approach in search of reliable climatic archives, *Sedimentology*, 47, 961–78.

Brierley, G.J. and Fryirs, K. 2000: River styles, a geomorphic approach to catchment characterisation: implications for river rehabilitation in Bega Catchment, New South Wales, Australia, *Environmental Management*, 25, 661–79.

Brown, A.V. and Brussock, P.P. 1991: Comparisons of benthic invertebrates between riffles and pools, *Hydrobiologia*, 220, 99–108.

Drysdale, R.N. 1999: The sedimentological significance of hydropsychid caddis-fly larvae (order: Trichoptera) in a travertine-depositing stream: Louie Creek, northwest Queensland, Australia, *Journal of Sedimentary Research*, 69, 145–50.

Erskine, W.D., Turner, L.M., Terrazzolo, N. and Warner, R.F. 1999: Recovery of the Snowy River: politics and rehabilitation, *Australian Geographical Studies*, 37(3), 330–6.

Frissell, C.A., Liss, W.J., Warren, C.E. and Hurley, M.D. 1986: A hierarchical framework for stream habitat classification: viewing streams in watershed context, *Environmental Management*, 10, 199–214.

Gale, S.J. and Bainbridge, S. 1994: Megafloods in inland eastern Australia, April 1990, *Zeitschrift für Geomorphologie*, 38(1), 1–11.

Kozarski, S. and Rotnicki, K. 1977: Valley floors and changes of river channel patterns in the north Polish Plain after the late Wurm and Holocene, *Quaestiones Geographicae*, 4, 51–93.

Ladson, A.R., White, L.J., Doolan, J.A., Finlayson, B.L., Hart, B.T., Lake, P.S. and Tilleard, J.W. 1999: Development and testing of an Index of Stream Condition for waterway management in Australia, *Freshwater Biology*, 41, 453–68.

Macklin, M.G. 1999: Holocene river environments in prehistoric Britain: human interaction and impact, *Quaternary Proceedings, Journal of Quaternary Science*, 7, 521–30.

Macklin, M.G. and Lewin, J. 1993: Holocene river alluviation in Britain. In I. Douglas and J. Hagedorn (eds), Geomorphology and geoecology, fluvial geomorphology, *Zeitschrift für Geomorphologie*, (Supplement) 88, 109–22.

Nanson, G.C. and Croke, J.C. 1992: A genetic classification of floodplains. In G.R. Brakenridge and J. Hagedorn (eds), Floodplain evolution, *Geomorphology*, 4, 549–86.

Pentecost, A. and Viles, H.A. 1994: A review and assessment of travertine classification, *Géographie Physique et Quaternaire*, 48, 305–14.

Raven, P.J., Holmes, N.T.H., Dawson, F.H. and Everard, M. 1998: Quality assessment using river habitat survey data, *Aquatic Conservation: Marine and Freshwater Ecosystems*, 8, 477–99.

Rosgen, D.L. 1994: A classification of natural rivers, *Catena*, 22, 169–99.

Schumm, S.A., Mosley, M.P. and Weaver, W.E. 1987: *Experimental Fluvial Geomorphology*. New York: Wiley.

Thomson, J.R., Taylor, M.P., Fryirs, K.A. and Brierley, G.J. 2002: A geomorphic framework for river characterisation and habitat assessment, *Aquatic Conservation: Marine and Freshwater Ecosystems*, 11, 373–89.

Tipping, R. 2000: Accelerated geomorphic activity and human causation: problems in proving the links in proxy records. In R.A. Nicholson and T.P. O'Connor (eds), *People as an Agent of Environmental Change*, Oxford: Oxbow, 1–5.

Tooth, S. 1999: Floodouts in central Australia. In A. Miller and A. Gupta (eds), *Varieties of Fluvial Form*, Chichester: Wiley, 219–47.

Townsend, C.R. and Hildrew, A.G. 1994: Species traits in relation to a habitat templet for river systems, *Freshwater Biology*, 31, 265–75.

Further Reading

Knighton, D. 1998: *Fluvial Forms and Processes: A New Perspective*. London: Arnold.

Schumm, S.A. 1998: *To Interpret the Earth: Ten Ways to be Wrong*. Cambridge: Cambridge University Press.

9

Glacial and Mountain Environments: Glacial Retreat as an Agent of Landscape Change

Stephan Harrison

An important direction in geomorphological research has been the stress on the interconnectedness of the components of complex landscape systems. This has meant that analysis of such systems is not best achieved through reductionist in-depth treatment of the individual parts, but through understanding of the interactions between the variables. The success of this view can be seen in the ways in which mountain and glacier systems have been investigated. The key issue here is not necessarily to understand how mountains and glaciers behave in isolation but to examine how they interact, since this gives us a much deeper understanding of the behaviour of such multi-variable systems in the real world.

One of the most important recent findings in mountain and glacial geomorphology has been the recognition that landscapes respond very quickly to glacier retreat. It has always been known that glaciers are capable of considerable landscape modification during advance phases, but their role during retreat has often been viewed as essentially passive. However, work over the last 30 years or so has demonstrated that this view is in-

correct. Periods of glacier retreat are associated with profound changes in related geomorphological systems and this is especially true in mountainous regions where steep slopes provide the kinetic and potential energy inputs to drive slope, fluvial and glacial processes. Therefore, a major initiative in contemporary geomorphological research in glaciated mountains is to understand the processes by which change occurs and the rate of landscape modification that results. This research was initiated by investigations in landscapes which were glaciated during the Pleistocene; more recent work has considered the role of glacier retreat in influencing present landscape change and on the possible impact of future climate warming on glaciers and mountains.

Perhaps the original research that put forward this idea of rapid landscape change during and after deglaciation was developed by June Ryder in the early 1970s. Her work (Ryder 1971) was concentrated in the upland valleys of central British Columbia, Canada, which had last been glaciated during the Pleistocene. She identified a number of alluvial fans which contained thick se-

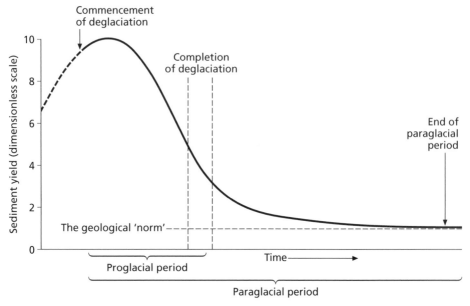

Figure 9.1   This is the Church and Ryder (1972) model of paraglacial response to deglaciation. Sediment supply to rivers increases markedly during glacier retreat and declines exponentially as sediment sources are exhausted. The period of glacier shrinkage is termed the 'proglacial' period. Modified from Church and Ryder (1972).

quences of reworked glacial material. At that time, and in the absence of accurate dating of these sediments, models of Holocene slope evolution suggested that these and similar deposits should have accumulated steadily throughout the period after glaciation. However, dating of a volcanic ash layer 2 m below the upper surfaces of these fans showed that the vast bulk of the material had accumulated by the early Holocene, some time before 6,600 years BP. The rest of the Holocene was only represented by a 2 m-thick layer of sediment. It was clear, then, that most fan development occurred shortly after glaciers had retreated from the area. This work resulted in a classic geomorphological paper which Ryder published with her husband Mike Church in 1972. This introduced a new term to geomorphology, 'paraglacial', which was used to describe non-glacial processes conditioned by glaciation and was originally employed to refer to fluvial processes. Working in Baffin Island and British Columbia,

Church and Ryder developed a conceptual model of sediment supply to valley bottoms and this is shown in figure 9.1 (Church and Ryder 1972). Here, sedimentation is highest immediately after deglaciation and slowly declines towards the level expected in non-glacial environments (the geological 'norm'). This phase of paraglaciation may be reset by renewed glacier advance and retreat so that sedimentation never reaches 'background levels'.

Whilst the processes operating during glacier retreat and paraglaciation also operate in non-glacial environments, numerous studies have shown that the intensity and rate of process are enhanced over non-glacial systems. There are a number of processes involved in paraglaciation (Evans and Clague 1994).

Rockfalls from glacially oversteepened valley sides are common after deglaciation and are caused by a combination of debuttressing (or the removal of the glacier 'support' from the cliff), stress release and fluctuating hydrostatic pressures within the joint system of the rock face (partly caused by changes in glacial meltwater pressures). Where glacier retreat is accompanied by the melting of permafrost in the valley sides, rockfalls may also be caused by reductions in frictional strength between the joint surfaces in the cliff as permafrost melts. In addition, unloading of the landscape as the glacier retreats means that small-scale seismic activity is common at this time and this serves to further destabilize steep cliff faces. All of these processes may mean that rockfalls continue, sporadically, for thousands of years after deglaciation.

Fluvial activity on steep valley sides means that debris flows are the main process by which material in lateral moraines and glacial tills are resedimented towards the valley bottoms during glacier retreat. The availability of water during this period may be enhanced by the melting of ice masses which were decoupled from the main mass of the glacier and buried under morainic material.

On many glacier forelands, the steep flanks of moraines are subject to periglacial processes such as solifluction, frost sorting and frost creep, and, over time, these reduce slope angles on landforms made of unconsolidated material. Strong katabatic winds from ice masses mean that aeolian processes are often very effective on glacier forelands and may winnow out fine material from tills and deposit these as sand sheets downvalley.

The rivers draining glaciated catchments also display 'paraglacial' behaviour. Glacier melting liberates vast amounts of water and sediments into the fluvial system and this may respond by rapid aggradation of the floodplain or incision. In addition, the rivers may also change their planform and respond to increased sediment supply by braiding or varying their gradient.

This period of rapid change can therefore be viewed as an adjustment of various geomorphological systems to changing stress levels, and a move towards equilibrium with newly imposed climatic and environmental conditions. It also is a good example of the idea of a 'complex response' whereby geomorphological systems may respond in very different ways and over different timescales to the same system perturbation.

From the research carried out on paraglaciation we can identify a number of important findings. First, the model of sediment supply to valley bottoms suggests that rapid readjustment of the system occurs immediately following deglaciation. Over time, the system reaches towards equilibrium and stability, either because all the available sediment has been reworked or the intensity and frequency of processes operating after deglaciation have declined. This period of time is known as the paraglacial period and may last from a few years to thousands of years. In certain high mountain temperate environments where glaciers are retreating, debris flow activity during paraglaciation ceases after ten years or so; in areas where steep bedrock cliffs remain the legacy of Pleistocene glacial activity, rockfalls associated with paraglacial readjustment may continue 20,000 years or so after glacial retreat.

Second, this model of paraglaciation is backed up by numerous studies from mountain environments where deglaciation is occurring or has recently occurred. An example of such a study is the one carried out on the flanks of the San Rafael glacier in southern Chile (Harrison and Winchester 1997). At the moment the glacier is flowing down a narrow, steep-sided valley and has been retreating and downwasting since the end of the nineteenth century. More recently, this retreat has accelerated and, since 1975, the surface of the glacier has downwasted some 100 m; this has exposed unstable lateral moraines to sub-aerial processes. The most important of these processes are debris flows, which have acted to redistribute the morainic sediments and deposit them downslope as debris flow cones. Dating of cone deposits along the flanks of this glacier by a combination of dendrochronology and lichenometry has shown that cone development, stabilization and fluvial incision occurs within 15 years or so of glacier retreat. This is a remarkably rapid readjustment of the geomorphological system and is achieved by a combination of the intense nature of fluvial activity associated with very high rainfall occurring in this region and by the fast rate of vegetation colonization which stabilizes cone surfaces.

Third, such studies which highlight the efficiency of paraglaciation have allowed geomorphologists to reinterpret landscapes that underwent glaciation and deglaciation during the Pleistocene. For instance, many of the large valley-bottom drift sheets and terraces that are common in the upland areas of Britain were originally interpreted as periglacial solifluction deposits which had accumulated following large-scale frost shattering of bedrock and the subsequent movement of the material downslope. In areas of the British uplands which had been covered by glacier ice during the last glaciation (around 26,000 to 18,000 years BP) these 'solifluction sheets' were assumed to have mainly formed during the short-lived Loch Lomond Stadial between 11,000 and 10,000 years BP, when permafrost was widespread in the British Isles. This interpretation therefore explains the geomorphological evolution of these landforms as being dominated by the slow accumulation and movement of material in valley bottoms; a process which must have taken several thousand years to accomplish. We can call this hypothesis the 'periglacial solifluction' hypothesis.

The paraglacial concept has allowed an alternative hypothesis to be developed (Harrison 1996) in which the sediments were laid down very rapidly by debris flow processes shortly after deglaciation. If this is correct, then the environment during which sedimentation occurred may have been relatively warm and the period of deposition occurred several thousand years before that implied by the periglacial solifluction hypothesis.

At the moment, it is not clear which of these two hypotheses is correct, although the available sedimentological and (limited) dating evidence that we have favours the paraglacial hypothesis.

Fourth, if the predictions of rapid future global warming are right, then we may experience a period of enhanced glacier retreat in many alpine and mountain regions. Our understanding of how mountain systems have responded to such retreat in the past allows us to predict with some accuracy their likely response in the future. The implications of renewed paraglaciation for economic and other human activity in mountainous regions are enormous. For example, increased sediment supply to mountain rivers by the destabilization of lateral and terminal moraines will adversely affect mountain hydroelectric power (HEP) schemes and silt up reservoirs. Instability of rock faces will make cable car stations vulnerable to collapse and catastrophic rockfalls, ice avalanches and rapid draining of moraine-dammed lakes will probably kill thousands of people. The last-named is a major hazard in the Peruvian Andes, where such outbursts are called *alluviones*, and their behaviour serves as a model and a forecast of future glacier behaviour in other mountain systems.

Conclusions

In conclusion, we can identify perhaps two key questions. First, given that renewed paraglaciation will accompany future glacier retreat, how will we accommodate such geomorphological instability into our models of economic and social use of mountain regions? Second, how far can the paraglacial model be used to interpret the geomorphology of mountain landscapes that were last glaciated during the Pleistocene? It may be that many of the landforms and sediments that were formerly interpreted as being periglacial in origin (e.g. rock glaciers, talus slopes, rockfalls) may be best interpreted as being the result of a period of intense paraglaciation. Such a reinterpretation has been attempted for upland 'solifluction sheets' as described earlier; the focus of research has shifted to these other landforms and, again, it will be accurate dating that may hold the key. The adoption of a new model of landscape development would be a radical departure and would help to downgrade the importance of periglaciation as a landscape-forming set of processes in previously glaciated areas.

References

Church, M. and Ryder, J.M. 1972: Paraglacial sedimentation: a consideration of fluvial processes conditioned by glaciation, *Geological Society of America, Bulletin*, 83, 3059–71.

Evans, S.G. and Clague, J.G. 1994: Recent climatic change and catastrophic geomorphic processes in mountain environments, *Geomorphology*, 10, 107–28.

Harrison, S. 1996: Paraglacial or periglacial? The sedimentology of slope deposits in upland Northumberland. In S. Brooks and M. Anderson (eds), *Advances in Hillslope Processes*, Chichester: Wiley, 1197–218.

Harrison, S. and Winchester, V. 1997: Age and nature of paraglacial debris cones along the margins of the San Rafael glacier, Chilean Patagonia, *The Holocene*, 7, 481–7.

Ryder, J.M. 1971: The stratigraphy and morphology of paraglacial alluvial fans in south-central British Columbia, *Canadian Journal of Earth Sciences*, 8, 279–98.

Further Reading

Ballantyne, C.K. and Harris, C. 1994: *The Periglaciation of Great Britain*. Cambridge: Cambridge University Press. This book provides a review of some of the landforms and sediments associated with Quaternary paraglacial and periglacial processes in Great Britain.

Benn, D.I. and Evans, D.G. 1999: *Glaciers and Glaciation*. London: Arnold. This is widely regarded as the most comprehensive textbook on glacial geomorphology currently available.

10

Coastal Environments: Geomorphological Contributions to Coastal Management

Peter W. French

There cannot be many people who have not visited the coast at some time in their lives. Coasts are, after all, the staple destination of many a package holiday. However, the perception of the coast as somewhere idyllic to live or spend leisure time is fundamental to a much wider contemporary issue in coastal management. People who visit the same stretch of coastline year after year may notice that it changes with each visit; for example, there may be more sediment on the beach in some years than others, the coast may have eroded back inland, or the beach profile may have got shallower or steeper. These changes reflect the dynamic nature of the coastline and highlight one of the fundamental conflicts between the coast as a natural, functioning physical environment and as a place where people want to live, play and work. The morphological processes causing these changes are becoming increasingly important to coastal scientists because understanding them can offer valuable information in the management of coastlines. Increasingly, coastal scientists including geomorphologists are contributing valuable knowledge to our overall understanding of how the coast 'works'. This

has been achieved by increased monitoring and modelling of coastal processes and change.

The terms 'coast' or 'coastline' cover a wide range of landscapes, including beaches, dunes, cliffs and salt marshes. The coastal geomorphologist may work on a few or all types, but some, such as beaches and cliffs, have received much attention in recent years because their dynamic nature has increasingly brought them into conflict with human activity and development. Coasts, like any physical environment, strive to achieve and maintain a state of dynamic equilibrium with the natural forces that shape them (typically, these may be referred to as 'forcing factors'). These forcing factors are many but include such processes as storms, rising sea level and changes in wave activity. If we look at a simple formula, we can start to understand how this occurs:

Total wave energy = energy used in erosion + energy used in sediment transport

This equation is much simplified. The purist would argue that waves can lose energy in other ways, such as through friction at the sediment–water interface, for example. However, for our purposes, consider a sandy beach backed by a cliff. If this beach receives a lot of sediment from up-drift, then waves hitting it would expend the majority of their energy transporting sediment along the beach and not reach the cliff. Thus, the erosion aspect of the above equation would be zero. Here we have a beach that is gaining sediment which compensates for that transported by waves. Hence, it does not pose any management problem all the time the beach is receiving new sediment and is in a dynamic equilibrium with the wave climate.

Now consider the same beach, but under conditions of increased wave activity and where sediment supply cannot compensate for that transported by waves. Here, waves have surplus energy after removing sediment. This is expended on erosion of the cliff. Put another way, the coast is not in dynamic equilibrium with its wave climate and so tries to adjust to a form which is stable. This adjustment process is commonly known as erosion.

Given that many of our coastlines today are examples of the latter situation, coastal managers are faced with the problem that many coasts are adjusting in response to factors such as increased wave activity and rising sea levels (Dolotov 1992 gives a good account of these issues). These processes are resulting in great changes in coastal form, such as cliff and salt marsh erosion. This then poses a key question:

If erosion is a case of unstable coastlines trying to adjust so as to regain a dynamic equilibrium, shouldn't this be allowed to happen?

The simple answer to this is yes, because once a coastline has achieved equilibrium, it will stop eroding, i.e., become stable. However, it is at this point that we return to our opening paragraph. The key problem is that because people like to use the coast as much as they do, this natural process of adjustment often impinges on human activity,

and so there is great pressure to prevent it from happening. In other words, we need to defend our coasts to oppose these erosive forces. The typical human response to stop erosion from happening is by some form of engineering, such as sea walls, offshore breakwaters, etc. However, given that coasts respond to changes in forcing factors, one sure way of exacerbating such changes is to build a structure along the coast that may change them still further.

A Major Decision: To Defend or Not to Defend a Coast?

One of the real issues in coastal management at present is formulating a response to the above observations. On the one hand, global warming is producing ice-cap melting and ocean expansion leading to rising sea levels, and also changes in atmospheric circulation and weather patterns that are leading to increased storminess. Thus, our coasts are trying to adjust to both of these factors (hence the prevalence of erosion). Logic tells us that we should allow coasts to adjust naturally and not try to stop it. Counter to this, however, is the view that in socio-economic terms, erosion means the loss of land which is owned by someone or has an income-generating potential.

In part, coastal defence planners have responded to this dilemma by adopting some of the 'softer' methods, such as beach feeding. Beach feeding is, in essence, artificially supplying sediment to a beach where waves are removing more than can naturally be replaced (see the example above). This has meant that as a method of coastal defence, beach feeding has become very popular, but it is still not without its problems or its detractors (see Davison et al. 1992 for a review).

Another possible response to a local population's calls for coastal managers to react to erosion issues is to 'do nothing'. Increasingly, doing nothing is a recognized response to erosion and is considered in every cost-benefit analysis (CBA) and environmental impact assessment (EIA) carried out in the planning stages of a coastal defence scheme. In effect, this is the baseline situation, i.e., what is predicted to happen to the coast if it is left alone. It may well be that following CBA and EIA, the preferred management solution is to opt for the baseline situation and leave the coast alone. This solution may be chosen even though it means that houses may fall into the sea. Such a decision is guaranteed to cause confrontation between coastal managers and the local population, particularly if people's homes and livelihoods are at stake.

Given the importance of decisions of this kind, therefore, recent developments have addressed the need for reliable ways of predicting a coastline's response. This increased attention has required two components. First is an increase in reliable monitoring relating to how a coast behaves. Each coast responds differently to the range of forcing factors described above, and so each needs to be assessed individually for its response to increased sea level and storm activity. Second, these data need to be applied to coastal models which predict how coastal landforms will respond. Taking historic trends in erosion, and by comparing these with changes in the forcing factors that caused them, it becomes possible to predict future trends. For example, French (1993) and French et al. (1995) have modelled changes in vertical accretion rates across the surface of salt marshes in response to sea-level rise, whilst European and other international initiatives have produced a range of models for predicting the response of beaches to sea-level rise and changes in sediment input.

A further development in coastal modelling gives us the ability to predict future coastal positions, and thus gain an insight into how a stretch of coastline is responding to erosive forces. By knowing historic rates of erosion and the conditions under which this occurred (i.e., a given rate of sea-level rise), it is possible to model the response of

a coastline to future sea-level rise. Bray and Hooke (1997) developed these ideas into a predictive model for soft cliff retreat, but it is also generally applicable to the retreat of any uniform lithology/sediment body, such as a dune front or salt marsh. In its simplest form, the model predicts future rates of erosion (R_2) on the basis of historical retreat (R_1) experienced under a given rate of sea-level rise (S_1) and extrapolated into the future using predictions of future sea-level rise (S_2), according to the formula:

$$R_2 = (R_1 / S_1) \times S_2$$

Given that sea-level rise estimates are generally made with reference to a time, i.e., 20 mm by 2020, then integration between the present and this future time will provide an estimate of total cliff recession between the present and, in this case, 2020. Furthermore, any date in the future up to that delineated by the future sea-level rise prediction may be used, thus giving a series of retreat figures at different time intervals. These can be plotted on a map to show the predicted coastal positions at preselected time intervals. Such an approach can prove invaluable in deciding the type and timing of future defence strategies.

Figure 10.1 provides a hypothetical example. Using the method described above, four predictive cliff positions have been plotted for 25, 50, 75 and 100 years into the future. Following calls from the local population for the erosion problem to be tackled, coastal managers can use such a map as the basis for making a decision. From the map in figure 10.1, it is possible to argue that based on predictions, none of the cliff-top development is likely to experience any effects of the erosion for at least 25 years, and so it is feasible to 'do nothing' for this period. This approach has the advantage of maintaining the sediment input derived from the erosion of the cliff to the coastal sediment budget. After 50 years, however, it is predicted that development (a) will have fallen into the sea and so at some time prior to this, a decision has to be made as to whether defences become necessary, or whether this development is expendable. Clearly, the advantages of this approach are that it allows a flexible and staggered response over time. For example, if development (a) is a factory, then it may not still be viable at this future time and is thus surplus to requirements. If development (a) is then left to fall into the sea, the next predicted problem occurs shortly prior to 75 years, when not only development (b) becomes threatened, but so does the main coastal road. Clearly, this raises more significant management issues because it is not only individual buildings that are at stake, but local infrastructure is threatened as well.

This approach is increasingly being used in the real world of coastal management. In the UK, Waveney District Council in Suf-folk has used this predictive method as part of its future coastal planning initiative. It has identified and marked on a planning map a predicted coastal position for 2068 (75 years after policy implementation) and uses it as a planning guideline with respect to future coastal development. Geographically, this lies between 150 and 190 m inland of the 1993 coastline. A similar approach was used on the Sussex coast, at Fairlight Cove. Penning-Rowsell et al. (1992) illustrate this example with the construction of seven-year retreat line intervals for the village of Fairlight. This example of predictive modelling was used to argue for the construction of coastal defences to protect the village from an increasing rate of erosion. In this case, it led to the construction of an offshore breakwater fronting the cliffs.

The limitations of this approach, however,

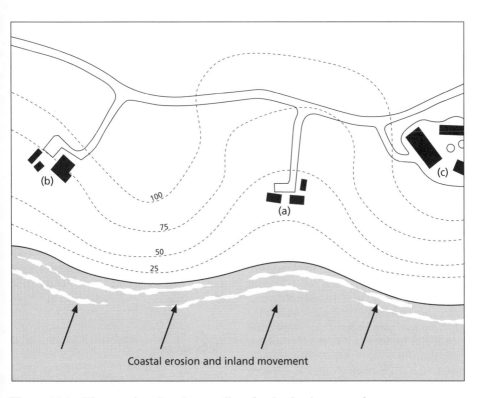

Figure 10.1 The use of predicted retreat lines for the development of a management strategy for cliffed coasts (French 2001).

need to be realized. It is critical to remember that we are dealing with predictions, and that these predictions are only as reliable as the model and data on which they are based. Clearly, the length of time covered by the dataset is critical as it forms the basis of predicting future response. One real issue with many coastlines is that these data are simply not available. This represents a major gap in our studies of coastal processes and geomorphological response to forcing factors. If we are to gain a better understanding of our coasts, it is vital for the future that there is accurate monitoring, and much more of it. Another factor is the model's reliability. The model used above is simplistic in its approach in that it assumes a constant lithology/sediment type and that sea level is the only driving force in coastal erosion. Clearly, other factors may be important, and Bray and Hooke (1997) do outline more complex models. However, we are still only just leaving the starting blocks in terms of accurate coastal modelling, and methods are developing rapidly. To achieve precise predictions, we need more monitoring data and more accurate models. This is the task facing the next generation of coastal geomorphologists.

Final Thoughts

Coastlines will continue to adjust to changes in forcing factors, regardless of human intervention. The operation of a coastal management policy that recognizes this position rather than being antagonistic to it will produce coastal protection with minimum impact elsewhere. The recent move towards soft engineering is a well-documented trend, which argues in favour of working with rather than against natural processes. Whilst this is a major step forward in the sympathetic management of the coastline, the techniques involved still produce a range of impacts. More recently, however, there has been a further shift to facilitate the desire to leave the coast alone whenever possible. This has

unfortunate consequences when it comes to arguing that homes should be allowed to fall into the sea. However, this should not detract from the management arguments for carrying out such a process, because leaving the coast to react to changing conditions naturally must be the best solution wherever the hinterland usage/value is not too prohibitive.

By using modelling to predict future coastline positions, it is possible to modify coastal defence practices and possibly to delay them for some time in the future. Such a process allows the free movement of sediment from the eroding part of the coastline for as long as possible.

In the preceding discussion we have considered cliff position, but equally such methods could be used to investigate other coastal problems such as landward dune migration, which could impinge on infrastructure, or salt marsh retreat, which might cause flood defence problems in future years. However, by recognizing and promoting the potential of abandoning land, it may well be that coastal planners and developers start to realize the implications of some of their work, and to adjust accordingly. This will, therefore, begin to remove the reason for coastal defence – inappropriate coastal development. Such a task is ideally suited to the geographer. It is, after all, the physical geographer who will continue to play a key role in the provision of baseline monitoring data, and in increasing our understanding of coastal behaviour and response to changing forcing factors. In addition, there could equally be a role for the human/social geographer to provide increased awareness of the human and social cost of some of the decisions made.

References

Bray, M.J. and Hooke, J.M. 1997: Prediction of soft cliff retreat with accelerating sea level rise, *Journal of Coastal Research*, 13, 453–67.
Davison, A.T., Nicholls, R.J. and Leatherman,

S.P. 1992: Beach nourishment as a coastal management tool: an annotated bibliography on the developments associated with artificial nourishment of beaches, *Journal of Coastal Research*, 8, 984–1022.

Dolotov, Y.S. 1992: Possible types of coastal evolution associated with the expected rise of the world's sea level caused by the greenhouse effect, *Journal of Coastal Research*, 8, 719–26.

French, J.R. 1993: Numerical simulation of vertical marsh growth and adjustment to accelerated sea level rise, North Norfolk, U.K., *Earth Surface Processes and Landforms*, 18, 63–81.

French, J.R., Spencer, T., Murray, A.L. and Arnold, N.S. 1995: Geostatistical analysis of sediment deposition in two small tidal wetlands, Norfolk, UK, *Journal of Coastal Research*, 11, 308–21.

Penning-Rowsell, E.C., Green, C.H., Thompson, P.M., Coker, A.M., Tunstall, S.M., Richards, C. and Parker, D.J. 1992: *The Economics of Coastal Management: A Manual of Benefit Assessment Techniques*. London: Belhaven Press.

Further Reading

Carter, R.W.G. 1988: *Coastal Environments: An Introduction to the Physical, Ecological and Cultural Systems of Coastlines*. London: Academic Press.

French, P.W. 1997: *Coastal and Estuarine Management*. London: Routledge.

French, P.W. 2001: *Coastal Defences: Processes, Problems and Solutions*. London: Routledge.

11
Dryland Environments: Changing Perceptions of Dynamic Landscapes

David J. Nash

Drylands are Important Environments

On the face of it, studying the geography of drylands may seem a rather, well, dry topic. This is despite the fact that drylands occupy over a third of the Earth's land area, making them the world's largest climato-vegetational zone. For much of history drylands have been considered areas to avoid because of their climatic extremes and the lack of surface water and food. This perception has been maintained to the present day, so that the popular image of the world's drylands is a largely negative one of vast, dry, empty expanses of sand, searing daytime temperatures, drought, poverty and famine, with people helpless in the face of runaway desertification. Indeed, this type of view appears regularly even in relatively recent geographical literature. Possibly the greatest challenge facing researchers today is to overcome some of these stereotypes and move away from the more traditional views of dryland regions. Far from being barren wastelands, drylands contain some of the Earth's most distinctive, beautiful, dynamic and regular landforms, and are host to a variety of intricate ecosystems. From a human perspective, drylands are not under-

populated wildernesses but are home to around one-fifth of the world's population, a high proportion of whom live in large urban settlements in addition to those who live in rural communities. A key goal of research in these areas needs to be to gain a better understanding of the operation of landscape processes so that we can understand how such processes impact upon human populations and propose more suitable sustainable policies to support dryland populations.

The Variety of Dryland Climatic Conditions

Some of the greatest shifts in our understanding of dryland environments have come in the past two decades, during which time the dynamic nature and diversity of environmental conditions, especially climatic conditions, in these areas have become recognized. The world's drylands are often classified on the basis of average annual rainfall levels into hyperarid (under 25 mm rainfall per year), arid (25–250 mm) and semi-arid (250–500 mm), but this is misleading as it masks the inherent climatic variability that we now know is a feature of these regions. Hyperarid areas, for example, have very high interannual rainfall variability, often receiving no precipitation for periods of more than 12 months, with the degree of variation decreasing in arid (less than 50 per cent interannual variability) and semi-arid (25 per cent variability) areas. The use of average rainfall figures also overlooks the importance of factors such as the timing and type of precipitation, and the levels of evapotranspiration, in determining moisture availability. Parts of the semi-arid Kalahari Desert, for example, receive up to 600 mm rainfall over the course of a year (wetter than many areas of the UK), but much of this occurs in distinct storms during summer months when potential evapotranspiration rates may exceed 1,500 mm. Furthermore, not all precipitation in arid regions occurs as rainfall, with sea fog and dew supplying much of the moisture to the dry coastal Atacama and Namib deserts, and snowmelt contributing water to high-altitude continental deserts such as the Chihuahuan.

Arid regions were once perceived as being hot all year round. However, we now know that this is not the case, with annual temperature regimes varying considerably due to altitude, latitude and location. Forty-three per cent of the world's deserts are hot deserts, where the warmest month has an average temperature of less than 30°C and the coldest of 10 to 30°C. Eighteen per cent are mild deserts (warmest month 10–30°C; coldest month 10–20°C), 15 per cent are cool deserts (warmest month 10–30°C; coldest month 0–10°C), with the remaining 24 per cent termed cold deserts (warmest month 10–30°C; coldest month under 0°C) (Thomas 1997). In the hot tropical and subtropical deserts, extreme temperatures of over 50°C may occur, giving rise to soil surface temperatures in excess of 80°C, whilst nighttime temperatures may drop below 10°C. In contrast, some high-altitude continental arid regions have winter temperatures considerably below freezing.

Recent Geographical Research in Drylands

The result of this increased understanding of the variability of climatic conditions, mainly arising from increased investment into drylands research and better instrumentation and monitoring, is reflected in much of the cutting-edge work being undertaken by geographers today. In physical geography, the recognition of spatial and temporal variations in the availability of moisture has had a major impact on our appreciation of the operation of geomorphological processes in drylands. Early studies of desert weathering processes, for example, tended to overemphasize the importance of temperature changes upon rock breakdown, mainly because of the perceived lack of moisture to drive chemical and biological weathering mechanisms. However, much research carried out today, both through laboratory simulation experimentation and field monitoring, places greater importance on the role of moisture in weathering environments. We now know that moisture can be available on a short-term basis even in some of the driest parts of the world as a result of inputs from dew or fog (Eckardt and Schemenauer 1998). The potential significance of this can be seen through a comparison of recent work on rates of slope retreat, largely controlled by the rate of weathering, on granitic inselbergs in the Namib Desert (Cockburn et al. 1999) and similar inselbergs in South Australia (Fleming et al. 1999). Rates of slope retreat are an order of magnitude higher in the Namib, most likely due to the frequent occurrence of salt-bearing coastal fog which increases the potential for the operation of salt weathering processes (Goudie and Parker 1998).

Similar shifts in perception have taken place in other areas of dryland geomorphology. The dynamics of desert fluvial systems were, until relatively recently, poorly understood in comparison with their temperate counterparts. This was due in part to the difficulties of measuring discharge and sediment transport in desert streams during flood events, but also to the chance of actually being in the right place at the right time when a flood occurred. What research was carried out was often qualitative or focused upon general catchment characteristics, with little detailed understanding of dryland floods despite the potential hazard they posed to infrastructure and human life. However, there are now a number of countries, including Israel and the United States, where detailed hydrological monitoring of dryland river catchments takes place using extensive networks of instrumentation. This monitoring has revealed exciting results about the controls upon flooding and the work done by desert fluvial systems. For example, the myth that all floods in drylands are enormous flash floods has been largely dispelled, with recent work in the Negev Desert revealing considerable variability of floods in terms of their 'flashiness' and magnitude (Reid et al. 1998). We now also know that dryland rivers, particularly those with gravel-beds, are capable of transporting considerably more bedload sediment for given ranges of shear stress on the river bed when compared to those from more humid regions (Tooth 2000). Studies in the Negev suggest that bedload transport rates at the threshold of entrainment when a flood passes over a dry river bed may be up to a million times higher than in a similar size temperate river (Laronne and Reid 1993).

Human and Physical Geographers Work Together in Drylands

The importance of climatic variability in drylands is also being increasingly recognized in areas of human geography, particularly those concerned with development issues. One of the most commonly cited problems said to be affecting the world's drylands is that of desertification. There are literally hundreds of definitions of desertification, but in its broadest sense it is the process of

land degradation in drylands, usually linked to the mismanagement of land and use of inappropriate technologies by local populations coupled with climatic variability. The term has almost reached the status of 'institutional fact' within organizations such as the United Nations, despite a considerable degree of scientific scepticism that desertification as a process actually exists (Thomas and Middleton 1994). The long-held notion that deserts are expanding and claiming large areas of potentially arable land has not yet been shown to be true. Studies of a ten-year time series of satellite imagery of the margins of the Sahara in Sudan, for example, show that vegetation cover (the lack of which is often, incorrectly, taken as an indicator of desertification) is very closely correlated with patterns of annual rainfall. Over the study period, the desert boundary oscillated north and south by up to three degrees of latitude, with the most extreme fluctuation in a single year occurring in 1984–5 when the boundary shifted by 110 km as a result of increased rainfall (Tucker et al. 1991). Desertification is usually associated with adverse human activities such as overgrazing, soil degradation and woodland clearance but, as this example shows, shifts in the distribution of the boundaries of the world's drylands can occur rapidly without the influence of humans. If the world's deserts were advancing at the rates proposed in the 1970s and 1980s, then the Sahara should have engulfed the entire Sahel by now. Similarly, if there has been overgrazing and soil degradation for several decades, we should have seen the complete collapse of dryland agricultural economies a long time ago. This is clearly not the case, even though it may appear to be so during severe drought episodes. Much of the ground-breaking human geography research into issues such as dryland management is now moving away from global-scale studies with an emphasis on the mapping and prediction of locations of environmental change (often based upon poor understanding of processes) towards a more local-scale focus (Mortimore 1998).

These local-level studies seek to involve local populations in management rather than imposing expert-led new technologies and systems upon communities to prevent land degradation (van Rooyen 1998). This change reflects the recognition that indigenous land-use practices are often much more specific and appropriate to local conditions than externally developed management systems, effectively reversing the previous view that all traditional approaches were inappropriate and damaging to the environment.

Overall, research into all aspects of the geography of dryland environments is undergoing a shift in perceptions. In physical geography this is a move away from traditionally held views that drylands are climatically static to a more holistic view of the importance of climatic variability on daily, seasonal, annual and decadal timescales for the operation of geomorphological processes. In human geography, perceptions of drylands as wilderness areas are being replaced by notions of drylands as landscapes under transition, with a greater need to consult local communities in their management. Perhaps most significantly, however, there has been an increased convergence of physical and human geography research activity in dryland regions. This is especially the case in relation to issues such as land degradation and desertification, which encompass both human and 'natural' environmental changes and require understanding of scientific and societal processes. This type of collaboration and cooperation is relatively unusual within the sphere of geography but represents the best way forward for the effective and sensitive management of such fragile environments. Long may it continue.

References

Cockburn, H.A.P., Seidl, M.A. and Summerfield, M.A. 1999: Quantifying denudation rates on inselbergs in the central Namib Desert using in situ-produced cosmogenic [10]Be and [26]Al, *Geology*, 27, 399–402.

Eckardt, F.D. and Schemenauer, R.S. 1998: Fog water chemistry in the Namib Desert, Namibia, *Atmospheric Environment*, 32, 2595–9.

Fleming, A., Summerfield, M.A., Stone, J.O., Fifield, L.K. and Cresswell, R.G. 1999: Denudation rates for the southern Drakensberg, SE Africa, derived from *in-situ*-produced cosmogenic [36]Cl: initial results, *Journal of the Geological Society*, 156, 209–12.

Goudie, A.S. and Parker, A.G. 1998: Experimental simulation of rapid block disintegration by sodium chloride in a foggy coastal desert, *Journal of Arid Environments*, 40, 347–55.

Laronne, J.B. and Reid, I. 1993: Very high bedload sediment transport in desert ephemeral rivers, *Nature*, 366, 148–50.

Reid, I., Laronne, J.B. and Powell, D.M. 1998: Flash-flood and bedload dynamics of desert gravel-bed streams, *Hydrological Processes*, 12, 543–57.

Tooth, S. 2000: Process, form and change in dryland rivers: a review of recent research, *Earth-Science Reviews*, 51, 67–107.

Tucker, C.J., Dregne, H.E. and Newcomb, W.W. 1991: Expansion and contraction of the Sahara Desert from 1980–1990, *Science*, 253, 299–301.

van Rooyen, A. 1998: Combating desertification in the southern Kalahari: connecting science with community action in South Africa, *Journal of Arid Environments*, 39, 285–97.

Further Reading

Mortimore, M. 1998: *Roots in the African Dust: Sustaining the Drylands*. Cambridge: Cambridge University Press.

Thomas, D.S.G. (ed.) 1997: *Arid Zone Geomorphology*, 2nd edition. Chichester: Wiley.

Thomas, D.S.G. and Middleton, N.J. 1994: *Desertification: Exploding the Myth*. Chichester: Wiley.

12

Environmental Modelling

Stuart Lane

In November 2000, many of the United Kingdom's floodplains were under water. This event was associated with an unusual weather pattern that resulted in a series of occluded fronts stalling across the country, and produced intense rainfall. The floods were identified by a range of organizations as symptomatic of a changing climate, attributed in this instance to the hypothesized effects of greenhouse gas emissions upon the climate system. How do we prove that this hypothesis is true? For a number of reasons, unfortunately, we cannot test this hypothesis using conventional experimentation. The complexity of the global environmental system, and the long timescale and large spatial scales over which human changes to the climate will be identified, mean that simple laboratory experiments based upon parameter manipulation (e.g. studying the energy balance in a closed box with varying greenhouse gas levels) are both meaningless and impossible. The second option is to investigate records of current climate during the period of human influence upon the environment. Again, this leaves us with uncertainty. The magnitude of observed temperature change, globally averaged, is still significantly less than temperature changes estimated from environmental reconstruction over the last million years. Observed changes are much lower than those associated with natural environmental variability (see chapter 10 by Peter French, this volume). By the time that a

greenhouse gas-induced climate 'signal' can be clearly disentangled from natural environmental 'noise', delays in climate response to greenhouse gas emissions are likely to have made the problem extremely serious. Environmental models represent the major tool for addressing the limitations of laboratory experimentation and environmental records in allowing us to extend the bounds of space and time in our environmental understanding.

Conceptual Models

The model can best be defined as some form of abstraction of reality: something that never contains the full complexity of the real world, but which has key characteristics that make the model, and model behaviour, representative of the real world. Some (e.g. Huggett 1993) would describe it as the process by which reality is reduced to manageable proportions. In practice, there is a hierarchy of models. The simplest type of model to understand, but sometimes the hardest model to develop, is a conceptual model. This is simply a statement of what the entities are in a model, and the way that they link with each other. An example is shown in figure 12.1 for the case of the ecology of a shallow eutrophic lake (after Scheffer 1998). This demonstrates the connections between key entities, and is the first step in deciding what entities, and hence

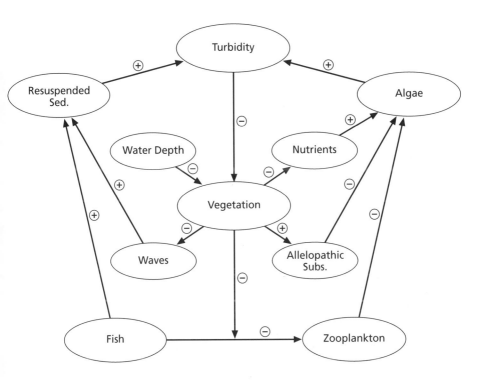

Figure 12.1   A conceptual model of interactions between nutrient loading and ecological processes for shallow lakes. The traditional emphasis upon eutrophication has involved the positive relationship between nutrient loading and algal growth. However, this diagram also illustrates how algal growth is influenced by both allelopathic substances (produced by aquatic vegetation) and zooplankton grazing, both of which reduce algal levels. It also shows how algal growth increases the effective turbidity of the lake. These links then involve feedbacks that can be traced around through these parameters. For instance, if algal growth results in more turbid water, the amount of bed vegetation growth in the lake will fall as light will no longer penetrate as far into the water column. Thus, more nutrients will be released as the binding effects of roots are lost, and allelopathy will be reduced. Both of these encourage more algal growth. This diagram provides a simple conceptual illustration of core interactions. Unfortunately, actual interactions are much more complex than this and are often ecosystem or even species specific, as a result of the evolution of biological organisms to cope with ecosystem interactions. These are reviewed in Lau and Lane (2001).

what processes, need to be included in the model. It is at this stage that the first assumptions are made in the modelling process: i.e., we decide what to include because it is important, and what can be assumed to be unimportant and hence not included.

Mathematical Models: Empirical

Mathematical models based upon empirical approaches involve making a set of observations of a number of phenomena and then using these observations to construct relationships amongst them. This commonly

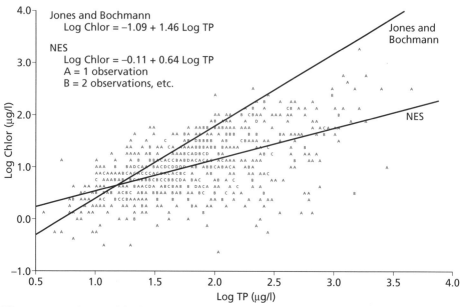

Figure 12.2   An empirical model based upon regression analysis: the relationship between chlorophyll content of a water body and nutrient loading (in this case, total phosphorus). This shows two regression relationships, demonstrating the fact that empirical models tend to be uncertain in terms of their generality. Do they apply beyond the time period of data upon which they are based? Do they apply for different locations? There is also considerable scatter about the relationship.

uses statistical methods and is based upon the assumption that one or more forcing (independent) variables cause changes in a response (dependent) variable. A good example of this is the eutrophication of shallow lakes. Eutrophication involves the progressive increase in primary productivity of a water body. Research has shown that primary productivity in lakes is limited by the availability of soluble phosphates, as these are required for photosynthesis. Thus, empirical relationships (e.g. Stefan 1994) between phosphate loading and level of eutrophication (e.g. chlorophyll-a concentration) have been developed, such that for a given phosphate loading it is possible to predict the probable level of eutrophication (figure 12.2). This relationship can be developed in one of two ways: (1) by observing many different lakes or rivers at the same point in time; or (2) by observing the relationship between phosphate concentration and eutrophication level at a lake through time. However the relationship is constructed, it is important to remember that empirical relationships involve statements of uncertainty. The scatter in the data in figure 12.2 means that for a given phosphate loading there is a range of possible levels of eutrophication. The greater the scatter, the greater the uncertainty. This is a very simple example of an empirical model. More sophisticated empirical models allow for greater than one forcing variable and more complex variable interactions.

The use of an empirical approach to modelling requires a number of assumptions. First, the empirical relationship must have a strong theoretical basis. Whilst the use of statistical methods allows assessment of how good the model is (e.g. by assessing the goodness of fit of the empirical relationship), this is not normally a sufficient test of a model: empirical relationships may be spurious, and hence have a poor predictive ability. Second, many empirical models perform poorly when used to predict beyond the range of observations upon which they are based. For instance, See and Openshaw (2000) report on the use of neural network modelling for flood forecasting. This is based upon using artificial intelligence methods to construct an empirical relationship between forcing parameters (e.g. upstream rainfall) and the key response variable (e.g. water level). Research has shown that these models are only good at predicting water levels associated with patterns of forcing parameters that have happened before. There are two important implications of this problem: empirical models don't always hold through time; and they often transfer poorly to different places. At the root of these problems is the poor generalizability of empirical models, and it is often argued that this is because they do not necessarily have a good physical basis, as they are grounded upon statistical interactions rather than fundamental physical processes.

Mathematical Models: Deterministic

Deterministic mathematical models deal directly with the accusation levelled against empirical models that they have a poor physical basis. They are based upon fundamental physical, chemical and occasionally biological principles. At the scale of the natural environment, we are fortunate in having a number of key principles, largely deriving from Newtonian mechanics: (1) rules of storage; (2) rules of transport; and (3) rules of transfer. Rules of storage are based upon the law of mass conservation: matter cannot be created or destroyed, but only transformed from one state into another. As with mass conservation, rules of transport may also be based upon Newtonian mechanics, e.g. every body continues in its same state of rest or uniform motion unless acted upon by a force. Rules of transfer allow for the possi-

bility that chemical reactions cause a change in the state of an entity. For instance, phosphate bound to aluminium or iron may become soluble in a eutrophic lake, and hence available for fuelling eutrophication, if the environment becomes reducing.

Stages in the development of a mathematical model are shown in figure 12.3a and illustrated in figure 12.3b for the case of a model of eutrophication processes in a shallow lake (after Lau 2000). Figure 12.3 introduces a number of important components in model building. First, it demonstrates the importance of proper conceptualization of the model. This process involves deciding what to include and what to exclude. If you compare figure 12.3b with figure 12.1, you will see that a very large number of components have been ignored. This process is commonly called 'closure' and involves defining the boundaries of the system that the model will address. Ideally, all relevant processes will be included, and the closure will not exclude the processes that might matter. In practice, processes are excluded for two reasons: (1) if they don't matter for the particular system being studied; or (2) if there are limitations on the possibility of including a particular process. In the case of (1), it is common to include a simpler version of a process's effects. For instance, in the case of lake eutrophication modelling, the system may be driven by both nutrient limitations and food chain interactions. The latter are controlled by the presence of bottom-growing vegetation which act as refugia for zooplankton that graze upon algae. It is not necessary to model plant life cycles in most shallow lakes as these are relatively straightforward functions of seasonality. Thus, they may be dealt with using simple parameterizations (see below). In this case, we dealt with the effects of high levels of algal growth upon turbidity and hence vegetation loss by a simple rule which said that if the algal concentrations reached a certain level, then vegetation die-back would occur and there would be a sudden increase in the rate of nutrient release from the bed. Veg-

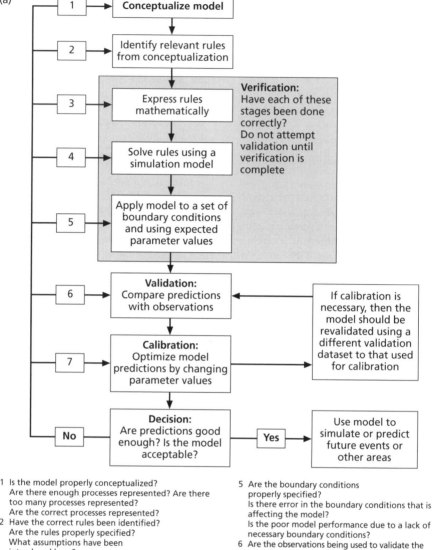

(a)

1   **Conceptualize model**

2   Identify relevant rules from conceptualization

3   Express rules mathematically

4   Solve rules using a simulation model

5   Apply model to a set of boundary conditions and using expected parameter values

Verification:
Have each of these stages been done correctly?
Do not attempt validation until verification is complete

6   **Validation:** Compare predictions with observations

7   **Calibration:** Optimize model predictions by changing parameter values

If calibration is necessary, then the model should be revalidated using a different validation dataset to that used for calibration

No   **Decision:** Are predictions good enough? Is the model acceptable?   Yes   Use model to simulate or predict future events or other areas

1 Is the model properly conceptualized?
Are there enough processes represented? Are there too many processes represented?
Are the correct processes represented?
2 Have the correct rules been identified?
Are the rules properly specified?
What assumptions have been introduced here?
Are they acceptable?
3 Is the mathematical expression of the rules correct?
Have any unsupportable assumptions been introduced during expression?
4 Are the rules solved correctly?
Is the computer programme free of programming error?
Is the numerical solver sufficiently accurate?
Is the model discretization robust?

5 Are the boundary conditions properly specified?
Is there error in the boundary conditions that is affecting the model?
Is the poor model performance due to a lack of necessary boundary conditions?
6 Are the observations being used to validate the model correct?
Are the observations equivalent to model predictions?
Are they representative of model predictions?
Where (in space and time) is the model going wrong when compared with observations and predictions?
7 Has the calibration process produced realistic parameter values?
Does the model produce more than one set of parameter values that are equally good at predicting when revalidating the model?
Is the model overly sensitive to parameterization?

Figure 12.3 A general approach to model development (a) and the conceptual model (b) applied to Barton Broad, Norfolk, England (Lau 2000). This shows that the model had three major components: the algal component (phytoplankton), a nutrient component (phosphorus) and an algal-grazing component (zooplankton). There were then interactions both within and between these three major components. For instance, phytoplankton are regenerated according to nutrient availability and die naturally. They are also grazed by zooplankton. As phytoplankton are generated, die and are grazed, so phosphorus is moved around the various components in which it can be stored.

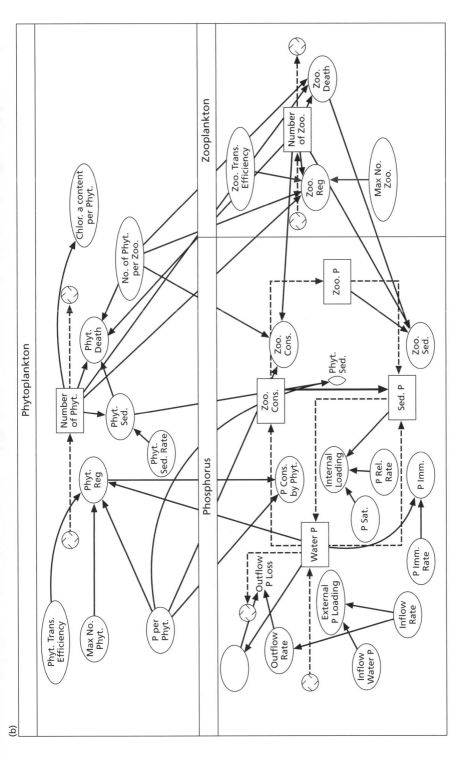

(b)

etation effects were not ignored but represented in a way that was appropriate to the parameters of interest within the model.

The second component of model building involves taking the conceptual model, identifying appropriate process rules and transforming these into a simulation model that can solve the equations. This can be the most difficult stage of model development as many equations do not lend themselves to easy solution. This is well illustrated by the case of predicting the routing of flood waters through a drainage network. The discharge entering the network varies as a function of time. In simple terms, the rate at which it moves through a part of the network depends upon the water surface slope in that part of the network (steeper slopes mean faster flows). Thus, there is a spatial dependence. Two problems emerge. First, the combined space-time dependence of flood routing means that the dominant equations are partial differentials (as they contain derivatives in time and space). This is common to almost all environmental models, as we are interested in how things move through space. Moving through space takes time, and hence all models should contain both space and time. Partial differentials are very difficult to solve. Second, all models require some form of initial conditions. In this case, we need starting values for water level and discharge throughout the drainage network. We will also need to know the inlet discharges. Hence, models have a crucial dependence upon the availability of data to initialize them.

Solving the governing equations commonly requires us to introduce parameters. This may be because during the conceptualization process, we chose to exclude certain processes, but we still need to represent their effects. In algal modelling, we considered the example of lake vegetation. In flood routing, we may choose to ignore lateral and vertical movements of water and flow turbulence, all of which affect the routing of discharge but whose inclusion in a model may make solution impossibly time-consuming if we are interested in drainage-network-scale flood routing. However, parameterization also results from the fact that whilst equations can have a good physical basis, this is insufficient to determine them directly. As the equations are solved, terms appear in the equation that cannot be determined directly (e.g. turbulence in a flood routing model) and parameters are introduced to represent these effects. Many of these parameters have a poor physical basis. They can be difficult to measure. Thus, parameter specification, or parameterization, is a central aspect of all modelling. The process of parameterization is rarely conducted independently from check data, in which the model is optimized by changing parameter values such that the difference between check data and model predictions is minimized.

Model assessment involves two important stages: verification; and validation. Verification is the process by which the model is checked to make sure that it is solving the equations correctly. This may involve debugging the computer code, doing checks upon the numerical solution process, and undertaking sensitivity analysis. The latter may be used to make sure that the model behaves sensibly in response to changes in boundary conditions or parameter values. Validation is the process by which a model is compared with reality. This normally involves the definition of a set of 'objective functions' that describe the extent to which model predictions match reality (see Lane and Richards 2001). A good example of an objective function is the root mean square error (εRMS):

$$\varepsilon RMS = \frac{\sum_{i=1}^{n} (p_i - o_i)^2}{n-1}$$

where p_i is the value predicted by the model; o_i is the observed value; and n is the total number of observations available.

Once an objective function has been determined, a model may be developed to reduce the magnitude of the error defined by the objective function. This may involve changing parameters or checking boundary conditions through the process of optimization described above. It may involve a more radical redevelopment of the model through the incorporation of new processes or alternative treatment of existing processes. Figure 12.4 shows the default (a), optimized (b) and predicted (c) values of chlorophylla concentrations obtained for the eutrophication model described above, as applied to Barton Broad, Norfolk, England (after Lau 2000). The model was optimized on data for 1983 to 1986, and then run with those optimized parameters for 1987 to 1993. This demonstrates the dependence upon optimization in this system (compare 12.4b with 12.4a). It also demonstrates a fundamental modelling problem: the optimized model performs progressively less well with time from when the optimizing period of data ended. As with the empirical models above, the parameterization appears to be only effective for a limited time period, questioning the extent to which the model may be used over longer time periods.

At this stage, the modeller must make a crucial decision: is the model, after optimization, sufficient to be able to use to simulate or to predict beyond the range of conditions for which it is formulated? As we noted in the introduction, this is the core purpose of a model: to extend the bounds of space and time. However, this decision causes us to look very critically at deterministic mathematical models.

Critical Perspectives on Deterministic Mathematical Models

Assumptions

For all modelling efforts, computer resources ultimately provide a trade-off between the amount of spatial resolution and time resolution that can be achieved and the degree of physical, chemical and biological processes that can be considered. This requires us to make assumptions about what processes to include, what model time steps to use and what spatial scales to address. This is a form of closure, so are our model results a reflection of reality or a product of the type of closure that we have used? In the case of the eutrophication model, we have excluded many processes (e.g. higher-order food chain interactions such as those associated with fish). If these matter, and they are either ignored or only partially represented, the model will not necessarily be a reliable decision-making tool.

Lumping and subgrid-scale processes

When we model, we have to specify the time and space scale at which the model is being applied. This is called discretization. However, we know that most processes operate at a range of time and space scales. Computational limitations mean that discretization is set at a scale that is only slightly smaller than that which is of interest. Scales smaller than the discretization scale are not modelled directly but must be represented in some way in the model, usually involving some sort of lumping together of their net effect at the scale being modelled. These are called subgrid-scale processes and may be spatial or temporal. Unfortunately, research suggests that subgrid-scale processes can have an unpredictable effect upon larger-scale processes. This is best illustrated by the question raised by Edward Lorenz (1963) in relation to chaos theory: '*How can the flap of a butterfly's wings* [a very small-scale process] *in Brazil set off a tornado* [a much larger-scale process] *in Texas?*' Some argue that chaos theory challenges the very assumption that the environment can be modelled deterministically.

Validation

The process of validation is highly problematic (Lane and Richards 2001). The basic

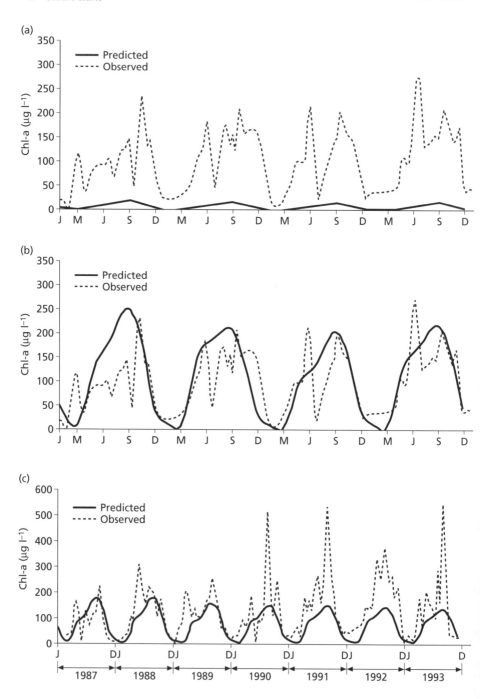

Figure 12.4 The default (a), optimized (b) and predicted (c) predictions of chlorophyll-a concentrations for Barton Broad (after Lau 2000).

assumption in validation is that check data are correct. Regardless of whether or not they are measured without error, and hence correct, these check data might not necessarily be sufficient. For instance, the spatial scale over which we measure in the field may be very different to the spatial scale at which models are used to predict. A good example is the measurement of soil moisture status, which is commonly at a point in the field. Models of hillslope runoff operate on a grid scale that is of the order of many metres and may so be predicting at a very different spatial scale. Similarly, a model often produces a range of predictions. A two-dimensional flood-routing model may provide predictions of water level through time at the downstream end of a river reach, as well as spatial patterns of inundation. These predictions may then be inspected in a number of ways (e.g. magnitude of flood peak, time to flood peak, duration of out-of-bank flow). Research suggests (e.g. Beven 2000) that whilst one of these predictions may be sufficiently accurate, this does not mean that all predictions are sufficiently accurate. Similarly, models of floodplain inundation may produce realistic estimates of water-level variation through time, but this does not mean that patterns of inundation within the floodplain upstream of the water-level recorder are also accurate. These sorts of issues demonstrate that model validation is not a straightforward exercise.

Parameterization and optimization

Parameterization and optimization are a necessary part of all models. However, they lead to a number of problems. First, the environmental system is normally always indeterminate: the number of model parameters is always greater than the number of equations describing the system. The large number of possible parameters that can be changed leads to the possibility that optimization produces the right results (i.e., accurate predictions) for wrong reasons (i.e., with the wrong parameter values). Beven (1989,

1996, 2000) has shown that this is a ubiquitous feature of hydrological models and leads to the conclusion that models suffer from equifinality: we can get a range of similar model predictions for very different sets of parameter values. Second, optimization is problematic if the quantity and/or quality of check data is poor. This leads to the possibility of obtaining the wrong results (i.e., predictions judged with respect to incorrect check data) for the wrong reasons (due to equifinality). Third, extensive parameterization and optimization will not be effective if there are fundamental inadequacies in the conceptual model upon which the mathematical has been developed. When there is a disagreement between models' predictions and observations, it is very tempting to focus upon optimization. However, this will be a futile activity if the fundamental conceptual basis of the model is not correct. The fact that many numerical models are heavily dependent upon parameterization and optimization has led some (e.g. Beven 1989) to question the physical basis of environmental models: whilst models have a physical basis, the predictions that they generate can often contain information that is largely as a result of parameterization. This is illustrated in figure 12.4, which demonstrates the parameter dependence for the algal modelling of Barton Broad. This means that they will only work for the time periods (e.g. figure 12.4a), locations and predictions for which they have been parameterized. In our discussion of mathematical models, we find that we have gone full circle, and we are forced to recognize that we have lost the very generality that we seek to gain by adapting a mathematical approach.

Conclusions

To finish on such a depressing note would do an injustice to the range of ways in which modellers are developing sophisticated solutions to these problems. Much of the

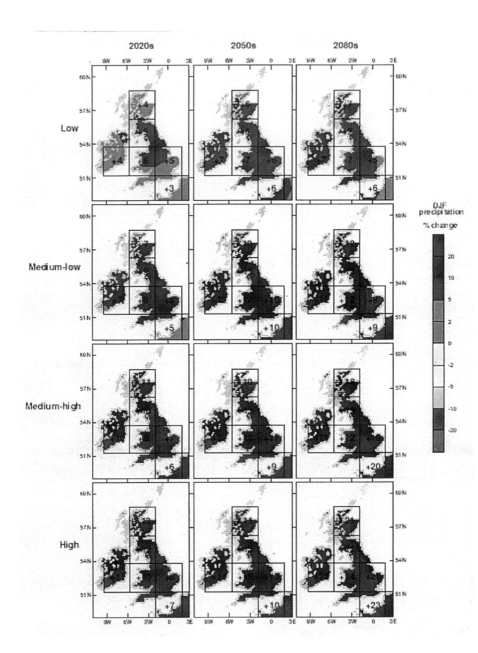

Figure 12.5　Predicted winter precipitation changes in per cent, from the UKCIP (Hulme and Jenkins 1998).

progress revolves around proper treatment and reporting of uncertainty in mathematical models. The possibility of error in parameter values and boundary conditions, as well as in check data, means that most models are associated with a range of possible predictions at a point in space and time. This allows the very problems of mathematical modelling to be turned to a core advantage: by undertaking a very large number of model simulations, and exploring changes in model predictions as a result of different combinations of boundary conditions and parameter values, we can put a range on our possible model predictions, which indicate the certainty (and hence uncertainty) associated with the model. Whilst this is not always straightforward (e.g. see Beven 2000), it makes us explicit in how confident we are that the model is correct by assigning statements of probability to our predictions that describe our level of confidence in how correct they are. Figure 12.5 shows such an analysis from the UK Climate Impacts Programme (UKCIP) (Hulme and Jenkins 1998). This encompasses uncertainties in both management of climate change (e.g. changes in greenhouse gas emissions) and in model parameters to provide a range of possible climate changes, classified as low, low–medium, medium–high and high. This reflects the natural state of predictions from numerical models: as with any crystal ball, they will be uncertain. Unfortunately, as a society, we are much less willing to accept these uncertainties. Much of the damage caused by the UK floods in November 2000 involved discrete events, where water levels in rivers exceeded the bankful threshold, and the consequences have real meaning for floodplain dwellers. Quantification and reporting of uncertainty is crucial. Wider understanding of the uncertainties associated with mathematical modelling is also critical. However, the ultimate challenge is the reduction of uncertainty so that we can increase the reliability of model predictions and so improve their suitability as decision-making tools.

References

Beven, K.J. 1989: Changing ideas in hydrology: the case of physically-based models, *Journal of Hydrology*, 105, 157–72.

Beven, K.J. 1996: Equifinality and uncertainty in geomorphological modelling. In B.L. Rhoads and C.E. Thorn (eds), *The Scientific Nature of Geomorphology*, Chichester: Wiley.

Beven, K.J. 2000: *Rainfall-Runoff Modelling*. Chichester: Wiley.

Huggett, R.J. 1993: *Modelling the Human Impact Upon Nature*. Oxford: Oxford University Press.

Hulme, M. and Jenkins, G.J. 1998: Climate change scenarios for the United Kingdom: summary report, *UKCIP Technical Report* 1, Norwich: Climatic Research Unit.

Lane, S.N. and Richards, K.S. 2001: The 'validation' of hydrodynamic models: some critical perspectives. In P.D. Bates and M.G. Anderson (eds), *Model Validation for Hydrological and Hydraulic Research*, Chichester: Wiley.

Lau, S.S.S. 2000: Statistical and dynamical systems investigation of eutrophication processes in shallow lake ecosystems. Thesis submitted in partial fulfilment of the requirements for the Ph.D. degree, University of Cambridge.

Lau, S.S.L. and Lane, S.N. 2001: Continuity and change in environmental systems: the case of shallow lakes, *Progress in Physical Geography*, 25, 178–202.

Lorenz, E.N. 1993 [1963]: *The Essence of Chaos*. London: UCL Press.

Scheffer, M. 1998: *Ecology of Shallow Lakes*. London: Chapman and Hall.

See, L. and Openshaw, S. 2000: A hybrid multi-model approach to river level forecasting, *Hydrological Sciences Journal*, 45, 523–36.

Stefan, H.G. 1994: Lake and reservoir eutrophication. In M. Hino (ed.), *Water Quality and its Control*, IAHR Hydraulic Structures Design Manual, 5, 45–76.

Further Reading

On conceptual models see R.J. Huggett, *Modelling the Human Impact Upon Nature* (Oxford: Oxford University Press, 1993). On mathematical modelling see M.J. Kirkby, P.S. Naden, T.P. Burt and D.P. Butcher, *Computer Simulation in Physical Geography* (Chichester: Wiley, 1993). Case studies of ecological modelling are provided by

M. Gillman and R. Hails, *An Introduction to Ecological Modelling: Putting Practice Into Theory* (Oxford: Blackwell, 1997). Case studies of hydrological modelling are given in K.J. Beven, *Rainfall-Runoff Modelling* (Chichester: Wiley, 2000).

13

Geocomposition

Rachael A. McDonnell

Capturing and exploring the spatial dynamics of form and process through numerical representations, whether these are patterns of economic activity or movements down a hillslope, continues to be a major focus of research for many geographers. Computer technology has been an important force to this end with ever more sophisticated capabilities for data collection, exploration and simulation, and the decline in cost of computer storage and processing in power enabling ever more complex problems to be addressed. This, allied with the ever-increasing availability of digital geographical data from a wide variety of different outlets, has meant that the potential for gaining new knowledge and understanding using computer technology has never been better. The result of all these developments has been that the computer has for many geographers become the experimental laboratory.

It is against this background that the term geocomputation has come into the geographical literature. It originated from a group of researchers at the University of Leeds School of Geography in the mid-1990s whose work focused on the development of a computational paradigm for doing human geography (Openshaw and Abrahart 2000). When it became obvious that the ideas and methodologies they were developing were equally applicable to physical geography problems, the term geocomputation was invented. Since then the term has begun to proliferate in the literature,

with impetus to the development and acceptance of the subject being given by the annual Geocomputation conference series, whose meetings have been held in Australasia, Europe and North America. These meetings have attracted researchers from many different disciplines such as computer science, geography, statistics, engineering, meteorology, economics and planning, who all wrestle with the complexities of spatial form and processes in their scientific explorations.

There have been many attempts to define exactly what geocomputation means, but as with all newly developing research areas, pinning down what is or is not part of it is difficult and no two people agree. Couclelis (1998) envisages it as the eclectic application of computational methods and techniques to portray spatial properties, to explain geographical phenomena, and to solve geographical problems. Longley (1998) summarizes the hallmarks of geocomputation as being research-led applications that emphasize process over form, dynamics over statics, and interaction over passive applications.

The use in the term of the word 'computation' as opposed to 'computing' is significant as it highlights the emphasis on seeking common methodologies rather than the application of a series of powerful techniques. Many researchers have found that existing software systems such as GIS have compromised their ideas by forcing the use of simplistic representations of the environment or the processes at work (see chapter 30 in this volume). The concern of geocomputation is to enrich our work with a toolbox of methods to model and analyse a range of highly complex, often non-deterministic problems (see Mark Gahegan's definition at the website listed in Internet resources below). This often results in combining different technologies such as statistical packages with GIS, with programmed models and with visualization software to help answer these questions.

Geocomputational Methods

So far this discussion has focused on ideas of geocomputation and these may seem rather intangible. This section will provide examples of three possible groupings of areas of development that have been given the geocomputation label to help illustrate what may be included under the term.

Exploratory spatial data analysis

Now that digital data are widely available, some researchers have adopted an inductive approach, which finds information from raw datasets without the constraints of pre-existing ideas (see Gahegan 2000). The sheer volumes of data now available for many areas defy conventional approaches to reporting and analysis. Exploratory spatial data analysis uses techniques for searching, pattern recognition and classification to determine the presence or absence of spatial heterogeneity, clustering, spatial correlation, and process and spatial structure. That is, by using various algorithms it is possible to discover whether there are associations between two different datasets such as high values in one bringing low values in another; or whether if one is spatially close to another then a particular reaction occurs; or if there are particular patterns in the occurrence of a particular phenomenon. Needless to say, this type of work requires high computer processing power.

There are an increasing number of examples of this type of research. Murray et al. (2001) used a combination of GIS and spatial analysis techniques to explore crime occurrence in Brisbane, Australia. Wise et al. (2001) describe how statistical data analysis software was linked to the ArcInfo GIS to query and analyse geographical data. Whilst both papers show the possibilities in such work, they also emphasize the challenges in exploratory spatial data analysis.

Visualization

An increasingly important way of exploring datasets is to use various methods for visualizing them. Obvious techniques such as graphing or mapping spring to mind, but today there are other methods that allow us to view the data differently to gain new insight. These are particularly useful where the phenomena are complex and multi-dimensional. For example, animation is useful where temporal dynamics are important, and watching a cartoon-like representation of, say, faulting in a landscape is much more comprehensible than statistics on rates of movement.

Visualization techniques are also useful when combining datasets to gain insight into interactions. By using a range of different visualizations such as animated maps and symbols, and different view directions, geographical areas may be more readily understood by both researcher and student alike and new understanding develop. For example, in figure 13.1 three views of a dataset are given, showing selected variables from the 1990 US Census, aggregated to state level. The first view (top left) is a conventional thematic map with residential rent values providing the shading scheme. The bottom image is a window onto the feature space defined by the census variables. It shows the relationships between 13 variables, depicted by a parallel coordinate plot, with each state providing a single 'string' through this space. The top right image is a

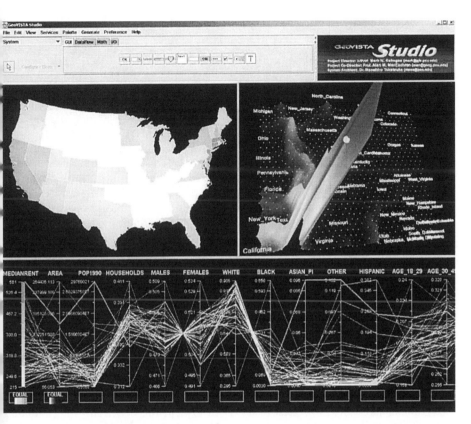

Figure 13.1 Three different linked visualizations of a dataset are given, showing selected variables from the 1990 US Census, aggregated to state level.

view 'inside' a neural classifier (a self-organizing map) as it attempts to cluster the states according to the census variables. The three views are linked; the user can explore the data from each perspective in a coordinated fashion. The visualizations were produced using GeoVISTA Studio (website listed in Internet resources below).

An interesting development has been the use of virtual reality environments in geographical studies. These can allow processes and forms of an environment to be explored using vision as well as other senses. Virtual reality field trips, for example, have been designed to help students understand landscapes (use a search engine to check the vari-

ous sites on the Internet). They are not meant to replace the valuable in-the-field experience, but they do help to bridge the gap between learning about individual landscape features in a somewhat isolated fashion in lectures and being faced with the overlapping, complex environment of reality.

Virtual reality systems are also used in commercial explorations. Shell International Exploration and Production (2001) show how immersive visualizations of three-dimensional structures are important in evaluation and management of oil and gas assets. The company now has a number of large-scale virtual reality centres around the world.

Modelling and prediction

Linking mathematically output and input variables to simulate or make predictions about an environment has been an important part of geography for many decades. Technology has revolutionized this and to-day computer models are a familiar tool in teaching and research. Developments in modelling languages and data collection techniques are allowing us to tackle more of the complexity of geographical systems. Lane and Bates (2000) give examples of the use of computational methods in the mod-elling of water flow to improve our under-standing of geomorphological and hydrological problems. In other work re-searchers have harnessed tools such as neu-ral networks and genetic algorithms from the artificial intelligence community, cellular automata rule-based methods, or probabil-ity-based functions to help with problems that cannot be realistically specified in the traditional 'A + B = C' deterministic ap-proach (see chapter 12 by Stuart Lane in this volume for some physical geography examples). These sophisticated tools enable geocomputation researchers to tackle issues such as the chaotic behaviour of systems and the uncertainty in data and process repre-sentation when developing predictive de-vices, whether this is for climate or stockmarket forecasting.

Geocomputation in Practice

An interesting case study of exploratory spa-tial data research was undertaken at the University of Leeds in the 1980s and 1990s with a study of the occurrence of childhood leukemia in northern England. The work involved searching for and locating clusters (that is, the occurrence of the same phenom-enon in close proximity) in a geographical dataset of point values. Whilst this sounds very simple it is computationally intensive, with many possible search methods and a vast amount of data to process, and the first

computer runs back in the 1980s took a month of central processor time on a main-frame. The search algorithms used found one cluster close to Sellafield and, surpris-ingly, an even bigger one in Gateshead thought to be associated with an incinerator (the work is summarized in Openshaw 1998). From these findings subsequent analysis tried to address the question of why children in these areas were more prone to leukemia. This involved the analysis of many different datasets to determine if there are any spatial associations between the clusters detected and any other phenomena. Whilst data on age group, gender etc. were rela-tively easy to access, one of the problems faced by the researchers was deriving datasets for variables that were difficult to measure, such as exposure to an allegedly harmful chemical.

The type of correlation and clustering analysis undertaken in the childhood leukemia project would have been impossi-ble with paper maps, or even using stand-ard spatial analytical packages. Imagine searching and combining all the different geographical data that could possibly be used to explain the occurrence of childhood leukemia. This type of geographical analy-sis was only possible using computational methods combined with super-computer processing capabilities.

The Future of Geocomputation

The place of geocomputation within geog-raphy and other disciplines is still to be con-firmed. Many of the problems we are facing are multi-dimensional and multivariate and there is no doubt that standard statistical and analytical methods are not enough. New spatially based approaches and methods, which may be borrowed from other disci-plines such as physics, computational sci-ence, engineering or economics, need to be applied to these diverse geographical ques-tions. Whether the researchers exploring these new areas will find enough theoretical, practical and institutional common ground for it to become a revolution or a separate field within geography (or even computational science) has yet to be established. For many researchers the limitations of current mainstream techniques for their particular area of interest make it worth the challenge.

In a wider context, the work that does emanate from this field must be scientifically acceptable with practitioners showing a clear appreciation of the assumptions and limitations of the computational techniques used, especially in their application to spatial problems (Couclelis 1998). Over the next decade, as the number of geocomputational studies grow, the credibility (or otherwise) of this research will become clear. The exciting nature of the work will make testing its validity and usefulness a highly enjoyable challenge.

References

Couclelis, H. 1998: Geocomputation in context. In P. Longley, S.M. Brooks, R.A. McDonnell and B. Macmillan (eds), *Geocomputation: A Primer*, Chichester: Wiley, 17–29.

Gahegan, M. 2000: On the application of inductive machine learning tools to geographical analysis, *Geographical Analysis*, 32, 113–39.

Lane, S.N. and Bates, P.D. (eds) 2000: *High Resolution Flow Modelling in Hydrology and Geomorphology*. Chichester: Wiley.

Longley, P.A. 1998: Foundations. In P. Longley, S.M. Brooks, R.A. McDonnell and B. Macmillan (eds), *Geocomputation: A Primer*, Chichester: Wiley, 3–15.

Murray, A.T., McGuffog, I., Western, J.S. and Mullins, P. 2001: Exploratory spatial data analysis techniques for examining urban crime, *British Journal of Criminology*, 41, 309–29.

Openshaw, S. 1998: Building automated geographical analysis and explanation machines. In P. Longley, S.M. Brooks, R.A. McDonnell and B. Macmillan (eds), *Geocomputation: A Primer*, Chichester: Wiley, 95–115.

Openshaw, S. and Abrahart, R.J. (eds) 2000: *GeoComputation*. London: Taylor and Francis.

Shell International Exploration and Production 2001: Visualise this, *GEOEurope*, 10, 34–5.

Wise, S., Haining, R. and Ma, J.S. 2001: Providing spatial statistical data analysis functionality for the GIS user: the SAGE project, *International Journal of Geographical Information Science*, 15, 239–54.

Further Reading

Burrough, P.A. and McDonnell, R.A. 1998: *Principles of Geographical Information Systems*. Oxford: Oxford University Press.

Cressie, N. 1993: *Statistics for Spatial Data*. New York: Wiley.

Longley, P.A., Goodchild, M.F., Maguire, D.J and Rhind, D.W. 2001: *Geographic Information Systems and Science*. Chichester: Wiley.

Internet Resources

- Details of the annual Geocomputation conferences: http://www.geocomp.org
- Mark Gahegan's definition of geocomputation: http://www.Ashville.demon.co.uk/geocomp/definition/definition.htm
- GeoVISTA Studio: http://www.geovista-studio.psu.edu/jsp/index.jsp
- The programmes used in the Leeds study of childhood leukemia are at: http://www.geog.leeds.ac.uk/research/geocomp.html

14

Strange Natures: Geography and the Study of Human–Environment Relationships

Noel Castree

Nature's not what it used to be. Consider the following. In 2001 alone, scientists and so-called 'lifescience' companies, like Monsanto, unveiled a seemingly non-stop array of new 'genetically modified' (GM) organisms. Perhaps the most notable of these were ANDi, the rhesus monkey – implanted with genes from a jellyfish! – and Bessie, an American cow, who gave birth to an Indian gaur, an ox-like animal threatened with extinction on the other side of the world. Not surprisingly, these remarkable feats of biological engineering have caused considerable controversy. In Britain, for example, government attempts to assess the safety of GM crops were sabotaged by environmental organizations like Greenpeace, whose UK leader, Lord Melchett, was famously arrested for his part in destroying field test sites. More notable still has been Prince Charles's outspoken opposition to GM foods. As he put it in a BBC Radio 4 lecture given in May 2000, producing these foods entails 'playing God with nature', which is why he has been a strong advocate of more natural, organic farming methods – methods he has used on his own farms in Cornwall and elsewhere.

Charles's desires to 'get back to nature' before we destroy it altogether was boosted

Figure 14.1 ANDi, the rhesus monkey, implanted with jellyfish genes, is one of a new generation of GM guinea pigs. But is it morally right to alter the biological make-up of other species?
Photograph: PA Photos/EPA.

in 2001 by a remarkable venture, again in Cornwall. The Eden Project, which cost £75 million, is an attempt to harness modern science and technology to preserve nature

Figure 14.2   The Eden Project, Cornwall – a vision of what nature *should* look like rather than an accurate portrayal of nature as it is?
Photograph: © Mick Hicks 2002/alamy.com

rather than tamper with it – in this respect contrasting with GM organisms. It was established by environmentalists concerned about species extinctions worldwide. Situated in two reclaimed clay pits, the Eden Project's giant geodesic greenhouses contain whole biomes from around the world, including those, like tropical humid forests, that are currently under threat from human activities such as logging. The name says it all: echoing the famous Garden in the Bible, the Eden Project attempts to be a storehouse for life on Earth. It literally transplants nature, recreating it in an artificial environment designed to ensure its continued existence.

Together, the proliferation of GM organisms and the Eden Project give us two glimpses of nature in the new millennium: the new 'engineered natures' of advanced societies and the fast-disappearing 'natural natures' bequeathed by evolution. Or do they? Despite their apparent differences, I would argue that in *both* cases nature is, in fact, a human fabrication. In other words, even the Eden Project – despite its intentions – serves up an 'unnatural' or 'strange nature'. This may seem a peculiar, even outrageous, claim, but I am not alone in making it. In recent years many human geographers have been arguing that nature today is no longer natural. They insist that it is *a human construction* through and through. In this chapter I want to explain what is meant by this and why it matters. I shall do so by saying more about GM organisms and nature-protection schemes like the Eden Project. But first it is necessary to say a little bit about how geographers have typically studied nature.

The 'Nature' of Geography

Since it was established as a school and university subject a century or so ago, geography has concerned itself with human–nature relationships. Indeed, in 1887 Halford Mackinder – one of Britain's first university geographers – defined the discipline as a 'bridging science' between the social and the natural sciences, one that would examine human–environment relationships in different places around the world. Since Mackinder's time, geographers have taken three broad approaches to understanding people–nature relationships. In the early twentieth century, a period when things like GM organisms would have been the stuff of science fiction, 'environmental determinism' was a popular doctrine. Environmental determinists believed that different environments determined, to varying degrees, what humans could and could not do, with more extreme environments (like the tropics and the poles) imposing severe constraints on human activity. By the 1940s and 1950s environmental determinism had been discredited. It had by that time become clear that humans had the ability to *adapt* to their environments far more than had previously been thought. Consequently, the idea of 'environmental possibilism' caught on in geography: that is, the idea that environments offered a set of possibilities for humans, not just constraints. However, no sooner had this idea become established than it was called into question. In the developed world, the mega-industrialization of the post-war years had, by the 1960s, started to have a visible effect upon the environment. Since then 'human impact' studies have been the mainstay of geographical research and teaching on people–nature relations. These studies – such as those of desertification, forest loss or water pollution – typically focus on how different societies are variously destroying, degrading or modifying the environments upon which they depend. And these studies typically try to figure out how best humans should, in future, 'manage' their activities so as to minimize environmental harm.

For all their differences, these three generations of approaches to people–nature relationships in geography draw upon a set of common ideas about nature – ideas that you will probably think are 'common sense'.

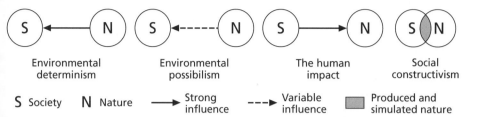

Figure 14.3   Geographical understandings of human–nature relationships.

First, nature is taken to be external to and different from people (for instance: the term environment, in many ways a synonym for the term nature, usually means the non-human world). Second, nature is taken to be something that is unchangeable. Humans might 'destroy' or 'alter' nature but, it is commonly thought, we can never create or make nature.

However, as GM organisms and (as I shall explain) the Eden Project show, these ideas of nature may have outlived their usefulness. Accordingly, in a kind of reversal of the environmental determinism of a century ago, many geographers are now interested in how humans manufacture nature (or, more precisely, how different societies manufacture different natures). This implies that nature is no longer external to humanity or unchangeable, such that the conventional distinction between people and environment collapses (see figure 14.3). As I shall now explain with reference to GM organisms and the Eden Project, geographers have argued that there are two main ways in which modern societies are 'constructing nature': physically and in the equally material realm of the imagination.

Producing Nature

GM organisms are a good example of the new power of humans to physically reconstitute nature 'all the way down'. The distinction between nature and humans dissolves, as companies like Monsanto are now able to 'make nature to order'. But why would humans want to materially remake nature? Nature has always presented opportunities and constraints to societies. Until recent decades, the best that societies could hope for was to exploit the opportunities and adapt to the constraints. Now, however, some societies – notably advanced industrial ones – have the technology to *overcome* these constraints altogether. We need not go into the details of how such technology (of which GM technology is an example) has been developed. More important, in the context of this chapter, is to understand some of the benefits of overcoming natural constraints to human action. Consider the cases of ANDi, Bessie the cow and GM crops. ANDi is one of a new generation of a GM guinea pig who, because of its genetic similarity to humans (primates are our closest biological relatives), can be used to develop cures for human illnesses. If scientists had inserted genes for serious human diseases into ANDi – rather than a jellyfish gene – then studying this rhesus monkey could help with developing new drugs and treatments for those diseases. GM technologies might also benefit non-humans. Bessie's mothering of an Indian gaur is part of a longer-term aim of protecting near-extinct species by breeding them through close relatives. Bessie, though a cow, is a genetic cousin of a gaur and therefore able to give birth to one if artificially inseminated with its genes. Finally, companies like Monsanto have argued that GM crops will have both human and environmental benefits. Because these crops can be 'genetically programmed' to be high-yielding or to resist natural pests and diseases, the argument is that they can help solve world hunger while requiring less land to grow on, in turn reducing the acreage of ploughed fields and exposed top-soil.

As these examples show, there are some important potential benefits to be gained by physically overcoming the constraints nature has traditionally imposed upon humans. However, as my earlier mention of Prince Charles and Lord Melchett indicated, there are potential problems we should recognize too. For example, is it morally right to alter the biological make-up of other species? More generally, do we risk losing nature forever if we go too far in tampering with its physical fabric, and does it matter? And do we risk creating serious new and unanticipated problems by 'playing God' with nature? These are the kinds of vital questions that many geographers are now asking in their studies of human–nature relationships. The answers are by no means straightforward and vary depending on who is doing the answering. Among politicians, environmentalists and ordinary people there are many who think that the physical reconstruction of nature has already gone too far. However, while intuitively appealing, is the argument that we should 'get back to nature' before it is too late actually a coherent one? Several contemporary geographers think not and a closer look at the Eden Project can help us understand why.

Simulating Nature

Where GM organisms are, if you like, 'unnatural', the Eden Project aims to deliver 'the real thing': that is, nature in the raw. At a time when nature seems to be fast-disappearing because of non-stop industrialization and population growth, the aim seems to be a vitally important one. However, it is arguably a misconceived aim because the experience of nature that millions of visitors to the Project have is anything but real. As

the geographers Rob Bartram and Sarah Shobrook (2000: 375) have suggested, 'the Project . . . [is] a simulation of nature . . . contriving a more-real-than-real, intense, and immediate experience'. What Bartram and Shobrook mean is that visitors to the Project are lulled into a false sense that they are communing with 'real nature'. Although there is no doubt that the soils, plants and trees in the geodesic domes are real, they are arranged in such a way that they are supposed to 'stand for' their wild relatives in the wider world. Thus the different biomes in the domes – tropical, subtropical, temperate, etc. – are supposed to be examples of much bigger, real-world biomes. This may seem unexceptional to you. However, this exercise in 'recreating nature' is arguably not as innocent as it seems. First, visitors to the Project have no way of knowing how accurate the things they see are as recreations of 'real nature'. For instance, since they cannot visit all the diverse real-world biomes depicted in miniature in the Project domes, the miniature becomes the reality for visitors. Second, the Eden Project tends to represent nature as if humans have had no involvement in its alteration over the centuries. But even the most seemingly 'natural' of the Project biomes – the tropical forest – has, in the real world, been inhabited by indigenous tribes for millennia. But this history of human settlement is written out of the exhibits in order to depict a 'pure nature' where humans have no place. In effect, then, a particular moral vision of what nature *should* look like – that is, a nature 'left in peace', free from 'human interference' – is being substituted for a more accurate portrayal of nature as it is. This moral vision arguably reflects the environmentalist viewpoints of the Project's conservationist, pro-nature founders.

The End of Nature and the End of Geography?

I have argued that we are witnessing the end of nature as we knew it. Nature is constructed both physically and in the imagination more than ever before. This should lead us, both as geographers and as citizens, to ask a key question: 'who is constructing nature and with what effects?' As the examples of GM organisms and the Eden Project show, there are no simple answers to this question but the stakes in getting the answers right are high. Geographical investigations into the 'social construction of nature' will play a role in delivering these answers as the twenty-first century unfolds. Far from heralding the end of geographers' interest in human–nature relationships, the end of nature marks the beginning of a new, exciting round of investigations. These investigations are as manifold as they are interesting. On the physical remaking of nature the key foci are science (since scientists study nature often in order to change it: see Demeritt 1998) and business (since, as GM foods show, business can make big profits by altering nature: see Castree 2001a); and on the changing representations of nature the foci are everything from the media – for instance, Gail Davies's analysis of wildlife films (Davies 1999) – to the metaphors competing groups use in struggles over the environment – for example, Bruce Braun's study of Canada's Clayoquot Sound (Braun 1997). In short, through their investigations into the ever-stranger natures that humans are making and remaking, geographers are ensuring that the discipline engages with some of the most momentous developments of our time.

References

Bartram, R. and Shobrook, S. 2000: Endless/endless natures, *Annals of the Association of American Geographers*, 90(2), 370–80.

Braun, B. 1997: Buried epistemologies: the politics of nature in (post)colonial British Columbia, *Annals of the Association of American Geographers*, 87(1), 3–31.

Castree, N. 2001a: Marxism, capitalism and the production of nature. In N. Castree and B. Braun (eds), *Social Nature*, Oxford: Blackwell, 1–21.

Davies, G. 1999: Exploiting the archive: and the animals came in two by two, *Area*, 31(1), 49–58.

Demeritt, D. 1998: Science, social constructivism, and nature. In B. Braun and N. Castree (eds), *Remaking Reality: Nature at the Millennium*, London: Routledge, 177–97.

Further Reading

Castree, N. 2001b: Social nature: theory, practice and politics. In N. Castree and B. Braun (eds), *Social Nature*, Oxford: Blackwell, 189–207.

Whatmore, S. 1999: Culture–nature. In P. Cloke, P. Crang and M. Goodwin (eds), *Introducing Human Geographies*, London: Arnold, 4–11.

15
Environmental Science, Knowledge and Policy

Sally Eden

Since Rio, the scientific argument for climate change has been ignored by many, accompanied by a sometimes sceptical media. But when people see and experience these ferocious storms, long summer droughts, torrential rains, more extreme and more frequent – they know something is wrong and that climate change now affects them. The people and our press are looking for a successful Hague agreement.

Thus spoke the UK's deputy prime minister, John Prescott, to the delegates at the sixth Conference of the Parties to the Framework Convention on Climate Change in the Hague in November 2000 (quoted from a press release). He was disappointed. The international talks at the Hague were a failure and had to be extended to July 2001 in Bonn when a diluted agreement was finally negotiated. Environmental problems like climate change are fraught with debates like these about scientific uncertainty, social acceptability and political bargaining. As John Prescott realized, what we know and believe about the environment is as important to policy as what the environment is like. Indeed, it is impossible to understand debates about the environment without understanding the complex ways in which we learn about the environment and why that learning does not always result in environmental policy that works.

The Role of Science

Science is crucial to how we find out about environmental change, especially globally.

In the climate change case, this crystallized when, in 1988, the United Nations Environment Programme and the World Meteorological Office set up the Intergovernmental Panel on Climate Change (IPCC) to research and advise governments on climate change. The IPCC is now the most authoritative voice on climate change science – and the most contested because, as John Prescott saw, science is not the end of the debate. Indeed, in recent years researchers and media commentators have suggested that, after decades of (relative) acceptance, science is having to deal with public criticism and distrust for the first time. In an influential argument, the German sociologist Ulrich Beck (1992) used environmental problems like nuclear fallout and water pollution to argue that we live in a 'risk society', where our lives are increasingly defined by risks that we often cannot see, solve or escape. In such a world, science becomes more and more involved in trying to manage environmental change and its inherent uncertainties. In the process, science becomes 'politicized' and politics becomes 'scientized' (see examples in O'Riordan and Jager 1996). Hence, the IPCC has been criticized for being politically biased in its judgements, because it is dominated by western academics, for failing to capture the huge complexities of climate processes, for providing a false consensus that excludes extreme views and for becoming a political football (e.g. Boehmer-Christiansen 1995).

Researchers question whether environmental science automatically has access to 'the truth' about environmental change. Science is never truly value-free because it is influenced by society in terms of the questions it asks to get funding, the credibility it gains, and even who does the research and where. This does not mean that science is wrong or biased, just that it is also a product of society and therefore can be criticized by society (and not just by other scientists). Hence, researchers have argued that what is needed is the democratization of science and of the environmental debate. This would require environmental science to become more reflexive in acknowledging its own uncertainties and ambiguities in public, and allow the environmental debate to include the views of many different groups and to open up discussion of uncertainties and risks (Beck 1992; O'Riordan and Rayner 1991). Appreciating this, the IPCC has tried to write about the uncertainties it encounters by stating the levels of confidence in climate change factors in its recent *Summary for Policy Makers* (the website is listed below).

The Construction of Knowledge

Governments and scientists have also tried to address rising public concern and distrust of 'experts' by providing information, supposing that distrust stems from misunderstanding of science. This is sometimes called the 'deficit model' because it assumes that providing information and 'educating' people will correct the problem through a one-way flow of information from the 'experts' to the public (Wynne 1995). But research shows that people do not simply accept in-

formation but reflexively judge and interpret it within the context of their daily lives, often against other information that contradicts it. What much of this research shares is some form of *social constructionism*. This contends that a person's knowledge or view of the environment is not a straightforward reflection of that environment but a social construct. We know about the environment through first-hand information (observing the weather, walking through the local park) and second-hand information (reading newspapers, talking to people) and we rethink our ideas all the time, in line with our own prejudices and morals. So cultural aspects and symbolism, not just information and facts, are critical to our environmental knowledge, and in consequence people's environmental understandings and attitudes are more complex than politicians or scientists often credit (see examples in Myers and Macnaghten 1998; Harrison et al. 1996; Irwin and Wynne 1996). And constructions link across geographical scales: since the early 1990s, the environment has increasingly been discussed in the media and science as a global entity and a global agenda for policy (Taylor and Buttel 1992), and people have struggled to connect these global constructions with their own, often more local, ones.

Furthermore, research shows that, whatever the message about environmental change, we tend to distrust the messenger more if they come from industry or government than if they are seen as 'independent' scientists or pressure groups. We are influenced by various 'claims-makers', such as pressure groups, scientists, industry spokespersons and journalists, who identify and define an environmental issue and bring it to our attention through the media (see examples in Hansen 1993). Often, a distinctive name or metaphor is coined, like 'the greenhouse effect' or 'Frankenstein foods', and this is a hugely effective way of conveying very complex environmental ideas through a simple and often highly emotional shorthand. How we construct the environment can change rapidly too. In a classic paper, Downs (1972) suggested that environmental issues go through a life cycle: they rise from relative obscurity to reach a high public profile and enthusiasm for action, but then decline from prominence to take a back-seat in people's minds. Mazur and Lee (1993) have charted such life cycles by measuring how US media coverage in the 1980s of different global environmental issues rose and fell. The construction of climate change is also fed by seasonal weather: the exceptionally wet months of 2000–1 in the UK were cited at the time in national newspapers (and by Prescott above) as evidence of climate change – but so was the hot summer of 1995.

So, our environmental ideas are heavily shaped by cultural assumptions and cues in particular places and times, and not simply by environmental change. This may seem obvious but has generated huge controversy. Critics have argued that social constructionism means that environmental problems are not 'real' and therefore we do not need policy to address them, or that every construction is equally valid and there is no way to judge who is right in a debate, turning it into a pointless exercise that devalues scientific judgements. Those who advocate seeing environmental knowledge as socially constructed do not go so far, but do consider that they are rebalancing the scales in favour of a more democratic environmental debate that will appreciate these diverse views instead of closing the debate to all but privileged groups – so science will not be excluded, but it will not have the floor to itself.

The Problems with Policy

These arguments also relate to policy. In the 1990s a new 'participation' agenda developed that emphasized the importance of involving 'ordinary' people in formulating and implementing environmental policy (Eden 1996). The argument was that, first, for en-

vironmental policy to be truly democratic, it should not be dictated top-down by government to the public but should reflect the concerns of everybody in a bottom-up or grassroots process. Second, for environmental policy to work in practice, it needs to be feasible for everybody to adopt or fit into their daily lives. This agenda was detailed in Agenda 21, a blueprint for worldwide sustainable development agreed at the Rio Earth Summit in 1992, and in the UK in the Local Agenda 21 strategies developed by local authorities.

But this is easier said than done. Public opinion polls have consistently shown concern about environmental change to be (relatively) high throughout the 1990s. Yet people's behaviour rarely matches their concerns, in terms of recycling, reducing emissions or other pro-environmental behaviours. This is referred to as attitude–behaviour dissonance in psychology – we believe in one thing but do another. Politicians often assume that providing information about environmental change and 'good' actions will correct the problem (and will win votes!), echoing the 'deficit model' described for science. The assumption here is that, if information is supplied to fill the deficit, people will change their views or behaviour, especially if they understand that some actions, like saving energy to reduce carbon dioxide emissions, may also save them money. It is strange that politicians still think this, when huge numbers of people continue to smoke despite decades of being told that smoking causes cancer. Researchers know that information does not necessarily change behaviour and that the reasons for this mismatch are hugely variable, relating to perceived moral norms, convenience and personal agency. So, understanding why policies are difficult in practice must include appreciating their cultural contexts.

And not only individually. Societies also have shared often unchallengeable assumptions and prejudices that confuse or even scupper international debate about environmental policy because it involves culturally unacceptable sacrifices. For example, the dominance of the car, the aspirant materialistic lifestyle and the distrust of government intervention, especially in energy use, make 'the US economy as dependent on fossil fuels as a heroin addict is on the needle' (Rayner 1993: 30), and about as able to reduce its consumption. The different approaches of European states to carbon dioxide controls likewise illustrate their different cultural styles and political identities. So, constructions of environmental change influence not only how we feel as individuals but how our representative governments argue for and against environmental policy in the international arena.

Governments are therefore beginning to recognize what social scientists have continually argued: that environmental problems are also social problems and therefore we can only understand them fully as such (Taylor and Buttel 1992). Social science is consequently catching up a little with natural science, which has dominated environmental change research funding and government attention so far. Some of the issues that I have only been able to touch on here, such as trust, reflexivity and complexity in our environmental views, will be essential areas for future research so that we can both understand and improve our relationship with the environment in the twenty-first century.

References

Beck, U. 1992: *Risk Society*. London: Sage.

Boehmer-Christiansen, S. 1995: Britain and the Intergovernmental Panel on Climate Change: the impacts of scientific advice on global warming, Part 1: integrated policy analysis and the global dimension, *Environmental Politics*, 4(1), 1–18.

Downs, A. 1972: Up and down with ecology: the 'issue-attention' cycle, *Public Interest*, 28, 38–50.

Eden, S. 1996: Public participation in environmental policy: considering scientific, non-scientific and counter-scientific contributions, *Public Understanding of Science*, 5, 183–204.

Hansen, A. (ed.) 1993: *The Mass Media and Environmental Issues*. Leicester: Leicester University Press.

Harrison, C.M., Burgess, J. and Filius, P. 1996: Rationalizing environmental responsibilities: a comparison of lay publics in the UK and the Netherlands, *Global Environmental Change*, 6, 215–34.

Irwin, A. and Wynne, B. (eds) 1996: *Misunderstanding Science?* Cambridge: Cambridge University Press.

Mazur, A. and Lee, J. 1993: Sounding the global alarm: environmental issues in the US national news, *Social Studies of Science*, 23, 681–720.

Myers, G. and Macnaghten, P. 1998: Rhetorics of environmental sustainability: commonplaces and places, *Environment and Planning A*, 30, 333–53.

O'Riordan, T. and Jager, J. (eds) 1996: *The Politics of Climate Change*. London: Routledge.

O'Riordan, T. and Rayner, S. 1991: Risk management for global environmental change, *Global Environmental Change*, 1(2), 91–108.

Rayner, S. 1993: Prospects for CO_2 emissions reduction policy in the USA, *Global Environmental Change*, 3(1), 12–31.

Taylor, P.J. and Buttel, F.H. 1992: How do we know we have global environmental problems? Science and the globalization of environmental discourse, *Geoforum*, 23, 405–16.

Wynne, B. 1995: Public understanding of science. In G.E. Markle, J.C. Petersen and T. Pinch (eds), *Handbook of Science and Technology Studies*, London: Sage, 361–88.

Internet Resources

- The IPCC *Summary for Policy Makers* can be found at: http://www.ipcc.ch 2001

16
Tourism, Environment and Sustainability: Everyday Worlds, Extra-ordinary Worlds

Tim Coles

Back to the Future: Tourism and Environmental Concern

In 1999 the British government published *Tomorrow's Tourism* (DCMS 1999). This was (and still is) intended to be a radical strategy to take tourism into the twenty-first century. Like many of its predecessors, *Tomorrow's Tourism* extols tourism for its contribution to economy and society. Tourism is to be nurtured by a partnership between the public and private sectors. Government aims to provide the appropriate infrastructure and regulatory environment, while the industry is urged to enhance the quality of products and experiences. Its advocates propose that *Tomorrow's Tourism* is different to previous strategies for its apparently innovative approach to and greater emphasis on the environment. In the first paragraph of his preface the prime minister, Tony Blair, proclaims that 'Britain is a wonderful country. Its people, landscapes, culture, character, history and traditions . . . These are the things which make Britain great and which make people – its own people and people from across the world – want to see Britain,

to know Britain and to understand Britain' (DCMS 1999: 1). Beyond the hyperbole Blair's statement recognizes the environment – in both its natural and physical as well as its human-produced, economic and socio-cultural manifestations – as the universal resource upon which all tourism is based. Tourism should be shaped by an attitude of 'wise growth' (discussed below). Rapacious, short-term (economic) benefits of tourism should not be accrued at the expense of irreversible environmental damage. Instead, tourism's principal asset should be respected, conserved and protected so that it will continue to generate benefits for future generations.

This is precisely the same sentiment expressed by Mathieson and Wall (1982: 97). They argued that 'in the absence of an attractive environment, there would be little tourism. Ranging from basic attractions of sun, sea and sand to the undoubted appeal of historic structures, the environment is the foundation of the tourist industry'. What followed was a discussion of the impacts and management implications of tourism development. Almost two decades separate these two pronouncements, but in microcosm they encapsulate the key themes in tourism geography: continuity, change and the centrality of environment. Through its development, exploitation and management as well as its interpretation and consumption by tourists, the environment has been the continued locus of research. Change stems from transformations in the organization of society and economy in the 1980s and 1990s, and its consequences for how environment is produced and consumed (see Ioannides 1995; Mowforth and Munt 1998). In turn, these have necessitated paradigm shifts which have altered the way we investigate as well as negotiate, produce and contest knowledge of tourism geographies.

The Sustainable Tourism Paradigm

By far the most heavily debated topic in the last decade has been sustainable tourism. Much of the discussion has taken its cue from the large body of work in the 1980s on tourism impacts and the future of the (then) dominant mode of consumption, mass tourism. Ever more systematic and sophisticated analyses of the impacts of tourism, in particular on the physical environment, resulted in normative, seemingly authoritative predictions about the future demise of tourism (Shaw and Williams 1994). Mass tourism was increasingly being viewed as untenable due to its heavy toll on host communities and environments. In many coastal resorts such problems as unplanned, uncontrolled development, highly seasonal demand, peak visitor flows often in excess of carrying capacities, high levels of leakages with commensurably low levels of multipliers, and a lack of investment conspired to challenge the wisdom of erstwhile conventional tourism development.

The End of Tourism As We Knew It: Alternative Tourism

The impetus for new forms of tourism came from two parallel developments. The shift towards a more flexible, sophisticated, IT-driven and market-oriented (post-Fordist) mode of accumulation induced changes in prosperity and flexible workplace practices, which in turn catalysed less regimented patterns of leisure, recreation and tourism (Shaw and Williams 1994). Longer-stay holidays of one to two weeks were replaced by more frequent, shorter breaks of several nights, often in midweek or at the weekend. Instead of the traditional 'works fortnight' on the coast, consumers were looking for new destinations, products and experiences compatible with their revised commitments. Enhanced choice resulted from the annihilation of space through faster, cheaper transport as access to previously unthinkable long-haul destinations was improved.

Mowforth and Munt (1998: 100) argue that increased consumer expectations and sophistication led to a panoply of so-called 'alternative tourism' forms, including: adventure, agri-, culture, eco-, ethnic, green, nature, risk, safari and wilderness tourism. Each aimed to exploit the many niches in an increasingly fragmented marketplace. Whereas the mass tourism formula had become somewhat stale and predictable, alternative tourism forms once more reintroduced the exotic and unknown. Tourism was to be cast again as a journey from the everyday to the extraordinary world. In addition to the experiential component, alternative tourism was organized quite differently to mass tourism. Cater (1993: 85) characterized alternative tourism as 'activities [that] are likely to be small-scale, locally owned with consequently low impact, leakages and a high proportion of the profits retained locally'.

Concurrently, environmental concern was centre stage of the global political agenda. The 1987 World Commission on Environment and Development (Brundtland) Report advocated sustainable development that must meet the needs of the present without compromising the ability of future generations to meet their own needs. As France (1999: 323) demonstrates, tourism was not explicitly mentioned in the report (only later in 1992 as Agenda 21), but many of the principles it espoused, such as environmental awareness and protection, local community empowerment and socio-cultural responsibility, were highly appropriate to tourism. In equal measure the principles of sustainable development, and hence by implication of sustainable tourism, appeared to represent the antithesis of and antidote to the dominant mass tourism paradigm. The pursuit of sustainable tourism began and alternative tourism forms became practically synonymous with the former. According to Wheeller (1993), sustainable tourism enjoyed considerable initial acceptance because it appeased its advocates' consciences. Precisely because it was not mass tourism it had to be less damaging, a more responsible way of behaving, and hence more viable in the long term. Later investigations have since exposed that not all forms of all alternative tourism can lay stake to such a claim. Some forms of alternative tourism are inherently unsustainable, while some forms of mass tourism evidently are intrinsically sustainable (see Fennell 1999).

Towards Sustainable Tourism

The linkage between sustainable tourism and development has been awkward to mediate and its logic often questioned (see Butler 1999). Rather than a type of tourist behaviour or activity, sustainable tourism is an ethos, an approach to the management of the relationship between tourism, tourists, the environment and the local community. In fact, the term has been extremely problematic to define and hotly contested (see Butler 1999; France 1999). Most salient features are evident in Bramwell and Lane's (1993: 2) definition. For them, 'sus-

tainable tourism is a positive approach intended to reduce the tensions and frictions created by the complex interactions between the tourism industry, visitors, the environment and the communities which are host to holidaymakers. It is an approach which involves working for the long-term viability and quality of both natural and human resources. It is not anti-growth, but acknowledges that there are limits to growth'. Four basic principles guide a sustainable tourism approach: holistic planning and strategy making; the importance of preserving ecological processes; the need to protect both human heritage and biodiversity; and the key requirement: to develop in such a way that productivity can be sustained over the long term for future generations (Bramwell and Lane 1993: 2). In the view of Hetzer (in Fennell 1999: 31), these principles translate into four pragmatic aspirations of responsible tourism: minimum environmental impact; minimum impact on – and maximum respect for – host cultures; maximum economic benefits for the host country's grassroots; and maximum 'recreational' satisfaction to participating tourists.

The academic literature is heavily punctuated with case studies and examples of sustainable tourism in action. These include whole-destination solutions, implementation by stakeholder groups, and tourist reactions to sustainable tourism (see Hall and Lew 1998). There have also been protracted critical discourses in which the nature and, indeed, existence *per se* of the theorized shift from mass tourism (associated with mass production or Fordism) to alternative tourism (accompanying the fragmented, highly niched markets characteristic of post-Fordism) have been challenged. Early theorizations of binary oppositions between the two paradigms have been rejected, as has the idea of a 'sustainability spectrum' between the polar conditions (Clarke 1997). The conflicting objectives of sustainable tourism and different aspirations of stakeholders are intensely difficult to resolve satisfactorily, especially since research on sustainable tourism lacks a political economy dimension (Williams and Shaw 1998). Finally, perhaps some of the most damning insights have centred on the notion that mass tourism does have a role to play in the tourism of the new millennium. As unthinkable as it may have once seemed, mass tourism does deliver benefits to host communities and environments. In one poignant exploration, well-meaning trekkers are demonstrated to have an environmental tariff greater or at least equal to conventional mass packaged tours (Mowforth and Munt 1998: 198), while mass tourism may offer environmental management possibilities by acting practically akin to a pollution sink by channelling harmful impacts into concentrated spaces (Becker 1995).

Tomorrow's Tourism Tomorrow

The jury is still out on the intellectual credentials of sustainable tourism. In spite of these deliberations, sustainable tourism has become the guiding principle of tourism management for producers and governors throughout the globe. In both the developed and developing world, the notion of responsible tourism development has found great favour among a multitude of local, regional and state governments, non-government organizations (NGOs) and supranational bodies (see Mowforth and Munt 1998). For example, in recent years the British government has developed its 'wise growth' approach. According to *Tomorrow's Tourism*, 'a "wise" growth strategy for tourism is one which integrates the economic, social and environmental implications of tourism and which spreads the benefits throughout society as widely as possible' (DCMS 1999: 48); that is to say, sustainable tourism by another name but avoiding the intractability of the 'sustainable' tag.

'Wise growth', and other state-led strategies like it, points to two very clear directions for further academic research. The first concerns the actions once a sustainable tour-

ism strategy has been agreed. Initiatives, policies and programmes have to be matched with careful monitoring of the nature and rate of transition. To date, monitoring has been relatively neglected, although it generates important debates (see Butler 1999). The English Tourism Council's 20 indicators of sustainability are a pertinent example (ETC 2001). They recognize that change is engineered gradually over the long term. Further research is required to benchmark, measure and evaluate progress over the medium and long term; test the probity of the selected indices; and ensure that the indicators remain robust and relevant to the changing times. Moreover, given governmental commitment to attain greater sustainability, questions must be answered of whether they are the most relevant measures for England, whether their selection was motivated by wider politico-environmental agendas, and whether the data are manipulated for propaganda purposes.

Heavy reliance on secondary data alone to examine the impact of tourism on the environment is also dubious. Rather, progress is required in a second area to address one of the greatest paradoxes of tourism research: namely, that the natural environmental impacts of tourism are most frequently gauged and evaluated by social scientists, not physical scientists as one may expect. Many 'experts' often lack the detailed knowledge and training to be able to judge the full environmental consequences of tourism development. For example, according to Butler (1999), this has manifested itself in the failure of many commentators to embrace the concepts of carrying capacities adequately. To extend the boundaries of knowledge in this manner is not going to be straightforward. Concerted research programmes are necessary, in which there are closer links between 'soft' and 'hard' scientists (human and physical geographers), as well as researchers trained in inter-tradition knowledge and skills. The schism between physical and human geography born of the increased specialization of the sub-disciplines is not insurmountable. However, for greater integration to occur several considerable differences in research traditions, perhaps most strikingly in the production of essentially normative versus intrinsically contestable knowledge, will have to be reconciled. The willingness to search for hard facts and to establish universal 'truths' so much evident in physical geography is greatly at odds with the philosophical developments in human geography during the 1990s. Human geographers have been grappling with the notion that all knowledge is contestable and hence dependent on the background (so-called 'positionality') of the investigator, especially the philosophical perspectives and methodological approaches s/he brings to a study. As is the case with monitoring, it may be a long time before there are tangible benefits in this area.

References

Becker, C. 1995: Tourism and the environment. In A. Montanari and A.M. Williams (eds), *European Tourism: Regions, Spaces and Restructuring*, Chichester: Wiley, 208–20.

Bramwell, B. and Lane, B. 1993: Sustainable tourism: an evolving global approach, *Journal of Sustainable Tourism*, 1, 1–5.

Cater, E. 1993: Ecotourism in the Third World: problems for sustainable tourism development, *Tourism Management*, April, 85–90.

Clarke, J. 1997: A framework of approaches to sustainable tourism, *Journal of Sustainable Tourism*, 5(3), 224–33.

Department of Culture, Media and Sport, Tourism Division (DCMS) 1999: *Tomorrow's Tourism: A Growth Industry for the New Millennium*. London: DCMS.

English Tourism Council (ETC) 2001: *National Sustainable Tourism Indicators: Monitoring Progress towards Sustainable Tourism in England*. London: ETC.

Fennell, D.A. 1999: *Ecotourism: An Introduction*. London: Routledge.

France, L. 1999: Sustainable tourism. In M. Pacione (ed.), *Applied Geography: Principles and Practice*, London: Routledge, 321–32.

Hall, C.M. and Lew, A.A. (eds) 1998: *Sustainable Tourism: A Geographical Perspective*. Harlow: Longman.

Ioannides, D. 1995: Strengthening the ties between tourism and economic geography: a theoretical agenda, *Professional Geographer*, 47, 49–60.

Mathieson, A. and Wall, G. 1982: *Tourism: Economic, Physical and Social Impacts*. Harlow: Longman.

Shaw, G. and Williams, A.M. 1994: *Critical Issues in Tourism: A Geographical Perspective*. Oxford: Blackwell.

Wheeller, B. 1993: Sustaining the ego, *Journal of Sustainable Tourism*, 1, 121–9.

Williams, A.M. and Shaw, G. 1998: Tourism and the environment: sustainability and economic restructuring. In C.M. Hall and A.A. Lew (eds), *Sustainable Tourism: A Geographical Perspective*, Harlow: Longman, 49–59.

Further Reading

Butler, R. 1999: Sustainable tourism: a state-of-the-art review, *Tourism Geographies*, 1, 7–25. A brief and very approachable summary of cutting-edge research on sustainable tourism.

Mowforth, M. and Munt, I. 1998: *Tourism and Sustainability: New Tourism in the Third World*. London: Routledge. A more developed critique of the emergence of 'sustainable tourism' which considers many of the intricacies of the critical debates surrounding tourism-based development revealed through the lens of Third World cases and evidence.

Internet Resources

- The University of Exeter is a centre of excellence in tourism research and teaching. The website of the Tourism Research Group within the School of Geography and Archaeology includes details of all its activities: its Tourism Associates consultancy arm; its new M.Sc. course in tourism, development and policy; and its work towards promoting the recently established Geography of Leisure and Tourism Research Group of the RGS (with IBG). http://www.ex.ac.uk/geography/tourism/tourism.html

17

Critical Geography and the Study of Development: Showers of Blessing?

Ben Page

> Rich nations now benevolently impose a straitjacket of traffic jams, hospital confinements and class rooms on the poor nations, and by international agreement call this 'development'.
>
> (Ivan Illich 1971: 162)

Imagine you're watching television. In the break a charity starts to make an appeal for its latest aid campaign in the developing world. A serious male voice tells you that 'In the year 2000 it is incredible to believe that 1.123 billion people in Africa, Asia and Latin America don't have access to an improved supply of water to drink'. As he speaks you watch a black woman walk to collect water. She has to struggle across awkward terrain with a baby on her back and a water container on her head. After waiting in a queue she edges through the mud to an unprotected spring to collect her water. The voice continues. 'In March 2000 the signatories at the Second World Water Forum in the Hague agreed to reduce this number by half by the year 2015, and to provide affordable, safe water for all people by the year 2025.' The woman on the film has collected her water and the camera is zooming in on her face. When the appeal to your generosity comes you hear the exasperated voice of male reason explaining how straightforward it is to solve this engineering problem, and you see the exhausted face

of the silent woman appealing to your emotions, challenging you to ring through a donation on your credit card. 'With your help . . .'

This is a parody, of course. Most development charities are alert to cliched visual imagery these days, and some are alert to gendered and racial stereotypes and even try to subvert them in their literature, but the essence of the idea remains the same. Material inequalities exist in the world, but they can be corrected by a process called 'development'. In some ways this is incontrovertible. It is undoubtedly true that vast numbers of people could live longer, easier lives if they had good access to safe, cheap water. So, because development seems such a worthy goal, it carries a tremendous sense of moral authority and, therefore, can be used to justify a wide range of political projects. In recent years geographers have been thinking about the development process more critically. This is not an academic exercise of taking something straightforward and turning it into something ambiguous. It is a willingness to accept that 'development' can be questioned. This chapter describes a number of ways in which contemporary geographers are thinking about the development process. In each case the issue of water supply is brought to the fore to show how it is possible for a critical development geography to enrich our understanding of an important issue without abandoning global humanitarian sentiments.

Big ideas are not very fashionable in geography at the moment and few projects come bigger than development. Since the Second World War vast areas of the globe have made it their goal to become more like Europe and America. Big ideas (or 'meta-narratives') are distrusted because they suggest that the whole world, with its diverse range of peoples and cultures, can be understood using only a small number of concepts. The people who get to choose which concepts are used tend to come from wealthy areas and they describe the world according to their own vision of how it should be. Every place is understood in relation to the history of Europe or America. When writers talk about a 'totalizing, homogenizing discourse', this is part of what they mean. The language and concepts that make up the 'development discourse' reflect relationships of power. They do not merely describe the world, they actively shape it. This is why some writers have even started to talk about entering a 'post-development' era.

How does the homogenizing character of development discourse impact on the provision of drinking water supplies? Throughout Africa, Asia, Latin America and the Caribbean the same western ideas are shaping water policy. A 'low-income country' in Asia is somehow the same as a 'low-income country' in Africa. The same policies can be followed in each place to produce development. For example, at the moment numerous international institutions are campaigning for the privatization of water supplies in the developing world. Government-run water supplies are currently being privatized in Buenos Aires, Accra and Shanghai. In each place the same multina-tional water companies from France, the UK and the United States are competing to win similarly phrased contracts to become the operator of water supply networks with similar technology (Bayliss 2001; Hall 1999). Advocates of privatization argue that water supplies organized along business lines are more efficient and are more popular because the old supplies operated solely by governments have failed (Uitto and Biswas 2000). Privatization is portrayed as progress towards a more developed form of water system. When geographers have advocated more local control of privatization (Marvin and Laurie 1999), or questioned the consequences of orienting supply around economic rather than social goals, they appear to be opposing progress. Greg Ruiters and Patrick Bond are two geographers who have been actively involved in resistance to water privatization (Ruiters and Bond 1999) in South Africa, but advocates of liberalization accuse them of being unrealistic and holding back development.

Despite the fact that geographers today claim they are anxious to avoid organizing their studies around meta-narratives, development remains a prominent and widely used concept. Almost every university geography department will have undergraduate courses in development geography and lecturers who are labelled specialists in the field of development studies. There are also many academic journals devoted to development studies to which geographers regularly contribute. *Third World Quarterly*, *Progress in Development Studies* and *Third World Planning Review* are just three that geographers often publish in and that provide lots of interesting case studies. Any claim that geography has entered a 'post-development' era is rightly treated with some scepticism (Power 1998). Development geography appears to be a flourishing sub-discipline despite a widespread questioning of the central concept around which it is organized. How do development geographers reconcile this apparent contradiction?

Geographical Histories of Development

Some, especially those who are sceptical about the claims of the development institutions, have begun to look at the history of development from a new and more critical perspective. Unsurprisingly, it was historians who began this approach, but geographers have a particular contribution to make because of their long history of close technical involvement with the planning of development projects and because development is such an obviously spatial process (Crush 1994). Much of the work of geographers in the nineteenth and early twentieth centuries was connected to the exploration and administration of overseas colonies, as Michael Heffernan's work on French geographical societies and French imperialism can illustrate (Heffernan 1994). As colonial policies were rescripted into the language of welfare and development after the Second World War, geography retained its role as a body of expert knowledge used to advise the new policy makers on aspects of land use, town planning and transport. This historical research project sets out to uncover the role that geographical ideas, language and institutions played in the history of the late colonial and early post-colonial periods when development was institutionalized. For example, Garth Myers traced the career of a British colonial official called Eric Dutton who was a major force behind urban planning in Africa in the first half of the twentieth century and also the author of a number of geographical books (Myers 1998). In relation to the development of water supplies, for example, geographers had a key role in assessing available resources and advising on water management for drinking and irrigation. Understanding this role could entail asking questions about who was allowed access to water, who the scientific knowledge about water resources was shared with, and what the water was used for.

Post-colonialism and Development Geography

The same antipathy to development's relationship with colonialism and with more contemporary asymmetries of geo-political power fuels another vein of critical development geography. This draws on a body of ideas, known as post-colonial theory, which emerged from literary and cultural studies. These geographers try to counter the totalizing character of development discourse by recovering the histories of particular places and people that were ignored in earlier writing about the developing world. Two pieces of work about South Africa illustrate this approach: Jennifer Robinson has looked at the way women were involved in managing urban housing programmes from the 1930s and Fareida Khan has looked at the role played by black South Africans in the history of conservation (Robinson 1998; Khan 1997). Geographers have also analysed the discursive tactics employed by the advocates of development (Klak and Myers 1997). In relation to the provision of drinking water, such research could focus on the forgotten history of indigenous technology and pre-colonial knowledge about water and water resources, such as the work of Bill Adams and his colleagues on African irrigation systems in Kenya (Adams et al. 1997). However, it could also ask questions about the strategies by which the language of development is used to persuade citizens in Third World states of the merits of privatizing urban water supply.

Geography and Alternative Development

Other geographers, who are less sceptical of development as an idea but are still critical of the way development works in practice, are hunting for alternative forms of development (Bebbington and Bebbington 2001). These geographers are often inspired by the work of anthropologists and various intriguing ethnographic research methods have been used to record the views of people whose voices in the past went unheard, such as street children (Young and Barrett 2001). Much of this work places a strong emphasis on new social movements and community development (Peet and Watts 1996). These grassroots initiatives, which sometimes become concretized as more formal non-governmental organizations, are seen as an institutional vehicle for producing a new form of development with more local integrity. Chasca Twyman's work on the ways that local communities are incorporated into conservation projects in Botswana or Claire Mercer's studies of the role played by NGOs in contemporary Tanzania illustrate this line of research (Twyman 2000; Mercer 1999). Such community-centred movements have been central to the development of water supplies for many years. In Cameroon, for example, community development has been integral to expanding access to rural water supplies since the late 1950s. But questions arise over the capacity of these non-governmental organizations to expand supplies, over their accountability and over the political consequences of their increasing influence.

Critiques of Development Practice

Despite an increasing academic hostility towards development as an idea, the apparatus of development (the development agencies, banks and NGOs) continues to wield a huge influence over the policies governments adopt across Asia, Africa and Latin America. Many geographers devote their research to commentating critically on current development projects. In particular, much of the research in the 1990s was dedicated to studying the consequences of structural adjustment plans – the bundle of neo-liberal economic policies (privatization, reduced protectionism, currency devaluation) advocated by major lenders such as the

World Bank as a recipe for economic growth and development. Bringing attention to the failures and negative consequences associated with these policies has been a vital task (Mohan et al. 2000; Robson 2000), particularly because those who advocate the plans back them up with an elaborate research machine of their own setting out to demonstrate their success. Under the banner of structural adjustment, water authorities across the globe have been seeking cost recovery from water users – more people have been paying more to get access to their drinking water. The key question becomes: is it fair to make everybody pay for their water according to the volume they consume regardless of their income?

Geography, Environment and Development

Geographers are not the only researchers analysing current development initiatives, but there are areas (urban studies, migration, gender issues) where they have particular expertise that makes their work amongst the most exciting. However, it is the studies of the politics of environmental change in the Third World that have gained most prominence. In the last few years an area of study known as 'political ecology' has emerged at the point where environment and development meet, and it is here that geographers have made their most substantial recent contributions to the field of development studies (Batterbury 2001; Bryant 1997). Philippe Le Billon's work illustrates the way in which natural resources such as diamonds, oil and timber have become central to understanding the history of war and national development in Cambodia and Angola (Le Billon 2001). Because they are interested not only in political economy and culture but also in physical ecological processes, geographers are well placed to show how development is simultaneously a process of social change and ecological change. In the case of providing urban water supplies, for example, patterns of global trade might influence the pattern of investment in the water supply infrastructure, but the physical availability of water and its quality are no less important.

Think back to the development charity campaigning for funds to run its water project. The international solidarity on which it is relying is easy to empathize with. For many people the attraction of geography is that it addresses issues that really matter, such as worldwide access to clean water. However, just because these issues are tremendously important does not mean that analysts cannot be critical of institutions and the policies that have been put into action in the past. Indeed, if development is to be regulated it needs to be scrutinized. Current geographical work sets out to disturb the simplicity of development propaganda in order to deepen our understanding of different places. Development geography is at an exciting moment precisely because it has dispensed with the view that anything can be justified as long as it is labelled progressive.

References

Adams, W., Watson, E. and Mutiso, S. 1997: Water, rules and gender: water rights in an indigenous irrigation system, Marakwet, Kenya, *Development and Change*, 28, 707–30.

Batterbury, S. 2001: Landscapes of diversity: a local political ecology of livelihood diversification, SW Niger, *Ecumene*, 8, 437–64.

Bayliss, K. 2001: *Water Privatization in Africa: Lessons from Three Case Studies*. PSIRU report [online], http://www.psiru.org.

Bebbington, A. and Bebbington, D. 2001: Development alternatives: practice, dilemmas and theory, *Area*, 33, 7–17.

Bryant, R. 1997: Beyond the impasse: the power of political ecology in Third World environmental research, *Area*, 29, 1–15.

Crush, J. 1994: Post-colonialism, de-colonization and geography. In A. Godlewska and N. Smith (eds), *Geography and Empire*, Oxford: Blackwell, 333–50.

Hall, D. 1999: *The Water Multinationals*. PSIRU

report [online], http://www.psiru.org.

Heffernan, M. 1994: The science of empire: the French geographical movement and the forms of French imperialism, 1870–1920. In A. Godlewska and N. Smith (eds), *Geography and Empire*, Oxford: Blackwell, 92–114.

Illich, I. 1971: *Celebration of Awareness*. London: Calder and Boyars.

Khan, F. 1997: Soil wars: the role of the African National Soil Conservation Association in South Africa, 1953–1959, *Environmental History*, 4, 439–59.

Klak, T. and Myers, G. 1997: The discursive tactics of neoliberal development in small Third World countries, *Geoforum*, 28, 133–49.

Le Billon, P. 2001: The political ecology of war: natural resources and armed conflicts, *Political Geography*, 20, 561–84.

Marvin, S. and Laurie, N. 1999: An emerging logic of urban water management, Cochabamba, Bolivia, *Urban Studies*, 36, 341–57.

Mercer, C. 1999: Reconceptualising state–society relations in Tanzania: are NGOs 'making a difference?', *Area*, 31, 247–58.

Mohan, G. 1999: Not so distant, not so strange: the personal and the political in participatory research, *Ethics, Place and Environment*, 2, 41–54.

Mohan, G., Brown, E., Milward, B. and Zack-Williams, A. 2000: *Structural Adjustment: Theory, Practice and Impacts*. London: Routledge.

Myers, G. 1998: Intellectual of empire: Eric Dutton and hegemony in British Africa, *Annals of the Association of American Geographers*, 88, 1–27.

Peet, R. and Watts, M. (eds) 1996: *Liberation Ecologies: Environment, Development, Social Movements*. London: Routledge.

Power, M. 1998: The dissemination of development, *Environment and Planning D: Society and Space*, 16, 577–98.

Robinson, J. 1998: Octavia Hill women housing managers in South Africa: femininity and urban government, *Journal of Historical Geography*, 24, 459–81.

Robson, E. 2000: Invisible carers: young people in Zimbabwe's home-based healthcare, *Area*, 32, 59–69.

Ruiters, G. and Bond, P. 1999: Contradictions in municipal transformation from apartheid to democracy: the battle over local water privatization in South Africa, *Working Papers in Local*

Governance and Democracy, 99(1), 69–79. Available online at http://qsilver.queensu.ca/~mspadmin/pages/Project_Publications/Papers.htm (accessed 12 September 2001).

Twyman, C. 2000: Participatory conservation? Community-based natural resource management in Botswana, *Geographical Journal*, 166, 323–5.

Uitto, J. and Biswas, A. (eds) 2000: *Water for Urban Areas: Challenges and Perspectives*. Tokyo: United Nations University Press.

Young, L. and Barrett, H. 2001: Adapting visual methods: action research with Kampala street children, *Area*, 33, 141–52.

Further Reading

Bryant, R. and Bailey, S. 1997: *Third World Political Ecology*. London: Routledge.

Crush, J. (ed.) 1995: *Power of Development*. London: Routledge.

Rahnema, M. and Bawtree, V. (eds) 1997: *The Post-development Reader*. London: Zed Books.

18

Globalization

Henry Wai-chung Yeung

'Geography is still important. Globalization has not diminished the economic significance of location.' So John Kay proudly declared in his fortnightly column in the world's famous pinkish newspaper, the *Financial Times* (10 January 2001: 14). The fact that business gurus like Kay write about globalization is absolutely nothing new; indeed, some geographers have associated the dreaming up of 'globalization' as a key word in the twenty-first century with such business gurus (Taylor et al. 2001). What is particularly interesting in Kay's *FT* column, however, is his unreserved defence of the importance of place and location in globalization. Coming from outside geography as an intellectual discipline, his views on globalization therefore provide significant legitimacy for the claim that geographers might have some very useful things to say about globalization.

Globalization as Both Material and Discursive Processes

Although the now influential text on globalization by economic geographer Peter Dicken – *Global Shift* – was published as early as 1986, geographers had not been very evident in the globalization debate until the early 1990s, when 'the end of geography' thesis became increasingly popularized in the media, policy circles and even academic worlds (see O'Brien 1992; Ohmae 1990).

In its essence, this thesis argues that the juggernaut of globalization, as a planetary force, is capable of penetrating all kinds of national boundaries and eroding any geographical differences inside them. The convergence effects of globalization are so strong that there is no place for geography in the processes of global change.

The critique of this strong convergence thesis of globalization has reverberated seriously in the social sciences (Held et al. 1999), and notably in human geography (Yeung 1998). In their laudable efforts to relate geography to globalization, Taylor et al. (2001: 1) noted that '[w]hatever your own opinion may be, any intellectual engagement with social change in the twenty-first century has to address this concept [globalization] seriously, and assess its capacity to explain the world we currently inhabit'. It is now generally agreed among human geographers that globalization should be viewed as both a set of material processes of transformation and resistance *and* a set of contested ideologies and discourses that operate across a variety of spatial scales. Globalization has seemingly led to significant material transformations in the global economy, most appropriately termed 'global shift' by Dicken (1998). These global transformations entail the rapid proliferation of cross-border trade and investments by global corporations and financial institutions, the ruthless penetration of global cultures epitomized by McDonald's, Hollywood films,

MTV and the Internet, and the reluctant power shift from nation-states to global governance.

These global transformations, nevertheless, are not geographically even, nor are they without resistance at different spatial scales and in different countries/regions. First, the global reach of corporate activities has failed to transform the world economy into a singular global production factory. In fact, what appears to be more convincing is the phenomenon of regionalization through which regions have emerged as the major motor of the global economy (Scott 1998). Global production seems to be taking place in such high-tech regions as Silicon Valley (northern California), Third Italy (Emilia-Romagna) and Baden-Württemberg (southern Germany). Global finance remains highly rooted in such existing global finance centres as the City of London, New York and Tokyo, and new international financial centres as Hong Kong and Singapore. Economic geographers have therefore argued for the examination of specific regions and territorial ensembles in order to appreciate the inherently geographical nature of globalization processes. Other geographical studies have shown that while economic globalization has been spearheaded by the cross-border operations of transnational corporations, the spatial transfer of business and industrial practices is by no means unproblematic (Leyshon and Pollard 2000). There remain significant place-based institutional limits to the globalization of business cultures and economic practices (Yeung 2000). For example, while capital can be transferred almost effortlessly across space, labour remains highly place-bound and locally embedded (Martin 2000). In other words, globalization encounters most 'friction' over space at the local and regional scales. It is clearly unable to erase spatial differentiation even in the realm of economic processes and governance. To cite the example of Singapore, even though it is often hailed as one of the most globalized economies of the world today, the state remains firmly the key driver behind Singapore's developmental trajectories. In fact, each round of economic liberalization and deregulation in Singapore is accompanied by another round of state re-regulation, albeit in different organizational and institutional forms (see also Yeung and Dicken 2000).

Second, the arrival of global cultures has further accentuated the awareness of local differences and cultural responses. While the global flows of information, images and artefacts are greatly facilitated by the advance in telecommunications and transport technologies, the recipients and consumption of these flows remain highly territorialized and embedded in specific geographical boundaries. Although few people in developed countries do not have access to global cultures in their embodied and commodified forms, a large segment of the population in developing countries is still highly dependent on localized products and cultural services. Most Filipinos and Brazilians, for example, are still drinking their local beverages instead of Coca-Cola and Pepsi. Furthermore, global cultures are equally resisted in many countries that view these cultural inflows as threatening to their national interests and political legitimacy. There are counter-movements to the globalization of cultures through the reassertion of local/national cultures, religions and other forms of social identities. Despite the dominance of American films in the English-speaking world, the local film industry in India and Hong Kong continues to go from strength to strength. These counter-movements, enmeshed in the global web of cultural flows, have created a world of cultural mosaics and differences. We are certainly still far away from a culturally homogeneous world of a global village.

Third, the political economy of globalization does not necessarily indicate the demise of nation-states as the primary locus of political governance. Some geographers have argued that the world is not becoming more unruly and disorderly in the governance sense (Herod et al. 1998). Rather, there is a reconfiguration of political governance and institutional power through which the nation-state is joined by supra-national institutions and other institutions (e.g. non-governmental organizations) to govern globalization. Just witness the immensely complex organization behind anti-globalization protests in Seattle, Melbourne and Genoa. Nation-states worldwide are experiencing a relative decline in their capacities to control the whereabouts of global corporations, to provide social welfare for their citizens, and to contain the post-Cold War world threats. These states are, however, not fixed entities to be overwhelmed by the unruly processes of globalization. Indeed, some nation-states are active agents in promoting globalization, for example through the liberalization of financial markets. Still other nation-states have globalized themselves by building interstate consensus and contributing to trans-state sources of power and authority. It is therefore premature to proclaim the death of nation-states.

Globalization is also as much a set of material processes as a set of contested ideologies and discourses (see Leyshon 1997). The ways through which globalization is represented can have an equally significant impact on material processes. While we cannot equate globalization with neo-liberalism *per se* (i.e., defined as a political economic ideology in favour of market mechanisms in lieu of state intervention), it is true that neo-liberal programmes of policies have greatly facilitated the advent of globalization. Just think of how liberalization and market-based economic policies have enabled certain giant corporations to emerge and dominate the global economy. In other words, globalization cannot proceed smoothly without its supporters and champions. As expected, business gurus, media pundits and policy makers (and even academics) are often the strongest supporters of globalization. In championing globalization and making it appear as a natural force, these people have essentially 'naturalized' globalization processes and portrayed them as 'necessary', 'inevitable' and 'beyond our control'. Powerless citizens have to choose either wholesale embracing of globalization or economic decline and social exclusion.

What's So Geographical about Globalization?

What geographers have plainly shown in their recent work is that this either/or ideological choice of globalization should be questioned (see Cox 1997). In fact, localities can reassert their power in the global economy through certain discursive practices and/or constructing alternative globalizations. These practices include the building up of social capital, experimentation with non-market forms of economic exchanges, the reorganization of local work practices and so on. To challenge the ideological supremacy of neo-liberalism and its much broader processes of globalization, some localities and communities have actively demystified globalization as a necessary and inevitable phenomenon. The rise of local civil societies in many parts of western Europe, for example, is a direct answer by localities to regulate the apparently unruly effects of globalization.

To sum up this discussion so far, globalization is understood in geography as a set of material processes and discursive practices that operate across different spatial scales. The differences that these spatial scales make are that globalization is about changing relationships between geographical scales and that these scales are socially constructed through ideologies and discourses (Taylor et al. 2001: 3). Geography's greatest contribution to the globalization debate is the recognition of how spatial scales matter in our understanding of the complex operations of globalization processes. Globalization is sometimes highly localized, whereas at other times it is the region that contains most of the effects of globalization. Viewed in this geographical perspective, globalization is *not* necessarily an essential global force that homogenizes national and local differences. It has as many local and regional dimensions as being a global force.

Bringing Geography Back into the Globalization Agenda

Now that we have a better appreciation of the geographical specificity of globalization, it is perhaps time to move beyond the simplistic notion of globalization as merely a set of end-state phenomena. There are at least three important geographical questions for the study of globalization by future researchers. First, we need more balanced empirical assessment of globalization processes. At both research and policy levels, it is really not helpful just to embrace globalization wholeheartedly or to condemn it unreservedly. We know from previous studies that globalization, like the two-headed Roman god Janus, has two faces and can harm as well as benefit localities. It is imperative for future researchers to focus on the uneven geographical outcomes of globalization processes and to evaluate critically these outcomes in order to arrive at better-informed policies. Second, there seem to be too many top-down studies of globalization. There is clearly a role for bottom-up studies of globalization by taking an agency approach, i.e., paying more attention to what people do and say. We need to understand not only the strategies of global corporations and international organizations. Equally important is the demand on us to appreciate how globalization is contested by social actors like you and me and those anti-globalization protesters (and their executive counterparts in transnational firms and banks) at various geographical scales, whether at the level of the regional resurgence of religious activities or at the level of individual household decisions. Last but not least, future researchers need globally coordinated research that is well executed locally. In other words, globalization research inherently requires global cooperation in research initiatives and networks across various countries and re-

gions. Through these cross-national/regional efforts, we might have a better chance of coming to grips with globalization as a complex set of phenomena that are increasingly shaping the lives of most people on this planet.

References

Cox, K. (ed.) 1997: *Spaces of Globalization: Reasserting the Power of the Local.* New York: Guilford.

Herod, A., Ó Tuathail, G. and Roberts, S.M. (eds) 1998: *An Unruly World: Globalization, Governance and Geography.* London: Routledge.

Leyshon, A. 1997: True stories? Global dreams, global nightmares, and writing globalization. In R. Lee and J. Wills (eds), *Geographies of Economies,* London: Arnold, 133–46.

Leyshon, A. and Pollard, J. 2000: Geographies of industrial convergence: the case of retail banking, *Transactions of the Institute of British Geographers,* 25, 203–20.

Martin, R. 2000: Local labour markets: their nature, performance, and regulation. In G.L. Clark, M.A. Feldman and M.S. Gertler (eds), *The Oxford Handbook of Economic Geography,* Oxford: Oxford University Press, 455–76.

O'Brien, R. 1992: *Global Financial Integration: The End of Geography.* New York: Council on Foreign Relations Press.

Ohmae, K. 1990: *The Borderless World: Power and Strategy in the Interlinked Economy.* London: Collins.

Scott, A.J. 1998: *Regions and the World Economy: The Coming Shape of Global Production, Competition and Political Order.* Oxford: Oxford University Press.

Taylor, P.J., Watts, M.J. and Johnston, R.J. 2001: Geography/globalization, *GaWC Research Bulletin,* 41, Department of Geography, Loughborough University.

Yeung, H.W.-C. 2000: The dynamics of Asian business systems in a globalising era, *Review of International Political Economy,* 7, 399–433.

Yeung, H.W.-C. and Dicken, P. (eds) 2000: Special issue on economic globalization and the tropical world in the new millennium, *Singapore Journal of Tropical Geography,* 21, 225–373.

Further Reading

Dicken, P. 1998: *Global Shift: Transforming the World Economy*, 3rd edition. London: Paul Chapman.

Held, D. (ed.) 2000: *A Globalizing World? Culture, Economics, Politics*. London: Routledge.

Held, D., McGrew, A., Goldblatt, D. and Perraton, J. 1999: *Global Transformations: Politics, Economics and Culture*. Cambridge: Polity.

Mittelman, J.H. 2000: *The Globalization Syndrome: Transformation and Resistance*. Princeton, NJ: Princeton University Press.

Yeung, H.W.-C. 1998: Capital, state and space: contesting the borderless world, *Transactions of the Institute of British Geographers*, 23, 291–309.

19

Historical Geography: Making the Modern World

Catherine Nash and Miles Ogborn

In many ways geography is a profoundly historical discipline. Rather than simply describing places, human and physical geographers are concerned with understanding the processes that have shaped places and landscapes in the present, whether these are patterns of climate change that are registered in the palaeoecological archives of a sediment core or the patterns of migration that can be traced in census records, oral histories and maps of urban growth. In human geography this interest in the making of places is also about tracing the lives of people in the past, and the social, political, economic and cultural institutions, systems and processes that shaped those lives and places, their relationships with each other, and the ideas that people used to make sense of the world and which, in turn, shaped the making of places and social relations. Sometimes this historical focus is on the recent past, sometimes on more distant periods; sometimes on reconstructing the geography of specific places in specific periods, or on the large-scale processes of change that have shaped the modern world. What geographers bring to this historical work is their spatial focus on distribution and scale, their attention to the specificity of places and their interconnections, and their interests in the cultural meanings as well as the political, economic and social characteristics of places (Graham and Nash 2000).

Key Themes and Sources

Historical geography today is a very diverse subject (Ogborn 1997, 1999), but its enduring themes include land use and human–environmental relations, settlement, migration and demography, and key periods of economic, environmental, political and social change and their new modes and relations of production, distribution and consumption (Dodgson and Butlin 1990; Graham and Proudfoot 1993). These have included the early modern shift in Britain from feudal to capitalist forms of social, economic and political organization and the changes in the nature of the rural landscape, agriculture and labour that they entailed and enabled. New landscapes were engineered by enterprising agriculturalists and landowners as wetlands were drained, moorlands enclosed and pastures improved. The draining of the fenlands of East Anglia in the seventeenth century is a classic case study within British historical geography (Darby 1940). Improvement in this period involved the shift to new forms of wage labour and the consolidation of rural landownership through enclosure beginning in the fifteenth and sixteenth centuries, and later backed by government in the late eighteenth and early nineteenth centuries. Along with schemes of improving land and livestock, this practice of converting former communally worked open field systems to individually worked separate plots transformed the landscape as it transformed the nature of agricultural labour. Rises in productivity and enclosure of open field systems had social and ecological impacts as livelihoods were lost with the loss of common land. Rural landscapes were newly ordered, productive, capitalized and commodified.

Local transformations of land and lives were also linked to changing geographies elsewhere. In historical geography these changes are often explored through focusing on local cases and by locating them within wider economic, social and political networks. Early forms of industrialization in the countryside and towns, for example, were financed by the profits of increasingly wider networks of trade and investment and other early modern forms of globalization: the exploration and settlement by the West of parts of the globe previously unknown to Europeans, the early stages of a global trading network involving the large-scale production of raw materials, food, minerals and commodities for industries and markets elsewhere, new extensive forms of capital investment and the transfer of profits and resources from distant locations to Europe. From at least the fifteenth century, global interconnections were developing and deepening in tandem with modern European industrial and urban development. These European networks variously displaced, altered or were grafted onto pre-existing circuits of trade and travel (Ogborn 2000).

This focus on the profound impacts of modernization extends forward in time to the dramatic changes gaining pace in the eighteenth century and accelerating in the nineteenth century: rapid urbanization and industrialization, the development of new intensive forms of mechanized agriculture, improvements in transport, population growth and migration to sources of industrial labour in towns and cities, new forms of bureaucratic organization, new versions of the state and widening spheres of communication and contact which extended the

degrees of interconnection between places at a variety of scales. This new spatial integration meant that local or regional forms of governance were subordinated to the centralized state and networks of communication, trade and influence deepened and expanded, 'transforming the intimate geographies of everyday life and animating grand transformations on a national or global scale' (Ogborn 1998: 19).

This attention to the different and increasingly interconnected geographies of the past is matched by a greater awareness of the ways in which ideas of social differences were shaped alongside processes of modernization, and how these processes were themselves experienced differently by people with different degrees of wealth and power. Historical geographers today are concerned not only with questions of class but also with other dimensions of domination and identity such as 'race', sexuality, nationhood and gender. This involves a focus on the transformation of material landscapes and on questions of meaning, power and identity; on the making of places and on the less material but no less significant cultural beliefs, practices and identities that both informed and were shaped by the shifting geographies of the modern world. Historical geographers are interested in the knowledges, practices and representations through which people have attempted to make sense of the world and their place within it, from scientific systems of mapping, measurement and classification of things, animals and people to cultural representations in different kinds of texts and visual images. This means using a wide range of sources of evidence. In his discussion of the historical geographies of urbanization, Richard Dennis (2000) has argued that while modernization was taking place 'on the ground, in the creation of new spaces, new scales and new patterns of segregation and specialization, and new forms of technology', it was also 'taking place in the mind, in how cities were spoken about, how they were visualized, mapped, painted and photographed'. Urban historical geographers therefore, he suggests, should be eclectic in their sources and approaches and use both qualitative and quantitative sources that can illuminate each other. This focus on both material changes and on questions of identity, power and culture is evident in one of the key themes of recent historical geography: the historical geography of colonization.

Historical Geographies of Colonialism

One key form of globalization in the past was European colonization, which imposed systems of rule and other cultural norms and practices on subordinated people. Its impacts have been profound. The effects of European colonialism in all its forms still continue; they can be seen in global patterns of uneven development today and in the patterns of migration from former colonies to former colonial powers. Colonialism involved material processes, knowledges and modes of representation. It has involved the transformation of environments, the displacement of colonized peoples and settlement of European migrants in colonized lands. For colonized people it meant the imposition of new systems of rule and culture, and the disruption and destruction of lives, ways of life and livelihoods. Colonialism involved new ways of understanding, representing, ordering and imagining the world, including measurements of space, forms of classification, metaphors of 'progress' and 'development', and the whole concept of a world divided into the 'Old' and the 'New'. European exploration, travel and trade were cultural and political as well as economic processes that were shaped by and which shaped European imaginative geographies. They were also deeply implicated in unequal power relations within Europe and between Europe and other parts of the world. The movements of people, plants, animals, ideas, cultures, raw materials, commodities and capital that were set in train by the twinned processes of European capitalist development and overseas expansion transformed social relations and dramatically altered environments (Pawson 1990). Environmental historians and historical geographers have chronicled the massive environmental changes resulting from capitalist and colonial settlement and agricultural systems in white settler contexts in North America or the 'New World', including the 'ecosystem shocks' of the British and Spanish invasions of the Americas and the Caribbean, the effects of the introduction of new species of flora, fauna and bird life in the Antipodes, and the ecological transformations of cropland and grazing expansion in the New World. Local projects that transformed the nature of particular places often had far-reaching effects that illustrate the interconnections between power, money, identity and culture, and the entwined and often unequal relationships between people.

Making Egypt Modern

For example, understanding what lay behind the cutting of the Suez Canal between the Mediterranean and the Red Sea allows many of these different facets of historical geography to be brought together. The canal itself was a dramatic transformation of the natural landscape which was based upon ideas of scientific and technological progress held by its French engineer Ferdinand de Lesseps and the venture's European backers, and on the labour of the thousands of workers recruited from the villages of southern Egypt who did the digging and construction work. On a global scale, the canal (and the similar project in Panama) was a way 'to open up the world to the free movement of commodities' (Mitchell 1988: 16) by bringing the rest of the world and their raw materials and markets more fully within Europe's orbit. Within north Africa, it was the most dramatic part of a broader transformation of land and lives in Egypt in the 1860s as the Nile delta was given over to the cultivation of cotton for the Lancashire textile mills whose other sources of supply had been cut off by the American Civil War. New roads, railways, dams, telegraph lines, forms of landownership and conditions of agricultural work were imposed onto the Egyptian landscape and people by the European finance that was drawn by the profits to be made. This economic dependence was facilitated by representations of Europe in Egypt (as modern, progressive and powerful) and of Egypt in Europe (as traditional, backward and dependent). The opening of the canal coincided with the building of a new opera house and the commissioning of Guiseppe Verdi to write *Aïda*, a piece which presented a European version of Ancient Egypt for a predominantly European audience. Yet the recognition of Egyptian dependence upon Europe, particularly after British occupation in 1882, also led to a growing nationalism which sought to foster an alternative Egyptian identity. The eventual outcome was the seizing of the Suez Canal from the British in July 1956, which signalled a profound change in the political geography of the post-war world.

Conclusion

The construction of the Suez Canal was made possible by new technologies and hard labour, but it also depended upon the idea of the *modern*. Tracing historical geographies of modernization also involves exploring the historical geographies of this concept (Nash 2000). The idea of the 'modern' can be located within European discourses which have used ideas of the modern, traditional and primitive to make distinctions between huge areas of the world, countries, regions and between people. Describing non-western livelihoods as unproductive and primitive, for example, has justified long histories of western intervention in the name of modernization. Reconstructing pre-colonial environmental histories provides important evidence of pre-colonial productive and

complex agricultural systems, often sustaining dense populations, that challenge persistent colonial discourses of empty or ill-used lands awaiting European settlement and western versions of development. How history is understood clearly matters and historical geographers are increasingly interested in the ways in which the past is remembered and represented (Ogborn 1996). This entails both considering the history of geography as a discipline as well as the popular representation of the past in museums, heritage sites, commemorative events, film and other forms of public culture. The representation of the past is deeply tied to contemporary conflicts about collective identities. How the past is understood has profound implications for the ways in which individuals and groups are located within states and nations, and local, family or community histories.

References

Darby, H.C. 1940: *The Draining of the Fens.* London: Cambridge University Press.

Dennis, R. 2000: Historical geographies of urbanism. In B. Graham and C. Nash (eds), *Modern Historical Geographies*, Harlow: Pearson Education, 218–47.

Graham, B.J. and Proudfoot, L. (eds) 1993: *An Historical Geography of Ireland.* London: Academic Press.

Mitchell, T. 1988: *Colonising Egypt.* Berkeley, Los Angeles and London: University of California Press.

Nash, C. 2000: Historical geographies of modernity. In B. Graham and C. Nash (eds), *Modern Historical Geographies*, Harlow: Pearson Education, 13–40.

Ogborn, M. 1996: History, memory and the politics of landscape and space: work in historical geography from autumn 1994 to autumn 1995, *Progress in Human Geography*, 20, 222–9.

Ogborn, M. 1997: (Clock)work in historical geography: autumn 1995 to winter 1996, *Progress in Human Geography*, 21, 414–23.

Ogborn, M. 1998: *Spaces of Modernity: London's Geographies, 1680–1780.* New York and London: Guilford Press.

Ogborn, M. 1999: Relations between geography

and history: work in historical geography in 1997, *Progress in Human Geography*, 23, 95–106.

Ogborn, M. 2000: Historical geographies of globalisation. In B. Graham and C. Nash (eds), *Modern Historical Geographies*, Harlow: Pearson Education, 43–69.

Pawson, E. 1990: British expansion overseas, *c.* 1730–1914. In R.A. Dodgson and R.A. Butlin (eds), *An Historical Geography of England and Wales*, London: Academic Press, 521–44.

Further Reading

Dodgson, R.A. and Butlin, R.A. (eds) 1990: *An Historical Geography of England and Wales*, 2nd edition. London: Academic Press.

Graham, B. and Nash, C. (eds) 2000: *Modern Historical Geographies*. Harlow: Pearson Education.

20

New Political Geographies 'Twixt Places and Flows

Peter J. Taylor

The United States is a nuisance. I mean this in a loosely legal sense. In English law behaviour constitutes a nuisance when it causes serious discomfort or inconvenience to others: a factory emitting noxious fumes over a neighbourhood is a dangerous nuisance. Extrapolating to the global scale, the United States is by far the world's greatest polluter who, by refusing to ratify the Kyoto Agreement on restricting greenhouse gas emissions, condemns the rest of the world to suffer the effect of its excesses. The United States is able to conduct itself in this obnoxious way because of its political power. As 'lone superpower' it looks after its own supposed interests first, and only then considers the consequences for others.

Such selfish behaviour is normal in the traditional conduct of international relations. In the competitive world of rivalry between states, each has had to look to its own ends and use any means available to satisfy them. Hence the twentieth century was a century of many wars both great and small. On several occasions during the Cold War this nearly led to nuclear catastrophe, not just for the United States and USSR, but for the whole of humanity. That dangerous world is now past. A divided political world has been succeeded by an integrated economic world: the Cold War has been replaced by globalization (see chapter 18 by Yeung in this volume). In the new circumstances, the

behaviour of states is much more circumscribed by the world economy than heretofore. Hence, rather than using the term lone superpower, a throwback to Cold War terminology, the United States' nuisance behaviour might be better described as the actions of the last state with a large degree of national sovereignty (Martin and Schumann 1997: 216) in a new globally connected world. This is a different sort of world, one where we need to think about politics and space in new and exciting ways.

Beyond Mosaic Political Geography

The traditional study of political geography has focused on the spatial structure of states and their aggressive competitive struggle for territory. This was to view the world as a mosaic of political spaces, sovereign state territories, which the Cold War simplified into a three-part mosaic of 'first', 'second' and 'third' worlds. Such 'mosaic thinking' is no longer appropriate for understanding political geography under conditions of contemporary globalization (Taylor 2000, 2001).

A new spatial framework is required. This has been provided by Castells (1996) in his depiction of a new global 'network society'. His analysis depends upon a distinction between 'spaces of places' and 'spaces of flows'. The mosaic of territorial states is an example of the former, the global transactions of financial markets are an example of the latter. Geographically, globalization is a result of a change in the balance of importance between these two different types of space. Combining the technologies of communications and computers has facilitated massive growth in connections and linkages across the world and in the process has often undermined the territorial integrity of states. A good example is the Internet, where governments are finding it very difficult to censor material within their own territories. The rise of network society is built upon such spaces of flows.

It is important not to think that spaces of flows are replacing spaces of places. Both are, and always have been, necessary for each other's existence. The key word in the above discussion is, therefore, balance, reflecting the changing saliences of the two types of space. Thus network society does not mean 'the end of the state'. States remain important institutions embedded in our modern world. Globalization is forcing states to restructure their activities as they are seriously challenged in some key functions (e.g. regulating financial markets). The development and deepening of regional institutions such as the North American Free Trade Association and the European Union are the clearest examples of spatial restructuring, providing 'havens' for states in the threatening turmoil of the world economy. In this way states remain important, but they are no longer omnipotent in international relations. World cities have emerged as important rivals in many ways. As nodes in a world city network, the major cities of the world have found a new lease of life in network society. Little more than two decades ago great cities such as New York were seen as massive political and economic problems, losing population and jobs and going bankrupt. Today New York is an archetypal global city (Sassen 2001), one of the most important nodes in the world city network.

This remarkable turnaround in the fortunes of large cities is probably the key geographical effect of contemporary globalization.

New political geographies for the twenty-first century, therefore, will be about how power in the world economy is organized between spaces of places (notably states) and places of flows (notably cities). I will illustrate this by taking a fresh look at three topics that have been central to traditional political geography – boundaries, capital cities and federalism – to show how they fare in a political geography beyond simple mosaics.

Boundaries: Alternative Cross-border Organization

Sovereign boundaries define territorial states. They are where flows – people, commodities, money – into and out of the state are traditionally controlled. But with globalization, boundaries are being seen less as spatial barriers and more as places of contact. Thus communities that were once resigned to being at the edge of the all-important state are now developing the idea of being the meeting place of nations.

Such opportunities are particularly evident in the European Union (Murphy 1993) and one particularly interesting example is along the north-west Mediterranean shore that straddles France and Spain (Morata 1997). Here there are two rival organizations promoting cross-border cooperation. The first is a regional alliance between Catalonia and its two French neighbours, Languedoc-Roussillon and Midi-Pyrénées. This 'West Mediterranean Euro-region' defines a cross-border territory. The second is the C-6 network, an alliance of six cities, four in Spain (Barcelona, Palma de Mallorca, Valencia and Zaragoza) and two in France (Montpellier and Toulouse). This defines a cross-border network. Both arrangements are promoting cross-border economic development through political cooperation but they are based upon very different forms of spatial organization. The Euro-region is a traditional 'space of places' strategy operating through a defined area, the C-6 network is an innovative 'space of flows' strategy operating through local nodes in the European city network. This is probably the clearest example of these alternative organizations of space operating in the same region. And yet, the two organizations compete rather than cooperate, an expression of the places–flows tensions we can expect in the political geography of the twenty-first century.

Capital Cities: New Roles Under Globalization

The importance of capital cities grew with the increasing power of the nation-state. By the end of the nineteenth century the great cities of Europe were the 'imperial capitals' of the great powers of Europe: London, Berlin, Paris and Vienna. These cities totally dominated their national/imperial territories: they were firmly planted at the top of national city hierarchies with economic growth and cultural flowering complementing their political centrality. This all-encompassing model of the capital city contrasts with the political specialization model in which the capital city has few functions beyond its capital role. Such capital cities are chosen or developed specifically to promote the territorial integration of the state. Cities selected to replace old capitals are usually more central (e.g. to Moscow from St Petersburg, to Ankara from Istanbul, to New Delhi from Calcutta); new capital city developments are usually located as spatial compromises (e.g. Washington, DC between North and South, Canberra between Sydney and Melbourne, Ottawa between Ontario and Quebec, Brasilia between Amazon region and the rest of the country). How do these different capital city strategies fare under new conditions of globalization?

Specializing in national politics is not a good springboard from which to take advantage of the new opportunities of globaliza-tion. Locked into a territorial politics in a space of places, most specialist capitals have not fared well. Ankara is no match for Istanbul in the global stakes, and the same can be said for Brasilia (with respect to São Paulo), Canberra (with respect to Sydney), Ottawa (with respect to Toronto) and New Delhi (with respect to Mumbai). In contrast, the all-encompassing capital cities have been able to convert to world city status relatively easily. London, Tokyo and Paris are prime examples. Where this has not happened, specific reasons relate to outcomes of twentieth-century wars: Vienna as a victim of the defeat of Austria-Hungary in the First World War and Berlin as a victim of the defeat of Germany in the Second World War.

However, there are two specialist capital cities that have been able to make the transition to world city status. Both are unusual examples. First, Washington, DC has gradually broadened its functions and climbed the US urban hierarchy (Abbott 1996). Being the capital city of a twentieth-century super-power with by far the largest national economy in the world created a situation for the growth of a capital city economy for firms needing to mix domestic and international contacts (e.g. the lobbying activities of global law firms). Second, Brussels, the unofficial capital of the European Union as the home of its most important institutions, was chosen on compromise grounds (being in a small and relatively unimportant country, Belgium). However, as the decision-making centre of the largest economic market area in the world, Brussels has attracted many corporations who wish to operate their European activities from near the political centre of the continent (Elmhorn 2001). Although Washington does not rival New York, Chicago or Los Angeles, and Brussels does not rival London, Paris and Milan, both cities have developed world city status as distinctive nodes in the world city network where economic and political flows meet in an interaction of the national and the global.

Table 20.1   First and second city global connectivity ratios by type of state

Centralized states		Federated states	
Top two national pairs	*Ratio*	*Top two national pairs*	*Ratio*
London/Manchester	4.4	New York/Chicago	1.5
Paris/Lyon	3.0	Milan/Rome	1.7
Tokyo/Osaka	3.7	Madrid/Barcelona	1.4
Stockholm/Gothenburg	3.2	Toronto/Montreal	1.5
Copenhagen/Aarhus	5.4	Sydney/Melbourne	1.2
Oslo/Bergen	3.7	Frankfurt/Hamburg	1.4

Federalism: State Structures and World City Formation

Federalism has been termed the most 'geographical' form of state because sovereignty is split between the central state and its federal units. This allows for different territorial interests to be represented countering separatist tendencies within the state. This spatial strategy failed in the aftermath of the Cold War with the demise of federal USSR and the disintegration of Yugoslavia. However, these represent very special conditions – the collapse of communism – that put unsustainable pressure on the existing state structures. In the rest of the world federalism has been successful in maintaining the spatial integrity of most large states across all continents, including the United States, India, Brazil, Germany, Canada, Nigeria and Australia. Under conditions of contemporary globalization, however, there is an unintended consequence of this spatial flexibility in terms of the configuration of the world city network.

In a federal system there is decentralization of power to lower units, whereas in a centralized state the power is concentrated. These two alternative spatial arrangements will create different forms of national urban hierarchies: with the latter one city will be pre-eminent, whereas decentralization creates a situation in which many important cities can flourish. This difference might well be accentuated with globalization since the advantages of national power concentration in one city should translate into a special advantage in global competition. For instance, in highly centralized Japan, it has been argued that the main constraint on Osaka in its aspiration for world city status is its 'Tokyo problem' (Hill and Fujita 1995). In table 20.1 the ratio of the connectivity in the world city network (as measured in Taylor et al. 2001 using global-scale business services) between the first- and second-ranked cities is shown for a selection of federal and centralized states. The contrast between the two types of state is stark: in federal states the first city is about one and a half more connected than its closest rival; in centralized states this increases to ratios of over three. It seems that not only does Osaka have its 'Tokyo problem', but Manchester has its 'London problem' and other second cities in centralized states are similarly afflicted. The obverse is that federal states are more conducive to more widespread world city formation. The resulting 'unbalanced growth' is leading to demands for decentralization in some states, such as in the UK with its recent devolution and plans for regional assemblies, and in Indonesia with the provincial resistance to rule from Jakarta.

Conclusion

These arguments have introduced new ways of thinking in political geography whereby traditional spaces of places are intertwined with new spaces of flows. Under conditions of contemporary globalization, simple competition between territorial states has been combined with the more complex mutual relations between networked cities. This is a crucial agenda for political geographers in the twenty-first century, but there are others of importance. I have concentrated on just a few selected themes – boundaries, capital cities, federalism – in the context of globalization, and there are other exciting research agendas notably in quantitative electoral geography, critical geopolitics and post-colonial political geographies. These indicate the range of topics covered in contemporary political geography.

References

Abbott, C. 1996: The internationalization of Washington, D.C., *Urban Affairs Review*, 31, 571–94.

Castells, M. 1996: *The Rise of Network Society*. Oxford: Blackwell.

Elmhorn, C. 2001: *Brussels: A Reflexive World City*. Stockholm: Almqvist and Wiksell International.

Hill, R.C. and Fujita, K. 1995: Osaka's Tokyo problem, *International Journal of Urban and Regional Research*, 19, 181–94.

Martin, H.-P. and Schumann, H. 1997: *The Global Trap*. London: Zed Books.

Morata, F. 1997: The Euro-region and the C-6 network: the new politics of sub-national co-operation in the West-Mediterranean area. In M. Keating and J. Loughlin (eds), *The Political Economy of Regionalism*, London: Frank Cass, 292–305.

Murphy, A. 1993: Emerging regional linkages within the European Community: challenging the dominance of the state, *Tijdschrift voor Economische en Sociale Geografie*, 84, 103–18.

Sassen, S. 2001: *The Global City*, 2nd edition. Princeton, NJ: Princeton University Press.

Taylor, P.J. 2000: World cities and territorial states under conditions of contemporary globalization, *Political Geography*, 19, 5–32.

Taylor, P.J. 2001: World cities and territorial states under conditions of contemporary globalization: looking forward, looking ahead, *GeoJournal*, 52(2), 157–62.

Taylor, P.J., Catalano, G. and Walker, D.R.F. 2001: Measurement of the world-city network, *GaWC Research Bulletin*, 43 [online], http://www.lboro.ac.uk/gawc.

Further Reading

Agnew, J. 1997: *Political Geography: A Reader*. London: Arnold.

Taylor, P.J. and Flint, C. 2000: *Political Geography: World-Economy, Nation-State, Locality*, 4th edition. London: Prentice Hall.

Internet Resources

• Globalization and World Cities Study Group and Network: http://www.lboro.ac.uk/gawc

21

World on the Move: Migration and Transnational Communities

Alisdair Rogers

On 18 December 2000 the United Nations declared International Migrant's Day, recognizing that 150 million people lived outside the country of their birth. Although this figure represents less than 3 per cent of the world's population, it is double the level of 1965 and most projections are that the volume and spread of international migration will go on increasing. The UN Convention against Transnational Crime and two protocols against trafficking humans were also signed that day in Palermo, Sicily. The discovery at Dover (UK) docks the previous June of the bodies of 58 Chinese people in a shipping container, the indispensable tool of global trade, was a terrible illustration of both the powerful motivations behind global mobility and the personal costs involved. The study of world migration, by geographers among others, is now providing key insights into both macro-scale economic and technological change and individual experiences of sustaining identities and livelihoods across great distances and in unfamiliar environments.

Globalization of Migration

Long-distance migration is nothing new, and by some estimates a greater proportion of the world's people were involved in the nineteenth century than today. Families of settlers and lone workers from Europe, India and China journeyed to the Americas, Australia and Africa in record numbers. Between 1846 and 1939, around 59 million people emigrated from Europe alone. As many as 12 million Africans were forcibly transported to the Americas as slaves by the mid-1800s. Two World Wars and a worldwide economic depression reduced the international flow of people just as they did goods. But in the late twentieth century a combination of factors restored momentum, as outlined by Castles and Miller in *The Age of Migration* (1998). These included: growing economic inequality between countries; the end of the Cold War; political instability, e.g. in the Balkans and north-east Africa; demographic differences between youthful and ageing societies; environmental deterioration and its impact on agrarian livelihoods; and cheaper, more widespread air transport.

As the intensity of global migration has risen, so has its extent. The International Labour Organization reckons that between 1970 and 1990 the number of countries counted as major receivers of migrants increased from 39 to 67 and the number of major senders grew from 29 to 55. Among the new receivers are the Persian Gulf states, enriched after the 1973 oil crisis, and newly industrializing economies such as Malaysia, Hong Kong and Singapore. New senders include Russia and China, where the government lifted a ban on overseas travel in the 1980s. Although the United States still accounts for one in every six migrants, the old pattern of South–North moves is breaking down to include many more South–South flows.

Although the broad causes and patterns of international migration are generally understood, they conceal significant subnational variations. This is where many of the key geographical questions are now found (Gorter et al. 1998). Does international migration reduce or increase economic differences between places and people, and do remittances of money sent back home compensate for the loss of skilled and working-age people? Is it a cause or an effect of the economic restructuring of regions and cities? Are positive effects at one spatial scale offset by negative effects at others, for example in the balance of taxes and welfare costs between local and national governments? Why do migrants come from and go to so few places, even within countries and regions with apparently the same levels of development? Within the major cities, where most migrants settle, will their levels of spatial concentration decline over time or not, and how is this associated with racism and social exclusion or 'cultural preservation' (see the special issue of *Urban Studies* 1998)? Does the settlement in suburbs of newer, wealthier immigrants such as Chinese and Punjabis in Vancouver spell the end of older patterns of ethnic segregation (Li 1997)? These simple questions are turning out to be difficult to answer conclusively, as conditions are changing so quickly.

New Migrants

The new and evolving diversity of types of migrant and migration makes matters of classification and definition trickier. The standard distinction between voluntary migration (including labour migration and family reunification) and forced migration (refugees, asylum seekers, displaced people and environmental migrants) is being refashioned by other forms of stratification by class, gender, race and sexuality. Furthermore, in crossing a national boundary, a person may be subject to redefinition by the state. Borders are simultaneously barriers, gateways and filters where legal status is redetermined, where some may pass with ease and others are detained or abused. They are where asylum seekers are demonized on television and in the press, but business travellers are depicted as citizens of the world. Public myths about immigration and the representation of migrant bodies are significant areas for future inquiry.

The diversity of types of migrant and experiences of migration is opening up areas of geographical inquiry, not least because governments are eager for more information and understanding. Restrictions on immigration by the North forced many desperate individuals to enter illegally or place themselves in the hands of organized smugglers, popularly termed snakeheads in China or coyotes in Mexico. Law enforcement agencies are desperate for intelligence about these criminal networks. At the other end of the spectrum, there is a growing number of skilled transients, or high-salary professionals, working abroad for multinational corporations and non-governmental organizations (NGOs), sometimes called 'expatriates' (Beaverstock and Smith 1996). Aspirant world cities, such as Singapore and Hong Kong, are actively trying to lure them by fashioning cosmopolitan environments. Small numbers of wealthy migrants may make a visible impact on the urban landscape in world cities such as London, where one in nine workers is foreign-born. But the personal pressures of overseas assignments, not least for dual-career households, mean high and costly rates of failure. Skilled workers in particular industries, notably IT, but also services such as health care, are being actively sought by countries traditionally hostile to immigration, such as the UK and Germany. Finally, there are business migrants, given permission to enter a new country in order to start up a business and invest substantial sums. But are such high-status migrants no more than a quick-fix solution to labour shortages, and does their presence lessen opportunities for native-born workers? Are they a 'brain drain' from their home countries, or can the 'knowledge networks' formed among expatriates to give advice and expertise to development projects compensate?

One of the laws of the pioneer of migration research in the 1880s, E.G. Ravenstein, was that women were less likely to move across national borders than men. But since the 1980s women are increasingly migrating as individuals in their own right, not just as part of the 'baggage' of male workers. Perhaps half of all international migrants are now female. Women in the South are enjoying relatively higher levels of freedom and equality, including the ability to earn a living and make decisions for themselves. Peasant lives are being uprooted by urban industrial change. At the same time, there is a growing demand in the North for workers in industries and occupations that typically employ women: health care, domestic service, hotel and restaurant work, entertainment and the sex industry. Bridget Anderson's compelling study, *Doing the Dirty Work?* (2000), reveals the abuses by both employers and the state faced by female migrant domestic workers (see also Henshall Momsen 1999). 'Global care chains', in which a migrant woman is paid to raise her employer's children, and in turn pays another woman to look after her own children back home, represent a kind of globalization of emotional labour (Hochshild 2000).

Transnationalism: The Next Research Frontier

Another assumption guiding migration studies until recently is that an individual moved from country A to country B and settled there, more or less integrating over time. This often meant a one-way transition from a rural peasant context to an urban industrial one. These assumptions are under increasing scrutiny and challenge from the emerging field of transnational studies, on the cutting edge of migration research. Initially defined by social anthropologists as 'the process by which transmigrants, through their daily life activities forge and sustain multi-stranded social, economic and political relations that link together their societies of origin and settlement' (Basch et al. 1994: 7), transnationalism may also extend to other social networks spanning national borders and involving routine interaction across them: international NGOs, social movements, organized criminal gangs, corporate employees and spiritual movements. There has been notably important research by geographers on transnationalism in the Asia-Pacific, including overseas Chinese business networks (Olds and Yeung 1999) and the impact of high-status Chinese transnationals on the landscape and politics of Vancouver (Ley 1995; Mitchell 1997). In the UK, the Economic and Social Research Council (ESRC) funded a programme of 19 projects on transnational communities, in which the work of geographers figures prominently.

The idea of migration as a circuit or a border-crossing social field points towards a more refined and less rigid and binary understanding of home, identity and national belonging. It may also be viewed as part of the new field of diaspora study (Cohen 1997). Diasporas, classically describing Jews and Armenians, for example, but now being used for other groups such as Kurds, Lebanese or Hindus, have complex and often ambiguous or conflictual three-way relations between their homeland (which may be 'lost'), their state of residence and their co-ethnics in the rest of the diaspora. They exemplify the recomposition of relations between territory and identity at the heart of cultural and economic globalization.

Approaches to Migration

The study of migration by geographers and others draws upon many approaches and methodologies (see Silvey and Lawson 1999 for a good overview). Contrasting theoretical approaches focused on individual-level, behavioural and micro-economic models on the one hand, or more historical-structural concepts on the other (see Gorter et al. 1998 for a review). Some kind of resolution has been identified in the study of social networks and social capital, regarded as both context and resource for migration. The work of Douglas Massey in US–Mexico migration (Massey et al. 1987) shows how networks reduce the risks and costs of migrating, and Thomas Faist (2000) uses social network ideas to explain why migrants travelled from a particular village in Turkey to Germany but not one nearby. Migration studies have always been handicapped by the lack of adequate datasets, particularly longitudinal or panel surveys that enable researchers to trace the socio-economic and spatial paths of individual migrants and their children over time. Finer-scale census data, individual-level datasets such as the UK Samples of Anonymized Records, and the linking of census data to entry records or tax records promise greater scope for inquiry, especially in North America. Such data will greatly increase the value of spatial modelling of migration flows and impacts (Stilwell and Congdon 1991). In cultural geography, there are ethnographic and biographical accounts that focus less on the causes of migration and more on the experiences of migrants, for example the collection on migrant literature edited by King, Connell and White (1995) entitled *Writing Across Worlds*. To understand transnational communities requires research-ing in both origins and destinations, exemplified by Katy Gardiner's studies in Bangladesh and Britain (Gardiner 1993).

References

Anderson, B. 2000: *Doing the Dirty Work? The Global Politics of Domestic Labour*. London: Zed Books.

Basch, L., Glick Schiller, N. and Szanton Blanc, C. 1994: *Nations Unbound: Transnational Projects, Postcolonial Predicaments, and Deterritorialized Nation-States*. Amsterdam: Gordon and Breach.

Beaverstock, J. and Smith, R. 1996: Lending jobs to global cities: skilled international labour migration, investment banking and the City of London, *Urban Studies*, 33, 1377–94.

Faist, T. 2000: *The Volume and Dynamics of International Migration and Transnational Social Spaces*. Oxford: Oxford University Press.

Gardiner, K. 1993: Desh-bidesh: Sylheti images of home and away, *Man*, 28, 1–15.

Gorter, C., Nijkamp, P. and Poot, J. (eds) 1998: *Crossing Borders: Regional and Urban Perspectives on International Migration*. Aldershot: Ashgate.

Henshall Momsen, J. (ed.) 1999: *Gender, Migration and Domestic Service*. London: Routledge.

Hochshild, A.R. 2000: Global care chains and emotional surplus value. In W. Hutton and A. Giddens (eds), *On the Edge: Living With Global Capitalism*, London: Jonathan Cape, 130–46.

King, R., Connell, J. and White, P. (eds) 1995: *Writing Across Worlds: Literature and Migration*. London: Routledge.

Ley, D. 1995: Between Europe and Asia: the case of the missing sequoias, *Ecumene*, 2, 185–210.

Li, P. 1997: The changing face of the suburbs: issues of ethnicity and residential change in suburban Vancouver, *International Journal of Urban and Regional Research*, 21, 75–99.

Massey, D., Alarcón, R., Durand, J. and González, H. 1987: *Return to Aztlan: The Social Process of International Migration from Western Mexico*. Berkeley: University of California Press.

Mitchell, K. 1997: Transnational subjects: constituting the cultural citizen in an era of Pacific Rim capital. In A. Ong and D. Nonini (eds), *Ungrounded Empires: The Cultural Politics of Modern Chinese Transnationalism*, London: Routledge, 228–58.

Olds, K. and Yeung, H.W.-C. 1999: (Re)shaping 'Chinese' business networks in a globalizing era, *Environment and Planning D: Society and Space*, 17, 535–55.

Silvey, R. and Lawson, V. 1999: Placing the migrant, *Annals of the Association of American Geographers*, 89, 121–32.

Stilwell, J. and Congdon, P. (eds) 1991: *Migration Models: Macro and Micro Approaches*. London: Belhaven.

Urban Studies 1998: special issue on ethnic segregation, 35(3).

Further Reading

Castles, S. and Miller, M.J. 1998: *The Age of Migration*, 2nd edition. Basingstoke: Macmillan.

Cohen, R. 1997: *Global Diasporas: An Introduction*. London: UCL Press.

Internet Resources

- For news on migration issues around the world, updated monthly, visit *Migration News* at http://migration.ucdavis.edu/mn/
- For material on transnational communities, as well as links to other sources on migration, go to the ESRC Transnational Communities website at http://www.transcomm.ox.ac.uk

22

Urban Geography: The 'Death' of the City?

Loretta Lees

At the beginning of the twenty-first century more and more people live in cities and are affected by urban processes than ever before. In 1950 about 300 million people lived in cities. Today about 3 billion, or roughly half the world's population, live in cities. It seems ironic, then, that the rhetoric of contemporary urban geography and planning is inflected with notions of the 'death' of the city. Three 'stories' in particular stand out. What binds these three stories together is that each story of death is countered by a related story of (re)birth. First, a decayed and dying inner city is seen to be reborn through processes of urban renaissance or gentrification (see figure 22.1). Second, there are concerns that unchecked urban growth is destroying the city itself with sprawl, pollution and congestion. Many planners now insist that these environmental problems can only be alleviated if cities and metropolitan regions are rebuilt in much denser, more urbanized and therefore more sustainable forms. Finally, while futurists predict the end of the city as the Internet and other communication technologies promise even the remotest of places access to the same network of services and other amenities once available only in central business districts, urban geographers are devising new conceptual models to represent the postmodern city.

Urban Renaissance

The process of gentrification was initiated in the late 1960s and 1970s as middle-class gentrifiers began buying and renovating property in dilapidated inner-city neighbourhoods common then in North American and British cities. Post-war, rapid suburbanization had led to the 'ghosting' of the central city, to disinvestment and ghettoization. By contrast gentrification was a back to the city movement of both people (repopulation) and capital (reinvestment) that countered the supposed death of the city. The rhetoric surrounding gentrification latched onto the notion of the 'rebirth' or 'renaissance' of the central city: pro-gentrification groups gave their newsletters titles such as 'The Phoenix' (in Brooklyn, New York City), playing off images of a phoenix rising from the ashes of dead city neighbourhoods.

In the 1990s debates over the causes of gentrification dominated urban geography. Two camps emerged, one broadly economic, the other broadly cultural (see Hamnett 1991). Marxist geographer Neil Smith (1996) argued that gentrification was the result of the differentiation and equalization of capital over the urban environment, an uneven dynamic of capital circulation he termed 'the rent gap'. Smith's explanation had a class-based dimension, as it was the

Figure 22.1   Gentrification in progress: the Lower East Side, New York City, 1988.

middle classes with capital who took over decrepit inner-city working-class neighbourhoods and displaced the 'indigenous' populations. By contrast, humanist geographer David Ley (1996) argued that gentrification was the result of the social and political choices made by a new (middle) class of people who turned their backs on the supposed sterility of suburban life. In seeking to live in the dying central city, they revitalized it. Others such as Liz Bondi (1991) added that gender was an important aspect of the gentrification process, for single and professional women were turning their backs on the gendered and sexist suburbs for life in the more diverse central city (on gay gentrification see Lauria and Knopp 1985). Most geographers today would argue that gentrification is both an economic and a cultural process.

What marks out contemporary gentrification is the increased role of urban policy in the process (see Wyly and Hammel 1999). Recent urban policy statements in both the UK and the United States overtly promote gentrification (see DETR 1999, 2000; US Department of Housing and Urban Development 1999) through initiatives designed to plug the gap between successful and failing cities and, as we shall see in the next section, to counter urban environmental problems. In addition to the role of policy, Lees's (2000) review of the gentrification literature identifies some other key issues for future research – the particular relationships between gentrification and the global city, race and gentrification, etc. She concludes: 'a more detailed examination of the "geography of gentrification" would constitute a progressive research programme and lead us to rethink the "true" value of gentrification as a practical solution to urban decline in cities around the world' (2000: 405).

Urban Sustainability

Planners and environmentalists are increasingly concerned about the sustainability of contemporary urbanization. Wackernagel and Rees (1995) developed the idea of the 'ecological footprint' to represent the environmental impact of cities in terms of the amount of land required to sustain them. They estimate that if the entire Earth's population consumed resources at the rate of the typical resident of Los Angeles, it would require at least three planet Earths to provide all the material and energy they consumed (see also Fokkema and Nijkamp 1996).

The city is choking to death due to (sub)urban sprawl, congestion and air pollution. As such, low-density suburban development is no longer the remedy for urban ills. Planners now argue that the way to make cities more sustainable is to make them more compact and urbanized (see Blowers and Pain 1999). In contrast to suburban living, more compact settlement patterns conserve open space and enable residents to use energy-efficient mass transit systems. In this way the densification of urban settlement is said to provide a solution to local-scale environmental problems such as traffic and air pollution from automobile use, but also to help reduce greenhouse gas emissions and the impact of cities on the global climate. The idea of urban densification has been rapidly incorporated into urban policy documents in a variety of countries (see DETR 2000, on the UK). The 'compact city' is a 'livable city' connected to the ideals of urban regeneration and gentrification (see Lees and Demeritt 1998). Its advocates argue that the redevelopment of abandoned industrial 'brownfield' sites instead of suburban 'greenfields' provides a way not just to protect the natural environment, but also to promote urban regeneration. But in so far as policies to encourage urban densification are often coupled with so-called growth management policies to discourage suburbanization through zoning restrictions on land use and taxes on driving vehicles, they have often proved controversial, particularly in the United States.

The idea of the 'compact city' has been

criticized for being unpractical (it is impossible to halt suburbanization), unrealistic (planning is limited in its influence) and undesirable (people do not want to live in each others' pockets) (see Filion 1996). Alternative concepts have been put forward, for example, the 'MultipliCity'. This approach to urban sustainability proposes clusters of development that include urban infilling and the creation of new ex-urban settlements, the result being a sustainable city-region. Such ex-urban development is becoming common practice, as we shall see in the next section.

To date much of the work on sustainable cities has focused on the spatial and physical environmental impacts of the problem. Blowers and Pain (1999: 296) argue:

> There is a danger that a simplistic technological and physical determinism will continue to deflect attention from more complex and unwelcome understandings of the unsustainability of cities. Debates about how to reconstruct the spatial and social elements of cities within an institutional framework that emphasizes social equality and political participation remain as relevant in the conditions of contemporary cities as they did a century ago.

The Postmodern Urban Condition

The emergence of the Internet has highlighted the fact that cities no longer enjoy the same monopoly on urbanity and urban functions that they once did. Traditionally, notions of the 'urban' have been defined by population density, and the concentration in cities of both economic functions, such as higher-order services, and social and cultural amenities, like art galleries, gives them a distinctly 'urban' feel. New forms of communication technology potentially challenge this definition by offering people in the countryside access to the same services and amenities as city dwellers (see Graham and Marvin 1996). While rural geographers have

debated the end of the 'rural' as urban cultural forms become so common as to challenge the distinctiveness of rural areas and traditions, the corollary is also true of cities. However, geographers have responded to bold declarations about the 'death of distance' by noting that the Internet, far from spelling the end of geography and equalizing the spatial differences between city and country, has also reinforced them as well. Fibre-optic cabling and broad-band access is most common in cities where the density of potential customers and the existing infrastructure make investments in these facilities most profitable (Graham and Marvin 1996, 2001).

The emergence of the Internet and the wired city is just one in a long series of social and economic changes encouraging urban decentralization and challenging the traditional concentric zone model of the city that has long been imprinted on the minds of geographers. The concentric zone model was conceived in the 1920s by the so-called Chicago School of urban sociologists, who based their ideas on the form of industrial Chicago at that time. They represented the spatial structure of the city through a series of concentric rings. Central to the city was the CBD, or central business district, surrounded by successive rings of industry, working-class housing, middle-class housing and commuter suburbs. By contrast, Dear (2000) argues that the 'new' postmodern city is increasingly centreless, multi-nucleated, disarticulated and polarized. In the post-Fordist metropolis industrial restructuring has given rise to a new geography of the city in which terms like edge city, exopolis, technoburbs and metropoles describe new urban patterns (see Garreau 1991; Soja 2000). In the UK, such developments can be seen along the M4 motorway corridor linking London with the old railway cities of Reading and Swindon and the port city of Bristol to the west. The 'new' city has become so unpredictable that Dear represents it as a centreless urban form. The blueprint is no longer the modern industrial city of Chicago but the postmodern post-industrial city of Los Angeles. Unlike Chicago, Los Angeles is seen to have no common narrative or story; as such it is very difficult to draw or represent.

In this 'new' city processes of decentralization coexist with central city processes of urban renaissance/gentrification. In the move from an industrial to a post-industrial era, central cities have shed their grimy industrial facades and replaced them with shopping malls, waterfront developments, restaurants and art galleries. City life has been commodified, repackaged and sold to city dwellers and tourists alike as they visit downtown attractions like New York City's South Street Seaport – once a working port, now a restaurant and shopping enclave. It seems that contemporary western cities are being pulled in two directions – towards the centre and towards the periphery. Advocates of urban sustainability have a complex polarity to manage.

Conclusion

Forty years ago Jane Jacobs condemned post-war city planning and urban renewal initiatives as 'the sacking of cities' in her bestselling *Death and Life of Great American Cities* (1961). Confident about what a city should be, she argued that 'there is nothing economically or socially inevitable about either the decay of old cities or the fresh-minted decadence of the new unurban urbanization' (1961: 16). At the beginning of the new millennium urban geographers are no longer so confident. In different ways each of the three stories highlighted here poses the question of what the city is and should be (see Paddison 2001b, Part I, on defining the city). For some gentrification is urban renaissance, while for others it represents the displacement and destruction of local urban communities. Debates about urban sustainability offer different normative prescriptions for what the city should be – the compact city is a quite different ideal

to the sustainable city-region. Unlike the Chicago School's concentric zone model, Dear's new postmodern city is almost impossible to represent – the best representation he manages to draw is a keno gameboard. Cities are more complex than ever before. The city is not dead but is being reborn as we speak. Contemporary planning is not about sacking cities but about sustaining and rethinking cities. The question on the lips of urban geographers today is, what is the city and what should a city be?

References

Blowers, A. and Pain, K. 1999: The unsustainable city? In S. Pile, C. Brook and G. Mooney (eds), *Unruly Cities*, London: Routledge, 247–98.

Bondi, L. 1991: Gender divisions and gentrification: a critique, *Transactions of the Institute of British Geographers*, 16, 190–8.

Dear, M. 2000: *The Postmodern Urban Condition*. Blackwell: Oxford.

Department of the Environment, Transport and the Regions (DETR) 1999: *Towards an Urban Renaissance: Sharing the Vision*. London: DETR.

DETR 2000: *Our Towns and Cities: The Future: Delivering an Urban Renaissance*. London: DETR.

Filion, P. 1996: Metropolitan planning objectives and implementation constraints: planning in a post-Fordist and postmodern age, *Environment and Planning A*, 28, 1637–60.

Fokkema, T. and Nijkamp, P. 1996: Large cities, large problems? *Urban Studies*, 33, 353–77.

Garreau, J. 1991: *Edge City: Life on the New Frontier*. New York: Doubleday.

Graham, S. and Marvin, S. 1996: *Telecommunications and the City: Electronic Spaces, Urban Places*. London: Routledge.

Graham, S. and Marvin, S. 2001: *Splintering Urbanism: Networked Infrastructures, Technological Mobilities and the Urban Condition*. London: Routledge.

Hamnett, C. 1991: The blind men and the elephant: the explanation of gentrification, *Transactions of the Institute of British Geographers*, 16, 259–79.

Jacobs, J. 1961: *The Death and Life of Great American Cities*. New York: Random House.

Lauria, M. and Knopp, L. 1985: Toward an analysis of the role of gay communities in the urban renaissance, *Urban Geography*, 6, 152–69.

Lees, L. 2000: A re-appraisal of gentrification: towards a 'geography of gentrification', *Progress in Human Geography*, 24, 389–408.

Lees, L. and Demeritt, D. 1998: Envisioning the livable city: the interplay of 'Sin City' and 'Sim City' in Vancouver's planning discourse, *Urban Geography*, 19, 332–59.

Ley, D. 1996: *The New Middle Class and the Remaking of the Central City*. Oxford: Oxford University Press.

Paddison, R. 2001a: Studying cities. In R. Paddison (ed.), *Handbook of Urban Studies*, London: Sage, 1–9.

Smith, N. 1996: *The New Urban Frontier: Gentrification and the Revanchist City*. London and New York: Routledge.

Soja, E. 2000: *Postmetropolis: Critical Studies of Cities and Regions*. Oxford: Blackwell.

US Department of Housing and Urban Development 1999: *The State of the Cities* [online], http://www.huduser.org.

Wackernagel, M. and Rees, W. 1995: *Our Ecological Footprint: Reducing Human Impact on the Earth*. Gabriola Island, BC: New Society Publishers.

Wyly, E. and Hammel, D. 1999: Islands of decay in seas of renewal: housing policy and the resurgence of gentrification, *Housing Policy Debate*, 10, 711–71.

Open University Understanding Cities series:

Allen, J., Massey, D. and Pryke, M. (eds) 1999: *Unsettling Cities*. London and New York: Routledge.

Massey, D., Allen, J. and Pile, S. (eds) 1999: *City Worlds*. London and New York: Routledge.

Pile, S., Brook, C. and Mooney, G. (eds) 1999: *Unruly Cities?* London and New York: Routledge.

Internet Resources

Websites

- US Department of Housing and Urban Development: http://www.huduser.org
- UK Department of the Environment, Transport and the Regions: http://www.detr.gov.uk
- The King's College London Gentrification Research website: http://www.gentrification.org
- Centre for Urban and Regional Studies, Newcastle University: http://www.ncl.ac.uk/curds
- The Urban Institute: http://www.urban.org
- Online planning journal: http://www.casa.ucl.ac.uk/olp

Listservs

- URBGEOG@LISTSERV.ARIZONA.EDU

Further Reading

Pacione, M. 2001: *Urban Geography: A Global Perspective*. London and New York: Routledge.

Paddison, R. (ed.) 2001b: *Handbook of Urban Studies*. London: Sage.

23
Feminist Geographies: Intersections of Space and Gender

Claire Dwyer

Why should geographers be interested in gender? When feminist geographers first answered this question in the late 1970s, their response was to point out that much of the geography then being written came from a male perspective. Although ostensibly 'gender blind', the figure of 'rational man' or 'mankind' which stalked the pages of geography textbooks was in fact very definitely a masculine figure – unencumbered by children to care for or domestic responsibilities, supported by a social network which included a wife at home! Writing in 1982, Monk and Hanson urged human geographers not to exclude 'the other half' in human geography. They pointed out that studies of agriculture in Africa which omitted to include the work of women, or studies of migration patterns that assumed everyone lived in households where only men were the breadwinners, gave only partial accounts of geographical patterns and processes. Writing in 2002, an answer to this question puts issues of gender even more centrally onto the geographical agenda. We might argue that an understanding of the gender relations between men and women in different parts of the world, or variations in ideas about masculinity or femininity in different places, is critical to how social, econ-

omic and political geographies are constructed and contested. However, our analysis would also go even further to suggest that gendered discourses and processes are implicated in the understandings of space and place we use in geography.

In this chapter I want to begin by discussing how we might think about gender and geography. I will outline some of the different ways in which feminist geographers have defined gender and how these ideas about gender have been important in reshaping geography. I will suggest that the project for feminist geographers is to explore the various ways in which ideas about gender – about masculinity and femininity – are being socially constructed in different times and places, and what impact these ideas have on the lives of men and women. This is illustrated by drawing on some of the work done by myself and others on the geographies of 'new femininities'. Finally, I will outline some of the new directions for geographies of gender.

Defining Gender

Perhaps the simplest way of thinking about gender is to define it in contrast to sex. While sex describes biological differences – being male or female – gender describes socially constructed characteristics – masculinity and femininity. Thus while individuals are born male or female, over time they acquire a gender identity, that is, an understanding of what it means to be a man or woman. This gender identity relies on the notion of biological difference and so is often assumed to be 'natural' rather than socially constructed. So 'natural' femininity might encompass, for example, motherhood, being nurturing, a desire for pretty clothes and the exhibition of emotions (Laurie et al. 1999: 3). Since these attributes are seen as 'natural', definitions of masculinity or femininity become fixed or normative, and those who are seen as acting outside these social norms risk condemnation or social sanction. Recognizing the ways in which gender identities – masculinities and femininities – are socially constructed, feminists have been able to analyse the (often highly inequitable) gender relations that are produced through material social practices. However, these analyses suggest that while dominant or hegemonic forms of gender relations can be recognized (often a form of patriarchal gender relations where men as a group are constructed as superior to women as a group and so assumed to have authority over them), gender relations and ideas about gender are historically and geographically differentiated. So at different times and in different places, particular social constructions of 'appropriate' masculinities and femininities emerge and become embedded within hegemonic gender relations.

It is also important to recognize that ideas about gender intersect with other aspects of identity, which include understandings of class, ethnicity, sexuality as well as (dis)ability or age. Indeed, much of the early feminist writing has been criticized for failing to recognize the diversity of experiences of women. Thus Leonore Davidoff and Catherine Hall (1987) illustrate how the 'moral order' of appropriate masculinities and femininities defined by the middle class

in Victorian England was achieved through ideas about class as well as gender. Similarly, Robert Connell (1995) argues that white men's construction of their gender identity is produced in relation to both an idealization of black masculinity as well as in opposition to white women. Embedded within normative social constructions of gender relations are also expectations about normative heterosexuality so that 'appropriate' gender identities usually mean heterosexual femininities or masculinities. This has led theorists to define dominant forms of gender relations as 'heteropatriarchy'. Gill Valentine's discussion of the time-space strategies of lesbians negotiating their sexual identities through different home and work spaces provides an excellent illustration of how such normative heterosexual femininities are constructed and resisted (Valentine 1993).

Before linking up these discussions about gender with geography, I want to mention recent work by post-structuralist feminists that has begun to challenge the idea of gender based upon biological sex which I outlined above. Influenced in particular by the work of Judith Butler (1993), some feminists have begun to question the assumption that the sexed body should be taken for granted. Instead, Butler argues that gender is itself a 'performance' maintained within what she calls the 'regulatory fiction of heterosexuality'. Her arguments about performance, and particularly the subversive potential of cross-dressing or drag, suggest that sexed bodies may themselves be socially constructed and implicated in our dichotomized ideas of gender. This perspective opens up new theoretical avenues for research on gender and embodiment, as biomedical technology increasingly challenges our ideas about the 'naturalness' of the body.

Gender and Geography

In her introduction to feminist geographies, Linda McDowell emphasizes that the goal of feminist geography is 'to investigate, make visible and challenge the relationships between gender divisions and spatial divisions, to uncover their mutual constitution and problematize their apparent naturalness' (McDowell 1999: 12). What this definition suggests is the many different and complex ways in which ideas about gender, space and place are entwined. On the one hand we might investigate how men and women experience spaces and places differently. McDowell herself has researched the different ways in which men and women working in merchant banking in London experience their work spaces (McDowell 1997), while Gill Valentine has considered women's experiences, and particularly their fears, within urban spaces (Valentine 1989). These different experiences highlight how our ideas about gender are actually constituted through different spaces. Thus women's experience of harassment within a macho working space like the trading floor of a merchant bank is a direct result of the assumption that such a place is not 'appropriate' for women. This entwining of ideas about gender and space has an important ideological underpinning. Feminists have shown that the social construction of a binary division between masculine and feminine is deeply entrenched in western thinking. Consider the following set of binary oppositions:

Masculine	Feminine
Culture	Nature
Mind	Body
Rational	Irrational
Public	Private
Production	Consumption
Work	Home

This mapping of binary distinctions within western thinking suggests the association of the feminine with nature and the body. It also emphasizes a powerful set of spatial divisions, so that the masculine becomes associated with the public sphere outside the home of work, production and knowledge,

while the feminine is associated with the home, and the domestic sphere of reproduction and consumption.

Feminist geographers have analysed the mutual constitution of these deep-seated spatial and gendered divisions in many different ways (WGSG 1984) and have examined how geographers have a deeply gendered approach to the understanding of space, landscape and nature (Rose 1993). At the same time geographers have emphasized the diversity of gender relations in different places as well as the specificities of gender within macro processes of globalization or economic restructuring (Laurie et al. 1999).

What these analyses suggest is the centrality of the study of gender to broader social, economic and political processes and the ways in which spatial concepts are profoundly gendered. Yet, as the quotation from McDowell above suggests, feminist work is also about challenging these associations. So we might want to ask whether there are ways of changing dominant ideas about gender through the reworking of different spaces and whether, at the beginning of the twenty-first century, new possibilities are emerging for the construction of masculinities and femininities. Such analyses require attention both to how men and women experience different spaces, and to the representation of space and place.

Geographies of 'New Femininities'

Our impetus for seeking to explore the geographies of new femininities (Laurie et al. 1999) came from both theoretical impulses and socio-economic realities. Increasingly, feminist geographers have recognized the diversity of different feminine identities (WGSG 1997), influenced particularly by post-structuralist perspectives on subjectivity. At the same time, rapid economic restructuring through the processes of globalization appears to be opening up 'new spaces' – of employment or social interaction – for women. Of course, such processes may not always be positive and the forces of globalization are equally likely to result in the remaking or reworking of existing gender relations (the restrictive policies of the Taliban in Afghanistan are perhaps the most extreme example). We sought to explore these questions through several different case studies that aimed to move beyond a Eurocentric celebration of 'new femininities' – so-called 'girl power' – to examine changes in a global context.

Nina Laurie discusses women's involvement in a workfare programme (PAIT) in Peru in the 1990s (in Laurie et al. 1999). The work, which was hard manual labour often in construction, was originally intended for men but was taken up by lower-class women as a means to provide for their families. Laurie argues that 'new social spaces' of work were created for women as they worked outside the home in a 'public' space – thus assumed gendered spatial divisions were challenged. Women gained confidence, developed new social networks and often transformed their domestic social relations. However, Laurie also illustrates the complex ways in which women negotiated these spaces so that their 'traditional' femininities were not compromised. Women used various strategies to ensure that their identities as 'good mothers' or as 'moral women' were retained, although they might be reworked in different ways. So a 'good mother' was no longer a woman who stayed at home but one who went out to provide for her family.

My own work on young British Muslim women (in Laurie et al. 1999) also considers the reworking of gender identities through different spaces. Drawing on the experiences of a range of young women, I illustrated how different sites – the home, the street, the school – were coded in particular ways for young British South Asian Muslim women. Thus 'the street' was a space within which young women needed to perform an 'appropriate femininity' (via clothing or behaviour) according to community-based discourses about femininity and ethnic or religious identity. Yet interviews with the young women revealed some of the different ways women were involved in strategically managing such expectations, and even subverting them, whether this was via adopting more orthodox forms of Islamic dress or by contesting the codes associated with so-called 'western' styles.

These brief examples suggest that gender identities are complex and unstable, and that they require constant reinforcing through dominant gender relations, since they are always open to possibilities for reworking or challenge. Space is central to these processes of contestation.

New Geographies of Gender?

Feminist geographers have emphasized that analysing gender relations is integral to understanding geographies. Thus the geography of gender is in many ways not a sub-discipline in itself but should be central within different sub-disciplinary fields. One measure of this is to look at the changes within economic geography, where the so-called cultural turn has ensured that issues of consumption (once relegated to the 'private' or 'domestic' sphere) are now central. An engendered approach has also been critical in histories of geography (Domosh 1991) and feminist approaches have contributed significantly to the development of more qualitative approaches to geographical research (McDowell 1992; Nast et al. 1994).

Within feminist geography increasing attention, as I have already suggested, is being paid to differences *between* men and women and the mutual constitution of gender identities with other axes of social difference. Thus the focus is on gender and sexuality (Bell and Valentine 1995), gender and age (Katz and Monk 1993) as well as gender and (dis)ability (Butler and Parr 1999) and gender and 'race' (Peake and Trotz 1999; Winddance Twine 1996). For too long masculinity has been given less attention than femininity; however, work is now being done on the social and spatial construction of masculinities (Bonnett 1996; Jackson 1991; Hopkins 2000). Interlinked with these perspectives is an increasing new focus on the intersection of gender identity and the body (Duncan 1996; Longhurst 2001). How are gender identities embodied? How do post-structuralist approaches to subjectivity change the ways in which we think about bodies as 'natural' or 'fixed'?

Notwithstanding such important critiques, it is worth concluding by emphasizing that as feminist theorists increasingly debate the significance of gender, recognizing its instability and multiplicity, geographers remain importantly placed to provide analysis of the specificities of place in understanding gender relations and the mutual constitution of gender divisions and spatial divisions.

References

Bell, D. and Valentine, G. 1995: *Mapping Desire*. London: Routledge.

Bonnett, A. 1996: The new primitives: identity, landscape and cultural appropriation in the mythopoetic men's movement, *Antipode*, 28, 273–91.

Butler, J. 1993: *Bodies that Matter*. London: Routledge.

Butler, R. and Parr, H. 1999: *Mind and Body Spaces*. London: Routledge.

Connell, R. 1995: *Masculinities*. Cambridge: Polity.

Davidoff, L. and Hall, C. 1987: *Family Fortunes: Men and Women of the English Middle Class*. London: Hutchinson.

Domosh, M. 1991: Towards a feminist historiography of geography, *Transactions of the Institute of British Geographers*, N.S. 16, 95–104.

Duncan, N. (ed.) 1996: *BodySpace*. London: Routledge.

Hopkins, J. 2000: Signs of masculinism in an 'uneasy place': advertising for 'Big Brothers', *Gender, Place and Culture*, 7(1), 31–56.

Jackson, P. 1991: The cultural politics of masculinity: towards a social geography, *Transactions of the Institute of British Geographers*, 16, 199–213.

Katz, C. and Monk, J. (eds) 1993: *Full Circles: Geographies of Women Over the Life Course*. London: Routledge.

Longhurst, R. 2001: *Bodies: Exploring Fluid Boundaries*. London: Routledge.

McDowell, L. 1992: Doing gender: feminism, feminists and research methods in human geography, *Transactions of the Institute of British Geographers*, 17, 399–416.

McDowell, L. 1997: *Capital Culture: Gender at Work in the City*. Oxford: Blackwell.

Monk, J. and Hanson, S. 1982: On not excluding the other half from human geography, *Professional Geographer*, 32, 11–23.

Nast, H. et al. 1994: Women in the field: special issue, *Professional Geographer*, 46, 54–66.

Peake, L. and Trotz, A. 1999: *Gender, Ethnicity and Place: Women and Identities in Guyana*. London: Routledge.

Valentine, G. 1989: The geography of women's fear, *Area*, 21, 385–90.

Valentine, G. 1993: Negotiating and managing multiple sexual identities: lesbian space-time strategies, *Transactions of the Institute of British Geographers*, 18, 237–48.

Winndance Twine, F. 1996: Brown skinned white girls: class, culture and the construction of whiteness, *Gender, Place and Culture*, 3, 205–24.

Further Reading

Domosh, M. and Seager, J. 2001: *Putting Women in Place*. New York: Guilford Press.

Jones, J.P., Nast, H. and Roberts, S. 1997: *Thresholds in Feminist Geography*. Lanham, MD: Rowman and Littlefield.

Laurie, N., Dwyer, C., Holloway, S. and Smith, F. 1999: *Geographies of New Femininities*. London: Longman.

McDowell, L. 1999: *Gender, Identity and Place*. Cambridge: Polity.

McDowell, L. and Sharpe, J. 1997: *Space, Gender, Knowledge: Feminist Readings*. London: Arnold.

Rose, G. 1993: *Feminism and Geography*. Cambridge: Polity.

Women and Geography Study Group (WGSG) 1984: *Geography and Gender: An Introduction to Feminist Geography*. London: Hutchinson and Explorations in Feminism Collective.

WGSG 1997: *Feminist Geographies: Explorations in Diversity and Difference*. London: Longman.

24
Mapping Culture

Peter Jackson

When the *Guardian* newspaper (4 November 1999) asked 'What is Britain's great contribution to world cuisine?', the answer was not roast beef and Yorkshire pudding, Cornish pasties or Bakewell tart, but chicken tikka masala. Whether bought as ready-meals from supermarkets, cooked at home or eaten at 'Indian' restaurants, Britain's apparently insatiable appetite for curry hints at the complexities of national identity, the legacy of empire, and the challenges and opportunities of living in a multicultural society. This chapter does not set out to trace the contours of Britain's changing culinary geography or the politics of contemporary consumption – subjects that have been addressed elsewhere (Crang and Jackson 2001). Instead, it suggests that underlying these anxieties over the shifting boundaries of Britain's culinary culture is a series of issues that lie at the heart of contemporary cultural geography – a field that has been transformed over recent years by its encounter with social theory and cultural politics.

Origins and New Directions

Cultural geography today encompasses questions of consumption and identity, race and place, geographies of gender and generation, sexuality and space, as well as more traditional concerns with the evolution of cultural landscapes. Cultural geography has a long history but in its modern form can be traced back to the work of Carl Sauer and his colleagues at the 'Berkeley school' in California. Sauer's colleagues and students were interested in the relationship between 'land and life' (Leighley 1963), in the domestication of plants and animals, the origins and diffusion of agricultural innovation and in what they called 'Man's role in changing the face of the Earth' (Thomas 1956). While an interest in cultural ecology still flourishes, present-day cultural geographers are as concerned with mapping contemporary metropolitan cultures as with tracing the evolution of rural landscapes in the historical past. Cultural geography focuses on the developed as well as the developing world and on the cultural processes and social practices that connect people and places at a range of scales from the local to the global. Contemporary cultural geographers also frequently venture beyond disciplinary boundaries, drawing from and contributing to debates in the arts and humanities as well as in the social and environmental sciences. In exploring the geographies of everyday life, traditional distinctions between the cultural and the economic, the social and the political, rapidly lose their meaning, especially in the context of a shrinking world of increasing transnational flows.

The transformation of cultural geography is in part, therefore, a reflection of changes in the material world – the advent of more flexible modes of capital accumulation, technological changes such as those associated with the Internet and the process of 'time-space compression' (Harvey 1989). But it also reflects a series of intellectual shifts associated with geography's engagement with social theory in general, and with the 'cultural turn' in particular (Cook et al. 2000). For tendencies towards cultural homogenization, driven by the forces of economic globalization, have been accompanied by a greater sense of cultural relativism, acknowledging the validity of different cultural perspectives rather than imposing a single 'universal' truth. Social movements associated with feminism and post-colonial studies, in particular, have contributed further to a general questioning of cultural authority (whether white, western or masculine).

The Cultural Turn

Since the mid-1980s geography, along with the rest of the human sciences, has experienced a series of changes that have come to be known as the 'cultural turn'. Following a period when political and economic questions were dominant, the cultural turn has brought cultural questions to the fore. It has also brought an increasing concern for the politics of representation, questioning social scientists' right to represent cultures other than their own. Influenced particularly by feminist criticisms of science, cultural theorists have been forced to recognize the politics of their own subjective position and to adopt a more reflexive (self-questioning) approach to the exploration of social difference and cultural diversity (Jackson 2000; WGSG 1997).

From feminism, too, has come a concern with our embodied identities as human beings. For our individual subjectivities cannot be understood as abstractions, divorced from our physical embodiment (see Dwyer, chapter 23 this volume). Cultural geographers have therefore begun to address the specificity of masculine and feminine identities, challenging the traditionally masculinist (and thoroughly disembodied)

conception of human rationality. Cultural geographers are also starting to explore the spatiality of different subjectivities associated with pregnancy, childhood and parenting, youth and age, health and illness.

From post-colonial studies has come an increasing interest in the way that our identities are constructed through the recognition of socially significant Others (whether articulated in terms of race or region, caste or colour, gender or generation). Edward Said's work on *Orientalism* (1978) is a landmark text in this connection, exploring the imaginative geographies through which we in the West have, over the years, reinforced our own identities through the invention of an exotic East. According to Said, 'the Orient' has provided Europeans with their deepest and most enduring sense of difference, a simultaneous source of desire and dread, fascination and fear. This work of the imagination has been underpinned by centuries of colonial exploitation and military adventure whose effects can be traced in a range of domains from the high arts of opera and museum culture to the popular world of film and fiction.

The cultural turn has also led to an emphasis on the cultural embeddedness of processes that were once regarded as unproblematically 'economic'. My earlier example concerning the British taste for curry is as much about (cultural) questions of identity and lifestyle as it is about (economic) issues of migration and marketing. For it is not just capital, labour and commodities that are increasingly mobile. As money, people and goods move around the world, so too are our ideas and imaginations reshaped in increasingly transnational ways. Indeed, recent work on the geographies of consumption and commercial culture (Jackson and Thrift 1995; Jackson et al. 2000) has called for a transcendence of all such dualisms (economic and cultural, production and consumption, global and local).

A Material World?

Where cultural geography was formerly dominated by textual approaches to the analysis of symbolic meaning and by the visual decoding of landscape iconography, questions of representation are now increasingly being joined by questions of practice, presentation and performance (Thrift 1996; see also Morgan, chapter 43 in this volume). This stems in part from dissatisfaction with existing theories of representation and the implication that visual or verbal texts can be 'read' in such a way as to reveal their 'real' underlying meaning. Similarly, the tendency to reduce the sensual world of human experience and social practice to a purely discursive world of language and text has come under increasing scrutiny. Recent developments in cultural geography have therefore witnessed a renewed interest in the material world, recognizing that material objects – and what anthropologists have called 'the social life of things' (Appadurai 1986) – also have a distinctive social geography. Emphasizing a renewed concern with human practice and the need to embrace the materiality of everyday life as well as its discursive formation, recent research has focused on the cultural geographies of material goods like food and clothing. Cultural geographers are now tracing specific commodity chains from the supply of raw materials through networks of marketing, distribution and retail to the sites of consumption (including circuits of use and reuse in the case of second-hand goods). Such research has focused on conventional retail spaces (like supermarkets, high street stores and shopping malls) and on 'alternative' consumption sites (like car-boot sales, charity shops and second-hand clothing stores).

Such research has been conducted via a range of methods including surveys and interviews, ethnographic work and textual deconstruction. Increasingly, too, researchers have sought to move beyond a politics of representation drawing on actor-network approaches and non-representational theory, posing some fundamental challenges to our understanding of the relationship between culture and nature (Whatmore 1999).

A Contested Field

As all of the previous examples suggest, contemporary cultural geography is a lively and contested field where much remains to be struggled over. Seeking to capture this sense of intellectual and political unrest, Don Mitchell's (2000) critical introduction to cultural geography refers to the field in terms of a series of 'culture wars', while Kay Anderson and Fay Gale (1999) employ a language of power and resistance, encompassing the clash of capital and culture and geographies of inclusion and exclusion. For the 'maps of meaning' through which we make sense of the world are not simply mental constructs that we carry around in our heads; they are made concrete through specific social acts and institutional forces. In this sense, Mike Crang's (1998) emphasis on cultural practices is a useful corrective to studies that emphasize discourse and representation to the exclusion of more material forces.

Making sense of this highly charged intellectual world can be a daunting experience for students. Cultural geography's agenda is vast and challenging, its intellectual sweep audaciously ambitious. Some might even question the wisdom of 'mapping culture' at all, as cartographic metaphors of this kind have often been associated with a colonial imagination and a desire to subordinate. But the openness of cultural geography's current agenda is part of its appeal. Underlying the metaphor of 'mapping culture' is a commitment to exploring contemporary issues of cultural diversity, insisting on a plurality of perspectives and on multiple ways of seeing the world.

References

Appadurai, A. (ed.) 1986: *The Social Life of Things: Commodities in Cultural Perspective.* Chicago: University of Chicago Press.

Cook, I., Crouch, D., Naylor, S. and Ryan, J. (eds) 2000: *Cultural Turns/Geographical Turns.* Harlow: Pearson Education.

Crang, P. and Jackson, P. 2001: Consuming geographies. In D. Morley and K. Robins (eds), *British Cultural Studies,* Oxford: Oxford University Press, 327–42.

Harvey, D. 1989: *The Condition of Postmodernity: An Enquiry into the Origins of Cultural Change.* Oxford: Blackwell.

Jackson, P. 2000: Cultures of difference. In V. Gardiner and H. Matthews (eds), *The Changing Geography of the United Kingdom,* London: Routledge, 276–95.

Jackson, P., Lowe, M., Miller, D. and Mort, F. (eds) 2000: *Commercial Cultures: Economies, Practices, Spaces.* Oxford: Berg.

Jackson, P. and Thrift, N. 1995: Geographies of consumption. In D. Miller (ed.), *Acknowledging Consumption,* London: Routledge, 204–37.

Leighley, J. (ed.) 1963: *Land and Life: A Selection from the Writings of Carl Ortwin Sauer.* Berkeley: University of California Press.

Said, E.W. 1978: *Orientalism.* New York: Pantheon.

Thomas, Jr, W.L. (ed.) 1956: *Man's Role in Changing the Face of the Earth.* Chicago: University of Chicago Press.

Thrift, N.J. 1996: *Spatial Formations.* London: Sage.

Whatmore, S. 1999: Culture–nature. In P. Cloke, P. Crang and M. Goodwin (eds), *Introducing Human Geographies,* London: Arnold, 4–11.

Women and Geography Study Group (WGSG) 1997: *Feminist Geographies: Explorations in Diversity and Difference.* London: Longman.

Further Reading

Anderson, K. and Gale, F. (eds) 1999: *Cultural Geographies.* Harlow: Longman.

Crang, M. 1998: *Cultural Geography.* London: Routledge.

Mitchell, D. 2000: *Cultural Geography: A Critical Introduction.* Oxford: Blackwell.

25
New Geographies of Disease: HIV/AIDS

Robin Kearns

Announcing a pledge by the United States for $50 million to support Uganda's AIDS programme, Secretary of State Colin Powell stated 'there is no war that is more serious . . . than the war we see here in sub-Sahara Africa against HIV/AIDS'. On the same day in May 2001, Reuters reported that Johannesburg graveyards were running out of space with the rising toll from the disease. With 50 years since the Second World War, it seems that in this new war on disease, all nations are allies.

However, news items such as these too easily suggest that AIDS is strictly a problem somewhere else. While African countries are undeniably the hardest hit by the pandemic, the reach of HIV/AIDS is global and a key contemporary vector of disease transmission is the Boeing 747. As human mobility increases, the networks people create enabling them to respond to, and cope with, change become a tangled web of influences opening pathways for disease transmission. The challenge for prevention programmes is to find health promotion messages that create behavioural barriers within such paths.

Health geographers have addressed HIV/ AIDS, as with other health issues, according to four dimensions: distribution, diffusion, determinants and delivery (Kearns 1996):

- illness and treatment facilities are seen to be *distributed* in space;
- disease and treatment innovations can *diffuse* (i.e. spread) through time and space;
- all diseases have underlying *determinants*; and
- various agencies and individuals offer health and social care *delivery* systems.

Over the last decade, and since the AIDS epidemic has taken hold, a fifth dimension can be added to the health geography formula: *difference*. A 'making space for difference' has involved an inclusiveness in not only health outcome or service style, but also in the very experience of well-being itself (Kearns 1995). The following sections sequentially survey work focusing on the virus, remedial services and the place-experiences of people most directly affected by HIV/AIDS.

The Scale and Character of the Disease

The emergence and spread of HIV/AIDS in recent decades has served as a warning that, notwithstanding general declines in mortality from infectious diseases over the last century, humankind cannot be complacent. The human immunodeficiency virus (HIV) which causes acquired immunodeficiency syndrome (AIDS) has brought a global epidemic far more extensive than anything predicted. The mortality figures make grim reading: an estimated 21.8 million deaths since the beginning of the epidemic, and 3 million AIDS deaths since the year 2000. Deaths, however, only tell part of the story. The United Nations reports that 36.1 million people are currently living with HIV/AIDS (http://www.unaids.org 2001). As with any disease, the experience of illness and its effects on family and community can be lost within stark mortality figures.

AIDS, as the abbreviation implies, is a syndrome (a collection of signs and symptoms) developing from an impaired ability to fight disease. This impairment involves a suppression of the immune system and an undermining of the body's defences against a range of viruses, infections and malignancies. The period between infection with HIV, the causative agent, and the onset of AIDS varies greatly and can be up to ten years. HIV is transmitted through exchange of bodily fluids. While in less developed nations AIDS is widespread throughout the heterosexual population, in western countries those who are HIV-positive are predominantly members of particular population groups such as gay men and IV drug users.

The Power of Mapping

Geographers have focused primarily on two related dimensions of HIV/AIDS: *distribution* and *diffusion*. With respect to both dimensions, it has been the virus that has tended to be of central interest rather than the experience of living with the disease. The application of spatial analysis techniques to studying AIDS has taken three forms. First, the global diffusion of AIDS has been a dominant focus (Gould 1993). Second, geographers have employed predictive modelling techniques to map future geographies of AIDS (e.g. Loyotonen 1991). Third, geographers have mapped the progression of the virus across nations and regions, especially focusing on the United States (Dutt et al. 1988). There has been general agreement that 'the HIV/AIDS epidemic is following a classical, but complex, spatial diffusion pattern' (Golub et al. 1993: 86).

The focus by medical geographers was initially, and understandably, on the disease agent – the Human Immunodeficiency Virus Type I. The complexity of this virus means that treatment for infection so far only delays the onset of AIDS. A 'hierarchical nodal pattern' of HIV was observed, spreading from central Africa to major urban centres either side of the Atlantic (Shannon and Pyle 1989). Later researchers analysed patterns of spread for the less prevalent HIV-2.

Given its interest in the virus rather than on people *per se*, the language of predictive modelling unfortunately distances itself from people experiencing the illness with such terminology as 'likely pairings of infectives and susceptibles' (Golub et al. 1993: 95). According to Michael Brown, this fixation on the virus in western countries is symptomatic of a tendency to create distances between geography and gay men and their everyday lives. Brown (1995) attributes the invisibility of (especially gay) experience of AIDS in geography to an excessive use of spatial analysis that tends to reduce incidence to dots on a map. According to Brown, the mapping and plotting of spatial analytic studies has focused on the *virus* to the neglect of the *people* dealing with the illness, with the result that 'gay men have been closeted by this spatial-scientific geography of AIDS' (1995: 3). The irony is that this 'erasure' of gay experience occurs at the same time as gains in visibility by the gay community in response to AIDS. Such gains include political representation, public parades and highly visible memorials such as AIDS quilts.

While issues of *distribution* and *diffusion* of the virus have received a good deal of attention, there have been fewer attempts to reveal socio-cultural and political implications of the disease. One response has been Rob Wilton's (1996) ethnographic study of the daily life experiences of Los Angeles men living with HIV/AIDS. Wilton disaggregates social, physical and psychological dimensions of daily life, leading him to discover the stark realities of disease that are inscribed into the human experience of place. He considers the proposition that people experience 'diminished worlds' after diagnosis, and through Wilton's analysis we become acquainted with the dynamics of respondents' life paths which involve community organizations, service agencies and personal social networks. By the end of Wilton's account, the respondents have become participants – not only in *his* research, but also in *our* understanding of HIV/AIDS.

Metaphors and Misconceptions

The strong interest in disease diffusion by geographers begs the question of where a disease travels from and to. In the case of AIDS, the African origins of the disease have played into stereotypes of fear, pestilence and the 'dark continent' associated with colonial times immortalized in Joseph Conrad's novel *Heart of Darkness*. Like afflictions throughout history ranging from plagues to influenza, mapping pathways of the epidemic has allowed AIDS to be attributed to 'them' rather than 'us'. High-profile legal cases brought against African foreigners infecting (white) locals have served to highlight possibilities of racism lurking beneath the veneer of disease prevention.

As Curtis and Taket (1996) point out, many of the terms used in association with HIV/AIDS act to foster stigmatizing myths. Terms such as 'AIDS carrier' and 'AIDS victim' are misleading for they confuse the phases of infection and imply subsequent helplessness, respectively. Such confusion in terminology adds to stigma, a socio-spatial process involving people who are identified as different being shunned and often literally set apart. In the case of HIV/AIDS, the

capacity for stigma is heightened with AIDS itself having become a metaphor, or stand-in term, for a range of anxieties in western thinking. According to Altman (1986: 194), 'the link with sexuality and blood makes AIDS particularly susceptible to metaphorical use'. Myths such as AIDS as 'the gay plague' or an 'African problem' have been given momentum by media attempts to popularize medical research. Behind the brevity of such accounts, however, lies the capacity to heighten homophobia, xenophobia and racism. In resistance to these processes, in western countries, the reality of AIDS has increasingly become a rallying point for the creation of new communities and service networks such as those documented by Law and Takahashi (1997) in West Hollywood.

From Geographies of Risk to Places of Respite

Given the fatality of the disease, an understanding of the mechanisms for its transmission are crucially important in establishing intervention strategies to control the spread of HIV. In developing countries, many of those people most vulnerable to infection are migrants and mobile workers whose beliefs and behaviours have come under the scrutiny of health researchers. The concepts of 'risk group' and 'risk behaviour' are well established in public health research and have been used by geographers to identify the determinants of healthy lifestyles. Although the 'risk group' idea moves in the direction of peopling the virus, the behavioural roots of the concept threaten to conflate risky (sexual) practices with particular 'Others' in society. Some researchers advocate a shift in research emphasis from high-risk *behaviours* to high-risk *situations*. Situations involve human interactions anchored in place, and Sheena Asthana's (1998) investigation of the commercial sex industry in Madras, and Ford and Koetsawang's (1991) work in Thailand, exemplify attempts to avert the generalization of behaviour to risk groups and, instead, to define the place-bound contexts of HIV transmission.

In lieu of any *cure* for AIDS, geographers interested in places of treatment have necessarily focused on sites of *care*. Persons living with AIDS need not only adequate but also secure housing. However, sufferers must often seek 'packaged' care such as hospices because of failing health and an associated inability to meet housing costs.

Chiotti and Joseph (1995) focus on Casey House, a facility for persons living with AIDS in downtown Toronto. As a place of care for gay men in the final stages of a fearful illness, Casey House holds considerable potential for generating stigma. Elsewhere in North America, facilities for AIDS patients (and especially those who are also afflicted by homelessness and substance addictions) have generated strong opposition, especially from nearby residents expressing the 'Not in My Backyard!' (NIMBY) catchcry (Takahashi and Dear 1997). Casey House, however, met with few objections and this is accounted for by its location within a largely gay part of Toronto that was already host to a range of other social service agencies. The challenge for planners seeking an inclusive city will be to develop an understanding of the relationship between public attitudes and the characteristics of potential sites of service provision. The operation of facilities by local people known to local residents holds potential to allay fears of being invaded by institutions and agencies.

Geographies Beyond the Virus

The global, regional and national scales adopted by geographers dealing with the distribution and diffusion of HIV/AIDS have allowed geographers to make important contributions to public health knowledge. However, if this is the benefit, the cost has been the relative invisibility of human experience. This brief review has indicated that the grim facts of HIV/AIDS have led to two types of geographical knowledge: patterns of morbidity and mortality; and explorations of the experiences of place for people living with HIV/AIDS. The priority given to mapping the disease has led to criticism by fellow geographers aligned to groups experiencing the disease first-hand. This criticism has been because in its traditional forms, medical geography reduced the body to little more than a container of disease, whereas health has increasingly spoken to geographers of the larger reality of people's experiences of place and identity. The challenge for health geographers confronting HIV/AIDS and other diseases is to find ways to integrate the global, national and state-level analyses of the determinants of viral distribution and diffusion with the local specificity (i.e., difference) of organizational response, service delivery and individual life experience.

The HIV/AIDS theme chosen as a focus for this chapter has generated discussion of the broad field of health geography. Students are encouraged to consider other health and health-care issues through reference to the texts by Jones and Moon (1987), Curtis and Taket (1996), Meade and Earickson (2000) and Gesler and Kearns (2001) listed below.

References

Altman, D. 1986: *AIDS and the New Puritanism*. London: Pluto.

Asthana, S. 1998: The relevance of place in HIV transmission and prevention: the commercial sex industry in Madras. In R.A. Kearns and W.M. Gesler (eds), *Putting Health into Place: Landscape, Identity and Well-being*, Syracuse, NY: Syracuse University Press, 168–90.

Brown, M. 1995: Ironies of distance: an ongoing critique of the geographies of AIDS, *Environment and Planning D: Society and Space*, 13, 159–84.

Callwood, J. 1995: *Trial Without End: A Shocking Story of Women and AIDS*. Toronto: Knopf Canada.

Chiotti, Q.P. and Joseph, A.E. 1995: Casey House: interpreting the location of a Toronto

AIDS hospice, *Social Science and Medicine*, 41, 131–40.

Dutt, A.K., Monroe, C.B., Dutta, H.M. and Prince, B. 1988: Geographical patterns of AIDS in the United States, *Geographical Review*, 77, 456–71.

Ford, N. and Koetsawang, S. 1991: The sociocultural context of the transmission of HIV in Thailand, *Social Science and Medicine*, 33, 405–14.

Golub, A., Gorr, W.L. and Gould, P.R. 1993: Spatial diffusion of the HIV/AIDS epidemic: modelling implications and case study of AIDS incidence in Ohio, *Geographical Analysis*, 25, 85–100.

Gould, P. 1993: *The Slow Plague: A Geography of the AIDS Pandemic*. Oxford: Blackwell.

Kearns, R.A. 1995: Medical geography: making space for difference, *Progress in Human Geography*, 19, 249–57.

Kearns, R.A. 1996: AIDS and medical geography: embracing the Other, *Progress in Human Geography*, 20, 123–31.

Law, R. and Takahashi, L. 1997: HIV, AIDS and human services: exploring public attitudes in West Hollywood, California, *Health and Social Care in the Community*, 8, 90–108.

Loyotonen, M. 1991: The spatial diffusion of the human immunodeficiency virus type 1 in Finland 1982–1987, *Annals, Association of American Geographers*, 81, 127–51.

Shannon, G.W. and Pyle, G.F. 1989: The origin and diffusion of AIDS: a view from medical geography, *Annals of the Association of American Geographers*, 79, 1–24.

Takahashi, L.M. and Dear, M. 1997: The changing dynamics of community attitudes towards human services, *Journal of the American Planning Association*, 63, 79–93.

Wilton, R.D. 1996: Diminished worlds? The geography of everyday life with HIV/AIDS, *Health and Place*, 2, 69–84.

Further Reading

Curtis, S. and Taket, A. 1996: *Health and Societies*. London: Arnold.

Gesler, W. and Kearns, R. 2001: *Culture/Place/Health*. London: Routledge.

Jones, K. and Moon, G. 1987: *Health, Disease and Society*. London: Routledge and Kegan Paul.

Meade, M. and Earickson, R. 2000: *Medical Geography*, 2nd edition. London: Guilford Press.

Internet Resources

- United Nations Aids Program: http://www.unaids.org

26

Social Exclusion and Inequality

Chris Thomas and Stephen Williams

In America, between 1979 and 1997 the average income of the richest fifth of the population jumped from nine times the income of the poorest fifth to around 15 times. In 1999,
British income inequality reached its widest level in 40 years.

Economist, 16 June 2001: 11

In 1755 the French philosopher and writer
Jean-Jacques Rousseau observed that 'from
the moment one man began to stand in need
of the help of another; from the moment it
appeared advantageous to any one man to
have provisions for two, equality disappeared' (1966: 199). Rousseau and the
Economist make the same point two and a
half centuries apart: inequality is endemic
in modern society.

Inequality emerged as a key geographical
concern in the development of a radical and
relevant discipline in the 1970s. This marked
a reaction against the spatial science approach
that had dominated the discipline in the
1960s, but also reflected a recognition of a
new agenda for geographical enquiry that was
concerned with the social and economic
problems of urban and rural spaces. Issues
of poverty, crime, health, racial discrimination and the provision of services such as
housing and education all received considerable academic attention, as exemplified in the
welfare geography approach of David M.
Smith (1977). Most recently these concerns
have begun to be refocused through the lens
of social exclusion – a development that opens
new avenues for investigating key issues with
which human geographers should engage.

Social Exclusion, Policy and Geography

The term 'social exclusion' emerged in
France in the 1970s to describe the condition of the people living in massive suburban housing projects, whose lives were
blighted by a combination of low incomes,
high rates of crime and poor-quality housing. Social exclusion rapidly became popular political rhetoric across Europe following
the adoption of the Social Charter within
the European Union in the early 1990s
(Room 1995) and has figured on the British
political agenda since 1997. The definition
adopted by the British government uses 'exclusion' as another way of seeing disadvantage. It suggests that 'social exclusion is a
shorthand label for what can happen when
individuals or areas suffer from a combination of linked problems such as unemployment, poor skills, low incomes, poor
housing, high crime environments, bad
health and family breakdown' (SEU 2001a:
11).

Philo (2000) suggests, however, that academic geographical concern for socio-spatial exclusion predates the recent emergence
of policy interest (in the UK at least) and

may be traced to Sibley's innovative *Outsiders in Urban Societies* (1981). The continuing development of geographical inquiry into social exclusion reflects the fact that this concept offers human geographers a new way of examining established concerns for inequalities across space and in place. Furthermore, the growing diversity of topics that geographers have begun to address under the heading of 'exclusion' demonstrates that the concept is concerned with more than just the incidence of economic inequality. In particular, one of the most significant deviations from previous constructs of poverty, disadvantage and deprivation is the ability of the term to emphasize not only the outcome, but also the *processes* by which people come to be excluded. It is this dynamic quality that holds out the most promise for geographical research, because it takes us beyond simple questions of identification of excluded groups by raising the more important question of *how* people are excluded. In so doing, we soon uncover a significant truth: namely, that exclusion is shaped in a diversity of ways and that in addition to a social dimension, exclusion may be defined by economic, political and cultural dimensions too. Each of these dimensions is capable of offering varying interpretations of the problem and setting differing research agendas for geography.

Exploring the Dimensions of Social Exclusion

Social

The social dimension of exclusion provides the clearest links to traditional geographical concerns with inequality. It was the awareness of social problems that prompted the development of a critical social geography 30 years ago, and it is a similar set of concerns that has driven the development of the term, the concept and the policy focus of social exclusion. The British government's definition of social exclusion quoted above presents a picture of disadvantage that suc-

cinctly delineates the social dimension of exclusion, but critically, the associated analyses and debates have also highlighted the processes through which such problems appear to be created. This has encouraged a more reflective, nuanced understanding that has drawn attention to a range of key social issues, including the breakdown of traditional households and families; rises in levels of teenage pregnancies; the incidence of crime and the disaffection of youth.

Economic

In detailing the growth of what he terms the 'exclusive society', Young (1999: vi) parallels the social changes identified above with 'massive structural change: where there have been fundamental changes in the primary and secondary labour markets; where employment patterns of women have radically changed; where structural unemployment has been created on a vast scale; where communities have disintegrated'. The emergence of a globalized economy, post-Fordist in organization, with much less reliance on heavy industry and traditional male employment, has been closely interwoven with the social dimension. This linkage between the economic and the social is reflected in policy; first, by the use of economic indicators as a means of identification of *social* exclusion (such as poverty and financial exclusion, exclusion from the labour market; and casualization of employment); and second, by the centrality given to economic solutions to social exclusion, particularly through paid employment. Thus, although the problem is widely conceived in terms of exclusion from society, much of the present understanding is shaped by the economic dimensions that surround participation (or otherwise) in the productive process.

Political

This realization underscores the significance of the political dimensions of social exclusion which, understandably, shape the ways in which the problem is defined and addressed. As in France in the 1970s, the key concern in Britain today is that social exclusion removes individuals and groups from the norms of everyday life, and this reduces participation in civic society. Policy, therefore, presumes to include people – partly through principles of equity but also, more practically, in the interests of effective governance.

However, many authors have noted that the political presumption of including people who are currently deemed as 'excluded' is problematic. Levitas (1998: 7) summarizes the difficulty when she writes that social exclusion 'represents the primary significant division in society as one between an included majority and an excluded minority'. This has implications for how both included and excluded groups are understood, and for the implicit political model of society itself. It draws attention away from inequalities and differences amongst the included and instead represents exclusion as a peripheral problem, existing at the boundaries of society, 'rather than a feature of society which characteristically delivers massive inequalities across the board and chronic deprivation for a large minority' (Levitas 1998: 7). In this, she shares with Byrne (1999) the conviction that the political readings are essentially selective – suggesting the border between inclusion and exclusion is an easy one to cross, whilst allowing the inherent inequalities of late industrial capitalism to go unchallenged.

Cultural

Considering the cultural dimensions of exclusion may usefully challenge the political reading of exclusion as a problem at the margins of society. Here, geographical research suggests that exclusion is everywhere. Sibley (1995: ix) argues that 'the human landscape can be read as a landscape of exclusion' in which the cultural dimensions of exclusion exist in subtle, powerful and endemic ways. He outlines how extant social

geographies have implicitly highlighted some forms of exclusion (although not in those terms) when they consider the make-up of modern urban and rural societies and the spatial outcomes of access to resources. In *Geographies of Exclusion*, Sibley goes further, to reveal more opaque forms of exclusion – instances that 'are taken for granted as part of the routine of everyday life' (Sibley 1995: ix). Exclusion, in this latter sense, is the 'edging out' of individuals and groups through mainstream reactions to attributes the excluded are thought to possess; a process that both 'normalizes' the exclusion and encourages its acceptance by mainstream society. Hence, people become excluded when they are felt not to belong, either by others or, less obviously, by themselves. Clearly this happens to the unemployed and the poor (who are the central concern of social exclusion policies), but also selectively affects women, children, migrants, travellers, disabled people, people of colour and the elderly. Exclusion, rather than being a simple state of separation from society, evidently exists also *within* society.

An Example of Social Exclusion: Leisure and Elderly Women

To illustrate how these different dimensions may combine in everyday life to produce grounded instances of exclusion, we refer to recent work on the exclusion of elderly women from urban leisure spaces (Hague et al. 2000a,b). This study, working with groups of elderly women in peripheral residential locations in Stoke-on-Trent, revealed something of the complex process through which the women came to be excluded from city-centre leisure sites.

First, the social dimensions to exclusion were reflected strongly through the age and, in the case of one group, the social class of the women. The shopping malls, 'trendy' restaurants and 'fashionable' bars of the redeveloped city centre afforded spaces that the women perceived as intended for other social groups, especially younger and more affluent people. Furthermore, a fear of crime in places that were now perceived as dominated by 'others' – especially the young – reinforced the incidence of exclusion as the women 'chose' not to visit sites where the social environments appeared to conflict with their own needs and expectations.

Second, these socially based exclusions were widely reinforced by the economic dimensions of the women's daily lives. As pensioners, few could easily afford the costs of public transport to the city centre and, particularly, the price of admission to leisure sites such as theatres and cinemas. Although the women articulated strong desires to visit entertainment facilities in the city centre, such desires were typically frustrated by prices that were widely seen as literally prohibitive and, by extension, exclusive.

Third, the study revealed clear evidence of forms of political exclusion. The groups of elderly women constituted a peripheral voice – both in a spatial sense and, more fundamentally, within local political processes. Their views, they perceived, were not generally heeded in discussions of urban design or local transport provision, resulting in a resigned acceptance of their peripherality to local decisions that took no real account of their particular needs and aspirations.

Lastly, the study underlined a powerful sense of cultural exclusion that revolved around their identity as elderly women in a city engaged in processes of reinvention and rejuvenation. The women repeatedly articulated the view that the leisure sites of the city centre were 'not for them'. This cultural discomfort was linked closely to personal attributes such as age, class, gender and income level – the redevelopment of urban spaces was understood as alien to them. Not only did these women suffer from 'obvious' exclusions due to access or affordability, but they experienced an almost tangible 'cultural dislocation', which led to their self-exclusion from much that the urban centre had to offer.

It was further apparent that the specific exclusions above were combined and routinized into the everyday lives and mythologies of the groups. Hence, we might suggest that many forms of exclusion become 'normalized' and thus invisible as 'social exclusion' according to the criteria by which so much current policy and practice are defined.

Space, Place and Exclusion

How, then, might geographers usefully engage with debates on exclusion? The different dimensions of social exclusion outlined above may suggest strategies for such an engagement, but perhaps more importantly, they affirm that there *is* a geography to exclusion, and it is a geography that operates at different scales. Social exclusion occurs across space and in place. At the local scale, policy has already identified the 'neighbourhood' – a place-bounded community – as a focus (SEU 2001b), whereas at the global scale, borders become more fluid for finance, but increasingly impenetrable for displaced peoples. In between, and a key arena for geographical investigation, is the late industrial restructuring of urban and rural spaces.

Whilst much attention has already been paid to the geography of social exclusion, the overwhelming focus of this attention to date has been the urban spaces of the late/ post-industrial city (Byrne 1999; Hague et al. 2000b; Young 1999). Yet social exclusion is just as significant an issue in rural spaces, as has been revealed in recent discussions of deprivation, disadvantage and lifestyle (Cloke et al. 1994). There is, therefore, both the opportunity and the need to develop a more rounded view of the geographies of exclusion that exposes – in a truly holistic manner – the spatial incidence and the variable construction of exclusion in contemporary societies. From work that is already completed, it is evident that the practices of exclusion operate differently and have differing outcomes within households, neighbourhoods, districts, cities, regions and nations. When we consider all the geographical contexts where the social, economic, political and cultural dimensions of exclusion intersect with space and place, it is clear that human geography has much to contribute to understanding social exclusion as a new way of viewing inequalities.

References

Byrne, D. 1999: *Social Exclusion*. Buckingham: Open University Press.

Cloke, P., Milbourne, P. and Thomas, C. 1994: *Lifestyles in Rural England*. Salisbury: Rural Development Commission.

Hague, E., Thomas, C. and Williams, S. 2000a: Political constructions and social realities of exclusion in urban leisure: the case of elderly women in Stoke-on-Trent, England, *World Leisure Journal*, 4, 4–13.

Hague, E., Thomas, C. and Williams, S. 2000b: Equity or exclusion? Contemporary experiences in post-industrial urban leisure. In C. Brakenridge, D. Howe and F. Jordan (eds), *Just Leisure: Equity, Social Exclusion and Identity*, Brighton: Leisure Studies Association, 17–34.

Levitas, R. 1998: *The Inclusive Society? Social Exclusion and New Labour*. Basingstoke: Macmillan.

Philo, C. 2000: Social exclusion. In R.J. Johnston,

D. Gregory, G. Pratt and M. Watts (eds), *The Dictionary of Human Geography*, 4th edition. Oxford: Blackwell.

Room, G. (ed.) 1995: *Beyond the Threshold: The Measurement and Analysis of Social Exclusion*. Bristol: Policy Press.

Rousseau, J.-J. 1966: *The Social Contract and Discourses*. London: Dent/Everyman.

Smith, D.M. 1977: *Human Geography: A Welfare Approach*. London: Arnold.

Social Exclusion Unit (SEU) 2001a: *Preventing Social Exclusion*. London: Cabinet Office/Stationery Office.

SEU 2001b: *A New Commitment to Neighbourhood Renewal*. London: Cabinet Office/Stationery Office.

Young, J. 1999: *The Exclusive Society: Social Exclusion, Crime and Difference in Late Modernity*. London: Sage.

Further Reading

Sibley, D. 1981: *Outsiders in Urban Societies*. Oxford: Blackwell.

Sibley, D. 1995: *Geographies of Exclusion: Society and Difference in the West*. London: Routledge.

Internet Resources

• The Social Exclusion Unit of the Cabinet Office: http://www.cabinet-office.gov.uk/seu

Part III
Studying Geography

Studying geography offers you the opportunity to master a large and growing range of techniques and skills. Geography departments now place greater emphasis on the systematic and thorough training in a range of methods which, it is intended, will make you better geographers and serve you in later careers. To traditional techniques such as cartography and fieldwork have been added new skills. Some depend upon access to increasingly sophisticated yet accessible computer software, such as geographical information systems, mathematical modelling and visualization technologies. Others require more social and personal skills honed for analytical and interpretive use, such as ethnography and image interpretation. Alongside these technical abilities, however, you will acquire competence in less dramatic but no less important skills such as writing, listening and making oral presentations.

This section includes four chapters written by specialists in geographical education designed to help you get the best out of your learning environment. There are practical tips and advice on writing essays and assignments, taking exams, making presentations and making good use of lectures. The section also introduces you to some of the key technical fields of geography, cartography and visualization, remote sensing and spatial modelling. It provides a series of guides about how to approach personal research projects and how to get started on a range of possible research methods, from questionnaire surveys to laboratory analysis. You can learn about the potentials and pitfalls of conducting your projects abroad.

Doing your own independent research can present you with awkward situations, about such matters as confidentiality and privacy. Geography students are now expected to think about the ethics of their own conduct, and the ethics of geographical research in general. The chapter by Tim Unwin addresses these issues.

27

Cartography and Visualization

Scott Orford, Danny Dorling and Richard Harris

Cartography in the Twenty-first Century

Cartography has always been associated with geography. Indeed, in some people's minds geographers and maps are inseparable. In its traditional sense, cartography is viewed simply as the presentation of geographic information, using hand-drawn maps to express a sense of geography on paper to a wide audience. Such a view is valid but limited. It is challenged by the major technological and conceptual changes that have occurred in cartography during the past 15 years. These changes are associated with the rapid development of computer technology, the increased availability of digital data and a growing need to understand an ever more complex world. Today maps are very rarely drawn by hand – anyone with access to a computer and mapping software can now create a map digitally and with relative ease. More importantly, cartography is branching out from its traditional role of (passively) presenting geographical information and is currently moving towards a more cognitive role where maps are used in an exploratory, research capacity. Twenty-first-century cartography is not just about drawing maps to communicate information to others. It is increasingly about using computational meth-

ods and visual displays to increase our understanding of the world around us by means of interactive, learning environments (MacEachren and Kraak 1997). Cartography is now establishing links with disciplines such as computer science and statistics, and as a result, it is shifting its status within geography to be slowly integrated under the banner of 'geographic information science' (GIScience) or 'geocomputation' – which includes geographic information systems (GIS) and exploratory data analysis (EDA) (Unwin 1999).

The recent changes in cartography have occurred partly in response to the emergence of scientific visualization (McCormick et al. 1987). This emphasizes the use of visual methods (or 'ways of seeing') as an important part of the way knowledge is constructed. The motivation of scientific visualization is to 'see the unseen' in the increasingly large and complex digital datasets used by scientists, and in this respect computational mapping is regarded as being a potentially excellent device to be used in the exploration of spatial digital data. To help conceptualize these innovations, DiBiase (1990) and MacEachren (1994b) have developed conceptual models that explain the different uses that maps can have. In DiBiase's typology, maps can be used for visual thinking (*exploring* geographic information) or visual communication (*presenting* geographic information). These categories are not mutually exclusive, however. In visual thinking, maps are used as research tools to help investigators search for properties in the data – revealing patterns and relationships and flagging unusual events. They are used in the exploratory and confirmatory stages of research and are not intended as a 'finished publication' or the 'final result'. They are intended to help the researcher. In contrast, maps associated with visual communication are used to illustrate a point, present ideas or demonstrate relationships to a general audience. They are intended to help other people. MacEachren (1994b) extends this model by incorporating the interactive and dynamic element afforded by computer technology. His model of cartography, shown in figure 27.1, is referred to as **[CARTOGRAPHY]**[3], with the space within the cube distinguishing maps used for visualization (visual thinking) and maps used for communication (visual communication). Using this model MacEachren argues that cartographic visualization is a private activity in which unknown facts are revealed in a highly interactive environment. In contrast, cartographic communication involves the opposite: a public activity in which known facts are presented in a non-interactive environment (Slocum 1999). Both of these cartographic activities have an important role to play in our understanding and communication of geographical data. However, whereas we know a great deal about cartographic communication, our knowledge of cartographic visualization is still very much in its infancy.

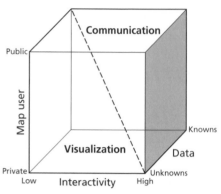

Figure 27.1   [CARTOGRAPHY][3] – a graphical representation of how maps are used. Axes relate to the user of the map, objectives of data use and the degree of interaction in the mapping environment. Within the cube, cartographic visualization is contrasted with cartographic communication along the three dimensions.

Adapted from MacEachren 1994b, fig. 1, 3, 6.

Cartographic Visualization: Seeing the Unseen

Although cartography has always been concerned with visual thinking, the growth of computational cartography and the interactivity that this affords has qualitatively changed what people can do with maps. Maps that allow a high degree of interaction allow brushing, panning, zooming, rotation, dynamic re-expression (where the map is automatically updated after a change has been made) and dynamic comparison (linked views) as part of their functionality (Dykes 1997). A linked view allows the user to select a point in one map to identify its location in another, for example. Such functions can be used in the exploratory analysis of spatial data and allow data to be analysed for unknown relationships, or answer questions such as 'what is the nature of the dataset?' and 'is there a spatial relationship between different features?'. Another important innovation in cartographic visualization has been the introduction of dynamic visual variables such as animation, multimedia and virtual reality (Orford et al. in press). Together, dynamic visual variables and interactive tools can allow greater insights in a data exploration environment. They are particularly useful at the start of a visualization session when the user is 'getting to grips' with the data.

Perhaps one of the most influential innovations to affect cartography during the past decade has been the World Wide Web (WWW). Its highly graphical nature and also its multimedia content mean that the WWW is an ideal medium for producing cartographic visualizations. Its interactivity and flexibility has enabled the growth of 'mapping on demand' (Cartwright 1997). Map presentation is rapidly incorporating the interactivity of the WWW to allow dynamic and interactive dissemination of mapped information. For a good introduction to Web-based cartography, students are referred to the book edited by Kraak and Brown (2001).

In 1990 Tufte wrote:

> The world is complex, dynamic, multidimensional; the paper is static, flat. How are we to represent the rich visual world of experience and measurement on mere flatland? (1990: 9)

As we have identified, technological innovations may yet allow cartographic visualization to break free from a two-dimensional and inert worldview. However, the research concerned with cartographic visualization is still very much in the development stages and many issues remain unresolved. For instance, unlike the conventional techniques used in cartographic communication, virtually nothing is known about good practice – when it is good to use cartographic visualization techniques and when it is not (Dorling 1998). For example, using animation to explore changes in the data is not necessarily easy, especially if many changes occur within a specified display time and all over the display area. Remaining blind to the most important trends or patterns is a frequent problem in visualization (MacEachren and Ganter 1990).

Cartographic Communication: Why Good Maps are Important and Why Getting it Right Requires Practice

Despite the changing research emphasis towards a more complex visualization, for the vast majority of users cartography is still about producing maps to present geographical information. Ironically, the development of computerized and Web-based cartography has enabled both a 'democratization' of map making and also a proliferation of rather inadequate maps! Thus Kraak and Brown (2001) have argued a case for a 'back to basics' in cartography. Since most geography students will still predominantly use cartography for visual communication (as opposed to visual thinking), an understanding of the principles of cartographic design remains essential.

The principles of cartographic design are a set of standard rules and procedures that have been developed to allow maps to be understood by general users. There are many good texts that explain these in detail (some are included in the reference list) and students are advised to refer to them for more detail. Briefly, however, cartographic design follows five basic stages. In the first stage the purpose of the map is identified together with the potential users. Usually geographers are interested in drawing 'thematic' maps that display the spatial variations of one or more variables, as opposed to 'topographic' maps that are used to emphasize the location of spatial phenomena. In the second stage, appropriate data are obtained and issues of planimetric position, projection system and scale are addressed. In stage three the data are simplified by identifying the spatial dimensions of the features being mapped – point, line, polygon (areal) or surface (volume) – and classified according to the scale of the data (nominal, ordinal, interval or ratio). In stage four the features are mapped using standardized graphics in a process called symbolization whilst other graphical elements, such as a legend, are also added. In the final stage, the map is assessed to determine whether users would find it useful and informative.

Of the five stages, stage four is perhaps the most important in determining the success of a map in communicating information. The process of symbolization is the critical step and it has been the focus of a

Table 27.1   The relationships between visual variables and the characteristics of the features to be mapped

	Nominal	Ordinal	Interval/Ratio	Point	Line	Polygon
Location	G	G	G	G	G	G
Size	P	G	G	G	G	M
Colour value	P	G	M	G	G	G
Colour hue	G	M	M	G	G	G
Texture	G	M	M	M	P	G
Orientation	G	M	M	G	P	G
Shape	G	P	P	G	M	G

G = good; M = marginally effective; P = poor.
Source: Adapted from MacEachren 1994a, fig. 2.28, 33.

great deal of research in cartographic design. There is no room to discuss these design issues in any depth here, but a few of them will be expanded upon. Although maps contain a lot of different types of graphics, these can all be classified into eight basic visual variables: size, colour value, texture, colour hue, orientation, shape and the two dimensions of the plane of a sheet of paper (Bertin 1983: 42). Table 27.1, adapted from MacEachren (1994a), shows how these eight visual variables relate to the spatial dimension and data scale of the particular feature to be mapped. Visual variables vary in their ability to display different types of information on a map and the skill of the cartographer is to know which one to choose to best display the information. A good example is the use of colour. Colour printing has become much more common in recent years, making it easier for students to use colour in their maps. But colour should be used carefully since although it can be effective at showing differences between nominal categories, it is not so good at showing differences in numerical data (Bertin 1983). It is also worth bearing in mind that traditional publication media may find reproduction of coloured maps difficult and costly. Bertin's framework has since been adapted and expanded to include new visual variables such as pattern, clarity (MacEachren 1995) and projection (Dorling

1992, 1994). These allow a greater degree of detail to be added to a map.

Once the data have been symbolized, other graphical elements are added so that the map communicates the information clearly and with as little ambiguity as possible. Although the principles of graphical design are contested (e.g. Tufte 1983, 1990, 1997; Cleveland 1985), it is generally accepted that the following five elements need to be included: title, legend, source of data, north arrow and scale. The title should explain the major theme being mapped, the geographic region and the date of the data and should be placed near the top of the map. The legend should include a clear explanation of what each symbol represents with the text written to the right of each legend symbol. Larger values should be shown at the top of the legend and smaller values at the bottom. The source of the data should be placed at the bottom of the map and include information on where the data were obtained, when they were collected and by whom. Other graphical elements may also be included such as labels indicating features on the map that may be of interest. However, it is bad practice to clutter the map with unnecessary information or 'chart-junk' (Tufte 1983) – decoration that may draw attention away from the important information on the map. When the map is finished it should appear visually balanced with no large empty spaces. For more information on issues of cartographic design see Slocum (1999), MacEachren (1995) or Robinson et al. (1984).

Final Comments

Cartography experienced rapid change during the 1990s as a response to swiftly evolving technologies and new concerns within science regarding data exploration and analysis. However, despite the fact that the majority of cartographic research today is geared towards the use of highly interactive map displays for visual thinking, the principal applications of cartography by students currently remain in the realm of visual communication. The growth of computerized and Web-based cartography means that more people than ever can create maps and hence there is a real need to foster 'back to basics' in cartography. The construction of a well-designed map that supports queries such as 'what is?' or 'where is?' and communicates facts effectively to the user is of increasing importance as the scientific and social communities demand speedier access to geographically referenced information.

References

Bertin, J. 1983: *The Semiology of Graphics*. Madison, WI: University of Wisconsin Press.

Cartwright, W. 1997: New media and their application to the production of map products, *Computers and GeoSciences*, 23, 447–56.

Cleveland, W.S. 1985: *The Elements of Graphing*. Monterey, CA: Wadsworth Advanced Books and Software.

DiBiase, D. 1990: Visualization in the earth sciences, *Earth and Mineral Sciences, Bulletin of the College of Earth and Mineral Sciences, PSU*, 59, 13–18.

Dorling, D. 1992: Visualizing people in space and time, *Environment and Planning B*, 19, 613–37.

Dorling, D. 1994: Cartograms for visualizing human geography. In D. Unwin and H. Hearnshaw (eds), *Visualization and GIS*, London: Belhaven, 85–102.

Dorling, D. 1998: Human geography: when it is good to map, *Environment and Planning A*, 30, 277–88.

Dykes, J.A. 1997: Exploring spatial data representation with dynamic graphics, *Computers and GeoSciences*, 23, 345–70.

Kraak, M.J. and Brown, A. (eds) 2001: *Web Cartography*. London: Taylor and Francis.

McCormick, B.H., DeFanti, T.A. and Brown, M.D. (eds) 1987: Visualization in scientific computing: a synopsis, *IEEE Computer Graphics and Applications*, 7, 61–70.

MacEachren, A.M. 1994a: *Some Truth With Maps: A Primer on Symbolization and Design*. Washington, DC: Association of American Geographers.

MacEachren, A.M. 1994b: Visualization in modern cartography: setting the agenda. In A.M. MacEachren and D.R.F. Taylor (eds), *Visualization in Modern Cartography*, Oxford: Pergamon, 1–12.

MacEachren, A.M. and Ganter, J.H. 1990: A pattern identification approach to cartographic visualization, *Cartographica*, 27, 64–81.

MacEachren, A.M. and Kraak, M.J. 1997: Exploratory cartographic visualization: advancing the agenda, *Computers and GeoSciences*, 23, 335–43.

Robinson, A.H., Sale, R.D., Morrison, J.L. and Muehrcke, P.L. 1984: *Elements of Cartography*, 5th edition. Chichester: Wiley.

Slocum, T.A. 1999: *Thematic Cartography and Visualization*. Upper Saddle River, NJ: Prentice Hall.

Tufte, E.R. 1983: *The Visual Display of Quantitative Information*. Cheshire, CT: Graphics Press.

Tufte, E.R. 1990: *Envisioning Information*. Cheshire, CT: Graphics Press.

Tufte, E.R. 1997: *Visual Explanations: Images and Quantities, Evidence and Narrative*. Cheshire, CT: Graphics Press.

Unwin, A. 1999: Requirements for interactive graphics software for exploratory data analysis, *Computational Statistics*, 14, 7–22.

Further Reading

Dorling, D. and Fairbairn, D. 1997: *Mapping: Ways of Representing the World*. London: Longman.

MacEachren, A.M. 1995: *How Maps Work: Representation, Visualization and Design*. New York: Guilford Press.

Orford, S., Dorling, D. and Harris, R. in press: *Introduction to Visualization in the Social Sciences.* Norwich: Environmental Publications.

Internet Resources

• The GeoVista Center: http://www.geovista. psu.edu/ This is a centre devoted to fundamental and applied scientific research on the visualization of georeferenced information and the development of geographic visualization technologies. It is a gateway site that introduces advanced visualization techniques associated with cartographic exploration. Includes demos, tutorials, further reading, etc.

• Map History/History of Cartography: http:// ihr.sas.ac.uk/maps/ THE gateway to the subject concerned with all aspects of mapping history, including information on conferences, books and articles, news, research and links to lots of images of maps on other websites.

• The British Cartographic Society: http:// www.cartography.org.uk/

• The Society of Cartographers: http://www. soc.org.uk/

28

Spatial and Locational Modelling in Human Geography

Michael Batty

What are Models? What is Modelling?

A model is a simplification of reality. In the last 50 years, as computers have swept the world, the idea of using mathematics to better our understanding and to gain some handle on making predictions has become commonplace. In essence, modelling is the process of building such simplifications. It usually consists of defining a limited number of variables, called inputs, which determine the phenomena to be explained, which are known as outputs. What connects the inputs and the outputs are relationships formulated as mathematical equations which embody the hypotheses that the scientist wishes to explain. The model builder basically 'tunes the model' to replicate some known data about the phenomena by 'calibrating' parameters – constants within the model – so that the model predicts the known phenomena as closely as possible. If the fit between the model and the data is good, this is taken as confirming the assumed hypothesis and at this point, the model builder engages in a debate as to whether the model is appropriate for making predictions in situations different from those in which it has been calibrated.

In one sense, this is no different from what goes on in the laboratory where the scientist sets up an experiment to test some hypothesis. In the case of mathematical modelling, the experiment is virtual and abstract, but just as the laboratory imposes strict controls on what is being tested, so does the computer model. This generic process has been developed in geography during the last half-century where we have been concerned with explaining how phenomena on the Earth's surface vary spatially at different scales. Spatial and locational modelling, however, has largely but not exclusively developed in the human domain, and to illustrate the process it is worth presenting a simple example.

Imagine that you wish to predict the percentage of households who own cars in different areas of a city. Our hypothesis is that car ownership is closely related to the numbers of households who own houses as this is indicative of greater income or wealth. The percentage of houses owned x within any area i called x_i is our independent or input variable and we wish to predict the percentage who own cars y, in equivalent areas i, as the output variable. The relationship between the input and the output is a simple linear equation which we can write as $y_i = \alpha + \beta x_i$. The two Greek symbols α and β are the parameters that the model builder has to 'fine-tune' to generate predictions of car ownership as close as possible to the observed data. This is usually done automatically by minimizing the sum of the squared differences between the predictions and observations. Once the process is completed, the model builder then assesses the quality of the model using various tests based on how well the model fits the data and how significant are the parameters.

For example, taking the percentage of car ownership and house ownership in each of the 33 London boroughs from the 1991 population census, we can fit the model as. $y'_i = 20.705 + 0.676 \, x_i$. The correlation between the observed and predicted car ownership is excellent at 97 per cent and the parameters are both highly significant. This model shows that in London, if you find an area with, say, 70 per cent of the housing owned by their occupants, then the model predicts that 68 per cent of these households will own one or more cars. Many spatial and locational models are much more complex than this in that elaborate systems of equations connecting one another are often specified. Constraints on what might happen, embodied as simple decision rules, are sometimes added. But at the end of the day, most models are little different in their purpose from the simple linear model that we have just presented where the focus is on explaining as much as possible from the simplest set of available data.

The Classical Tradition: Social Physics and Regional Science

Social science and physical science have been entwined since the Enlightenment, and before the ink was dry on Isaac Newton's great *Principia* scientists were speculating on how classical physics could be applied to human phenomena. In essence, the notion that there were human forces which could be explained by mechanical analogies gradually gained momentum. By the twentieth century, the idea that human movements ranging from migration to rush-hour traffic, international trade to shopping trips, could be modelled in analogy to gravitational force, held sway. For the last 50 years, these models have dominated spatial analysis, being variously referred to as 'social physics' or spatial interaction modelling. The basic idea is that interaction varies directly with the size of the objects generating or attracting the movement, and inversely with respect to some measure of spatial impedance usually based on distance between the origin and destination of that interaction. Location can be predicted as a function of such interaction if it is assumed that everyone who is located at some place is involved in some kind of movement. Such models have been used particularly to show how shoppers can be attracted to shopping centres, consistent with earlier theories of central places and their hinterlands. Predicting where people live has also been the subject of such models based on mechanisms within the housing market where transport cost and land rent are traded off in making the locational decision.

Let us consider a typical form which predicts the flow of expenditure from a residential location i to a shopping centre j, called S_{ij}. This flow is a function of the expenditure available for shopping at i, C_i, the size of the shopping centre at j measured by its floorspace F_j, and some index of the distance between i and j which we will measure as the travel cost t_{ij}. The model is specified as $S_{ij} = G \, C_i F_j / t_{ij}^{\alpha}$. As in our linear model, there are two parameters to fine-tune – G and α – which are adjusted until the predicted sales at the shopping centre S_j – calculated by summing S_{ij} over I ($S_j = \Sigma_i S_{ij}$) – are closest to the observed sales. This model in effect predicts the sales that are generated in each central place hinterland associated with each shopping centre, and these models are widely used by commercial retailers and developers. For example, large supermarket chains such as Tesco in the UK make predictions using these kinds of models at least once every week in their quest to capture market share by finding sites for new stores.

There are many other related models within this classical tradition. Industrial and economic location form the mainstream of what came to be called regional science in the 1950s. Models were built around ideas from regional economics based on macroeconomic theory originating from the Keynesian point of view. In particular, the idea of interdependence within the economy where some activities are more basic than others has been widely applied through multiplier and input–output analysis. A related theoretical tradition from economics involves models which show how land is allocated using market mechanisms within cities. Originating from Von Thünen's classic 1826 model of the way agricultural land use organizes itself around a market, analogous urban models underpinned by the micro-economic theory of the consumer and the firm have been widely developed as part of what is called the 'new urban economics'. These models too are consistent with social physics, based on interaction, density, potential and accessibility, and a link between them has been established through theories of discrete choice in economics and psychology.

Although the classical approach still dominates, there are many limitations and enough inconsistencies to have brought the general approach into mild disrepute within the last 20 years. The single biggest intellectual problem involves the notion that space is the dominant focus. In fact, it is now clear that thinking about geographical systems as being in some sort of spatial equilibrium deeply implicit in such models is only half the story. The idea that time, dynamics and change are all important is now on the agenda. The other problem is one of scale. The classical approach depends upon statistical considerations of large numbers, which has meant that models have been aggregative and that the intrinsic heterogeneity of people and places has been ignored. Current developments are seeking to change this focus.

Location Modelling and Complexity

The view that human systems are never in equilibrium and the explanation must be on how activities and peoples change through time is strongly tied up with the notion that change is non-linear. Non-linearity is driven by the fluctuations and feedback at the most micro level, and it is clear that creativity, innovation, history even, in human systems can only be explained in this way. Exponentially growing or declining populations represent the most obvious sources of such change, but when such systems co-evolve with others it is their non-linearity that leads to competition. The distinct structures that emerge can only be explained through models which generate discontinuities such as catastrophes, bifurcations, and sometimes chaos. For example, the spontaneous growth of 'edge cities' – large shopping and commercial malls on the periphery of urban areas – the gentrification of inner cities and the formation of ghettos can only be explained using these ideas about dynamics in which feedback is central.

Several different approaches have been developed. Nesting classical models within non-linear dynamic frameworks enables spontaneous development to be simulated, consistent with equilibrium interaction patterns. Variants of these models in which micro-changes set off new unforeseen developments in space have been developed for the growth of cities, and a new strain of urban economic geography is developing based on the way cities and regions compete with one another. New approaches based on morphology where the focus is on generating self-similar or fractal shapes have been applied to cities, linking these ideas to traditional social physics where scaling ideas predominate. Self-similar locational structures such as central places, street systems and concentric zones of development can be grown using reaction-diffusion models where local processes of change can be simulated using cellular automata which grow spatial structures from the bottom up, analogous to the way crystals grow in the laboratory.

These locational models are very different from those that came a generation or more ago. They require much more detailed data at the individual level, on the dynamics of change and decision-making processes. They require state-of-the-art computer systems, GIS, and visualization technologies. There are some impressive practical examples such as the TRANSIMS model, where every individual and car on a city network is being simulated, while several cellular automata models are being developed to simulate the impacts of urbanization and urban sprawl in US cities. There are also more conceptual models which illustrate how polarization can emerge in cities where locational preferences do not suggest that such segregation is at all obvious. But most of all these new ideas are changing the conceptual basis of prediction in human geography. In the past, we have assumed that the search is for 'parsimonious models'. Now the focus is on building models to explore our own conceptual hypothesis making rather than on exact prediction.

Further Reading

A clear introduction to the design and construction of a mathematical model is contained in N. Gilbert and K.G. Troitzsch, *Simulation for the Social Scientist* (Milton Keynes: Open University Press, 1999). To brush up on mathematical techniques (and useful for classical spatial interaction models), see A. Wilson and B. Bennett, *Mathematical Models in Human Geography and Planning* (Chichester: Wiley, 1985). A sense of the variety that forms the classical approach is given by A. Wilson, *Complex Spatial Systems: The Mod-*

elling Foundations of Urban and Regional Analysis (Harlow: Prentice Hall, 2000). A good if pedagogic introduction for building classical models from scratch is D. Foot, *Operational Urban Models* (London: Methuen, 1982).

An overview of models within regional science is contained in P. Nijkamp (ed.), *Handbook of Urban and Regional Economics*, vols 1–3 (Amsterdam: Elsevier, 1991). Many of these models are now linked to GIS and a useful compendium of articles is P. Longley and M. Batty (eds), *Spatial Analysis: Modelling in a GIS Environment* (Chichester: Wiley, 1996). Newer computational approaches are introduced in S. Openshaw and R.J. Abrahart (eds), *GeoComputation* (London: Taylor and Francis, 2000). Finally, for those interested in how these approaches relate to the wider theoretical context, see P. Krugman, *The Self-organizing Economy* (Oxford: Blackwell, 1996).

A Note on Data Sources

The public decennial population census provides data for small areas such as census tracts. Such data are now invariably digital and can be read using GIS software. A UK archive is at http://www.mimas.ac.uk/, while there are other, usually commercial, sources for employment and population data at finer scales such as post or zip codes. Software for large-scale spatial and locational modelling is usually purpose built, one-off and costly, but a series of plug-ins is gradually being developed for desktop GIS software (see http://www.esri.com/). Some freeware is available at sites such as http://www.geog.uu.nl/pcraster/ and http://www.flowmap.geog.uu.nl/, but it is surprising how far you can develop such models within standard software such as spreadsheets and desktop GIS as well as within statistical packages such as SPSS.

A series of example programs which can be run with software such as Excel is R. Klosterman, R.K. Brail and E.G. Bossard (eds), *Spreadsheet Models for Urban and Regional Analysis* (New Brunswick, NJ: Rutgers University Press, 1993).

29
Modelling in Physical Geography

Susan M. Brooks

What is a Model?

Modelling has been a recognized method for carrying out research in physical geography for a very long time. It can therefore be argued that there can be few, if any, physical geographers who have no experience of modelling whatsoever. Furthermore, it can be introduced to investigations at virtually any academic level, ranging from GCSE coursework to Ph.D. level and beyond. Modelling can provide tremendous insight into the way natural systems operate and, as a consequence, it has grown in popularity throughout the twentieth and twenty-first centuries. Of course, this has also been possible due to unprecedented advances in computer technology and, in particular, increasing versatility of the World Wide Web. These assertions largely depend on what we mean by 'modelling' and how we define what constitutes a model. As Kirkby et al. (1992) have said:

> There is a continuous range of models from strictly deterministic quantitative models with a unique outcome; through a range of stochastic models in which outcomes are more or less predictable; to entirely qualitative models, including those which have been traditionally successful in geomorphology, like W.M. Davis' Geographical Cycle of Erosion.

Over the past decade or so, modelling in physical geography has been revolutionized, driven by the fast pace of advancement in computer hardware and software capabilities. Not only is our capacity to design and implement state-of-the-art computerized models better than it has ever been, but the models available now reach a wide sector of the population worldwide. Running computer models of the physical world is no longer exclusively for the specialist researcher, but can be achieved by anyone who has an interest in the physical world. It is now possible to obtain a wide range of models over the Internet, many offering sophisticated capabilities, at very low cost.

However, there are many issues that need to be considered when using models to carry out research in physical geography (Anderson and Burt 1990), also outlined in chapter 12 by Lane in this volume. These have become increasingly overlooked as ever more detailed conceptual and mathematical models are used by a growing number of people. Simply because a model looks good and is capable of carrying out a number of functions, it does not necessarily mean that it works well or produces valuable information. Models have to be used with care and thought if they are to provide sound insight into the way physical systems work.

The Earliest Models: An Inductive Approach

Research generally proceeds along one of two lines, whatever academic level is adopted. First, there is the inductive approach. This approach involves researchers accumulating as much knowledge as possible about the way a system works. You may find yourself using this approach if you have no prior knowledge about what you expect to find. You simply gather up data (qualitative or quantitative) and then use the data to develop a generalized theory. The generalized theory is really a model.

The early models used in physical geography were straightforward representations of physical systems as simpler or smaller-scale versions of reality. Conceptual models, outlined by Lane (chapter 12 this volume), might fall into this category. This is possibly one of the easiest modelling strategies to comprehend, simply involving a series of observations which are used to form a general idea of the overall development of a system. They may involve entities, linked together through a series of concepts revealed by the initial observations. However, models such as this provide little insight into the way processes operate or into the complex interrelationships that often exist in physical systems. This is not to say that these modelling efforts are without value or pointless. They often initiate serious debate and spur thinking among scientists to improve upon the model.

All models make assumptions. No single model can ever be a perfect replica of the system it aims to describe and emulate. Conceptual models make many simplifying assumptions. They are also fairly restrictive in the ways they can be applied. Physical geography is often concerned with understanding systems that cannot be observed directly (too remote spatially or operating in the distant past or in future time periods). Because of this, as well as the simplifying assumptions, conceptual models have been heavily criticized and replaced by more detailed

models. Mathematical models represent a category of more detailed models. While they still make assumptions, it is argued that such models are more readily spatially and temporally transferable. Hence they do not only apply to the circumstances and locations under which they were developed. Included among these later models are more complex processes, and more realistic inclusion of the effects of phenomena such as climate change, making them more readily applicable to other regions of the world, as well as other time periods.

The increasing desire to include greater complexity in physical models, to allow fewer and more realistic assumptions, has spurred a move away from an inductive approach towards a deductive approach to modelling.

More Recent Models: A Deductive Approach to Modelling

The general trend in physical geography has been towards increasingly quantitative approaches, especially in the period from the 1950s to the 1990s. Numeric data have been gathered to try to understand a range of processes, normally over a very short period of time (one season is typical, but some longer-term projects have run over a decade or so). This has sparked enormous interest in a deductive approach. Unlike the inductive approach, where the research begins with observations that accumulate to form a general picture of how systems work, the deductive approach begins with a testable idea (a hypothesis). Data or information are then gathered and used to support or refute the theoretical idea that was suggested initially (Haines-Young and Petch 1986). Models that have been built around existing theory can be run over and over, using different datasets to investigate the conditions under which they work best. In this way, they can provide a robust test of the initial theory, highlight the circumstances in which the theories work best or, perhaps, do not work at all. The models can be used to examine multiple hypotheses in order to determine which is the more likely to be suited to the problem in hand. In this way, models can greatly assist in the understanding of system behaviour in a comparatively efficient way.

This deductive methodology differs somewhat from the original purpose of many quantitative models. Many models were originally designed to *predict* how a system might behave under particular circumstances. Good examples are rainfall-runoff models (Beven 2000), intent on predicting flooding extents, timings and heights. However, models have been far more widely used, with special strengths in helping to determine how systems behave and to explain this behaviour.

Many physical geographers use the deductive approach today. The starting point is an idea, or hypothesis, that can be tested against observations. You may well use this approach if you have a good idea of what you are looking for in a system, perhaps from lectures, lessons, tutorials or books you have read. You approach your research by testing your hypothesis after collecting only data that are appropriate to the problem. Then you examine the data to see whether they support or refute the original hypothesis. Many modelling approaches are like this because the model has often been constructed based on existing theory. Mathematical modellers must have a sound theoretical basis in which the modeller has already decided which elements of a system to include and which to omit as being irrelevant. The model therefore has very particular data requirements that are determined by the model structure. A good example in rainfall-runoff modelling is the prediction of the timing and quantity of runoff during rainstorms. The data required to drive such models are normally rainfall along with measurements of the infiltration capacity of soil. Hence the volume of runoff can be calculated from the balance between these two datasets. The timing of runoff reaching the

catchment outlet depends on topography and the configuration of the channel network. It is therefore sensible to collect only data actually required by the model, as opposed to collecting any data you can about the system in question. In effect the model structure determines the data that are collected (very different from the inductive approach).

You may also gather output data to test whether the model can simulate (or predict) the same behaviour as the real system. This is the validation procedure, discussed by Lane (chapter 12 this volume). In rainfall-runoff modelling you may collect data for runoff under different rainstorms and compare these against predictions made by the model. You may find the model performs very well and gives a close approximation to the data that have been measured. Equally possible is that the model comes nowhere close to simulating the amount and timing of runoff that you have actually measured. However, this is not necessarily a bad outcome and much can be learnt from a model failing to perform well.

What could be wrong? First, if you have ever tried to collect data for rainfall, infiltration or runoff you will be aware of the difficulties of doing so, and the errors that can result. Either the input data may be poor, in which case the model is being driven by unrealistic error-prone data, or the output data may be inaccurate, in which case you are comparing model predictions with inaccurate field observations. Second, have you considered exactly where you took your measurements and whether these locations and points in time are the best choice? Rainstorms are highly variable from place to place, and during a storm there is great temporal variability in the amount of rain falling. Similarly, soil infiltration rates are highly variable, often reflecting the compaction (or density) of the soil. For example, infiltration into soil may change during storms due to swelling of soil colloids. These are very specific points in relation to rainfall and infiltration measurements, but they reveal a general and very important point about modelling. Many models take little account of uncertainty in data, requiring you to input specific numbers and, in turn, providing specific numbers back in the form of outputs. These models are deterministic models, and often fail to simulate real-world behaviour. Some models do take uncertainty into account. They allow inputs for a range of values for each parameter (such as rainfall or infiltration) and provide a range of possible values for the output. These models are called stochastic models and they do not have a unique outcome. They are therefore quite difficult to interpret rigidly, but do not make any great claim to being able to pinpoint and predict unique outcomes.

Putting data matters aside, models can also fail to provide decent simulations because they are structurally flawed. One of the more serious flaws relates to the choice of theory to base the model upon, and the inclusion/omission of key processes. Models are based on known theories that are thought to determine system behaviour. Sometimes deliberate choices are made to exclude processes that are not thought to be important, to maintain simplicity in model structure and data requirements. If those processes play no part in the system output then it might well be better to exclude them. However, if we are applying a model to a new location or time period, then we really cannot be sure that these decisions are sound. There is a lot we do not know (or which may even be unknowable) about the real world and, hence, is impossible to include in models. This possibility must never be ruled out when interpreting the results of model simulations, even if the models provide reasonable simulations compared with measured data.

One central issue relates to the scale at which models are applied. When you take field measurements you will often choose a relatively small scale to work at. Again, using the example of runoff generation, the key input parameters are rainfall and infiltration. These may be measured using raingauges and infiltrometers, respectively. Each instrument comprises a cylinder with an approximate diameter of 10 cm. What you are doing when you collect rain in a gauge or use an infiltrometer is taking point measurements. It is important to consider whether these point measurements reflect the behaviour of the whole hillslope or catchment. For example, if you were to take 50 measurements of rainfall over a hillslope, is it reasonable to assume that the hillslope as a whole will receive the average of the 50 measurements over its entire extent? The answer may be debated, and measurements taken at the different scales can help to resolve the question. However, as with many issues to do with modelling, there is no right or wrong answer. It is just important to consider and discuss how this issue affects the accuracy of model simulations.

One of the most effective ways in which deductive models are used is to test a whole series of ideas to see which is the more likely to be true. A simple example might relate to the effect of vegetation removal and climate change on runoff generation. We may wish to assess whether it is the removal of vegetation or a change in climate that might be responsible for a change in runoff amounts. A model may be set up to assess each scenario in turn and the outputs inspected to provide clues as to which parameter is the more significant. This is a very simple example, but physical geographers have used such an approach to test a whole series of competing hypotheses (or theories/ideas). This example illustrates the importance of having a sound model structure as the model is being used to provide explanations of how systems behave. Traditionally, when models have been used for prediction the structure is of less importance. Often poorly structured models can provide predictions which are as good as a well-structured complex model. However, physical geographers have largely moved away from using models for prediction in favour of their use in providing explanations of system behaviour.

Conclusions: The Way Forward

There are many models available today. Looking back at the various ways they have been used by physical geographers offers many opportunities to consider using models for several different purposes. First, you may decide to use (or develop) a descriptive model that summarizes all your observations about a system. This conceptual model might then provide a highly useful generalization that applies to other similar systems. Second, you might choose to use an existing model to predict how the system you are investigating is likely to behave in different circumstances. You may be trying to predict how a river responds to rainfall, which storms might lead to flooding or how extensive such flooding might be. Third, you may use a complex model that includes many processes to try to explain how a system might behave under different scenarios.

Finally, given all the controversial issues raised above, as well as the tremendous capability of computers to handle large datasets, some physical geographers are reverting to the inductive approach of early times. In this they gather or generate huge datasets and then try to devise a model that best simulates these observations. This 'new' approach is termed exploratory data analysis, where research starts from many observations and then proceeds towards a model (Brooks and McDonnell 2000). In effect, we have turned a full circle from the earliest approach to modelling right up to the current day. Whichever approach you take, you should find much of value in adopting a modelling strategy for planning and carrying out your research. However, models should be applied 'thoughtfully' and with care, offering discussion and consideration of the many issues raised here (Beven 1989).

References

Anderson, M.G. and Burt, T.P. 1990: *Process Studies in Hillslope Hydrology*. Chichester: Wiley.

Beven, K.J. 1989: Changing ideas in hydrology: the case of physically-based distributed models, *Journal of Hydrology*, 105, 157–72.

Beven, K.J. 2000: *Historical Development of Rainfall-Runoff Modelling*. Chichester: Wiley.

Brooks, S.M. and McDonnell, R.A. 2000: Research advances in geocomputation for hydrological and geomorphological modelling towards the twenty-first century, *Hydrological Processes*, 14, 1899–1907.

Chorley, R.J. 1965: A re-evaluation of the geomorphic system of W.M. Davis. In R.J. Chorley and P. Haggett (eds), *Frontiers in Geographical Teaching*, London: Methuen.

Davis, W.M. 1899: The geographical cycle, *Geographical Journal*, 14, 481–504.

Haines-Young, R.H. and Petch, J.R. 1986: *Physical Geography: Its Nature and Methods*. London: Paul Chapman.

Kirkby, M.J., Naden, P.S., Burt, T.P. and Butcher, D.P. 1992: *Computer Simulation in Physical Geography*. Chichester: Wiley.

Internet Resources

There is a vast collection of software available over the Internet. Various conditions are attached to using software, and sites vary enormously in their provision for research. Some will charge for software and this can be expensive, but often evaluation copies of software are available if you simply want to gain an idea of how a piece of software works. This can be useful if you want to assess its likely suitability for your research. Some sites allow you to download software free of charge but impose a time limit on its validity. However, there are some sites that allow you to download and use software free of charge, in exchange for signing a licensing agreement and/or acknowledging the use of any software in publications and reports.

Some sites I have used recently and which are relevant to the above discussion are listed below.

- http://www.pcraster.nl
  Here you can download software that allows you to create a digital elevation model of a catchment, and to devise and run simple routines to model infiltration, runoff generation and water flow through the catchment. It has a wide range of tools for modelling different aspects of catchment behaviour and is an

exciting place to begin to explore the potential of catchment models. The software is free of charge provided you sign a licensing agreement.

- http://www.greenhat.com/
  This site is a must for anyone interested in studying interactions between plants, soils and the environment. Models available simulate water flux through structured soil columns, allow variation in rainfall inputs, enable simulation of the effects of plant growth and include many other features. Software is relatively inexpensive but can also be downloaded free of charge for a limited trial period.

- http://www.sc-dlo.nl/
  Along similar lines to the previous software, this site offers up-to-date versions of the model SWAP (Soil-Water-Atmosphere-Plants), and other environmental models.

- http://www.es.lancs.ac.uk/es/Freeware/Freeware.html
  One of the early conceptual models for catchment modelling is TOPMODEL. It has since been refined and extended over a period of almost 30 years. It can be downloaded free of charge from this website and is another good place to start exploring the scope of models.

- http://www.geo-slope.com
  If you are interested in slope stability modelling, then the geoslope office package is up-to-date, comprehensive and offers impressive capabilities. However, it is complex to get used to using and is comparatively expensive. At this site you can download an evaluation version that provides a good insight into the whole package, and there is also a student version with tutorials to get you started in slope stability modelling. This is well worth having a look at.

30

GISystems, GIScience and Remote Sensing

Rachael A. McDonnell

Geographical Data and Computers

We handle geographical data every day in our lives. Knowing what and where something is underlies many of our actions and interactions. Our spatial sense is based on the breaking down of a view into a series of objects, and then using descriptors such as 'beside', 'to the right of' or 'in front of' to express their relationships to each other.

Whilst this spatial reasoning is second nature to most living things, translating that into a computer-manageable form of bits and bytes of code is difficult because of the complexity of the basic information. Digital databases need to provide capabilities for storing the 'what and where' information as well as the spatial relationship data between individual phenomena. For example, if stream tributaries are represented as a series of lines, then it is important that information is included that states that these individual lines join together to form a whole entity of a river.

The software systems designed to help our spatial work are known as geographical information systems (GIS). These systems are powerful sets of tools for collecting, storing, retrieving at will, transforming and displaying spatial data from the real world (Burrough and McDonnell 1998). They are able to handle many forms of geographical data such as maps, field observations, tables and remotely sensed images, allowing us to analyse through a computer the distributions, interactions and processes of phenomena for an area, be it a street, country, continent or world. The main commercially available software systems such as ArcGIS (and its sister ArcView), GeoMedia, IDRISI and MapInfo offer to varying degrees the ability to query, analyse and combine large datasets. The results of all this work may be presented in a variety of forms ranging from tables and maps through to animated sequences.

The current systems support two main data structures for defining the location, properties and spatial links of an object of interest. These are called raster and vector data structures. The first may be thought of as a chequerboard in which each square (or other tessellated shape) is classified in terms of a particular phenomenon. A useful analogy is to think of the raster data structure as a board of the same-sized 'Lego' blocks with different colours representing different classes. The resolution, that is, the size of the pixels, will dictate how detailed the representation is.

The second data structure, known as vector, is more akin to our map-based ideas. The outline of each individual unit of phenomena is described using the basic building objects of points, lines and polygons with explicitly defined coordinates. These primitives are joined together to form complex spatial descriptions and are a sort of 'Kinex' version of geographical data. The all-important spatial relations of connectivity etc. are explicitly defined within the database. This information is known as the topology and its inclusion in the database is important for subsequent analysis. The GIS software chosen for use in a project will usually determine the data structure adopted. This will influence to a certain degree the subsequent steps involved with developing a GIS database and the analytical functions available (see Burrough and McDonnell 1998).

Developing a GIS Database

The conversion of our geographical information into digital form based on either the raster or vector data structure is the first step of any GIS project. Ten years ago this would almost certainly have meant the manual conversion of paper sources into digital form using equipment such as digitizers and scanner (for more detail see Heywood et al. 1998; DeMers 1999; Longley et al. 2001). This was always a tedious task during which errors were often introduced, so care and attention were paramount.

For many studies this is still the case, but today there are vast quantities of data already available in digital form that may be used in GIS. Many national mapping agencies are in the process of converting all their paper sheets into digital form, which is now accessible to outside users though often at a cost. Remotely sensed data are also predominantly available in digital form. Standard data formats have been agreed upon which allow digital files to be read directly into a GIS without awkward translations. The Internet has been an important stimulus for these developments and there are now online virtual data warehouses where users can search and download files, again often at a cost.

Whether the data used are digital or converted from paper sources, it is important to consider the characteristics of the information such as their scale and age, data collection techniques used, quality of data, size and the shape of individual mapping units. These have to be compatible with the application or the results will be poor. The final database will usually consist of a number of different thematic layers of information, which are all registered to the same map projection.

Analytical Capabilities of a GIS

GIS provide many different analytical functions that range from basic querying across one or more layers to more complex modelling of processes and interactions. The analysis may be applied to a particular location, a neighbourhood or a region. For example, one might need to know all reported incidences of crime at a particular address, within 3 km of an address, or in a city or county as a whole.

One of the most used capabilities of GIS is the combining of two or more data layers through a function known as overlay. This may be used to integrate maps or to study the co-occurrence of variables. For example, one may wish to examine the link between proximity to major roads and the occurrence of burglaries. By combining these two layers, it is possible to determine whether there is a *prima facie* case for supporting the hypothesis that houses close to fast roads are more likely to be burgled than ones further away.

Some more specific algorithms are also included in many GIS such as those used for analysing elevation datasets. For studies in many areas of geography, knowing steepness of slope, aspect etc. is important for interpreting processes, whether this is human settlement, water runoff or landslides. The starting dataset of altitude and location values is known as a digital terrain model, and to this may be applied pre-programmed functions to derive new maps to show, for

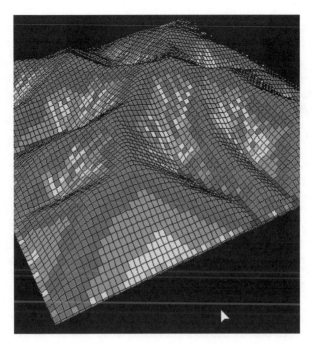

Figure 30.1   Digital elevation model of a landscape showing drainage lines derived by a GIS.

example, local slope or plan or profile curvature of a hillslope, as shown in figure 30.1. This type of information has been used extensively in the modelling of runoff to determine the route of water flow downslope.

With many studies there is a need to examine changes over time of a system. Current commercial GIS offer limited capabilities for handling data that have time as well as attribute and spatial components. This problem has been overcome by linking the GIS to a computer modelling language. The GIS then acts as a data-holding and displaying system and the computations of the model are undertaken in the separate programme. Data, usually held in raster format, are transferred between the two systems before and after the running of the model. Although a little messy, with the standardization in data formats and the increasing 'openness' of GIS this is becoming easier.

Geographical Information Science

Since the middle of the last decade there has been a shift in emphasis for much of the research undertaken in the field of spatial analysis and modelling. As GIS became more sophisticated, studies based on the development and use of technology have begun to be replaced by research on the fundamental ideas and theories underlying geographical information and its use. This new and growing area of research has been called geographical information science. It is characterized by the broadening of GIScience from an essentially technical subject to one that includes more cognitive, philosophical and methodological discussions.

At the first GIScience conference held in the United States in October 2000 (http://www.giscience2000.org), papers were given by researchers working in philosophy, psychology, computer science, engineering,

Figure 30.2   This 1-metre-resolution black-and-white image, collected by the IKONOS satellite, shows London around the London Eye.
Reproduced by kind permission of Space Imaging (http://www.spaceimaging.com).

information science, architecture, earth science as well as geography departments. The common theme of interest amongst all these different disciplines is how we define and represent space, whether this is at a building, a city or global scale, and how this may be translated into machine-readable format. The possibilities for the conceptual model are immense given the attribute, relationship and behavioural characteristics of phenomena present as well as our own differing experiences and cultural backgrounds. Representing the various dimensions of geographical information using formally defined schema is one of the many challenges being addressed by GIScientists.

Remote Sensing

For both GISystems and GIScience research, geographical data acquired using remote sensing techniques have been an immensely important input to the work. Not only are the data used directly, but also many geographical sources rely on these systems to provide information. For example, aerial photographs and now high-resolution satellite images of cities are an important input to large-scale mapping projects, as shown in figure 30.2. This type of image is useful for updating the mapped information of urban areas which change often.

Remote sensors measure reflected or emitted electromagnetic radiation without coming into contact with the phenomena. The Earth's surface absorbs, reflects and emits this energy, and the variations in output across the wavelengths of the electromagnetic spectrum may be used to detect the occurrences of different material. The patterns of energy amount and wavelength can, for example, be used to distinguish between different types of woodland in an area, or different cloud densities and temperatures to help in the forecasting of rainfall.

The sensors used vary in terms of the area captured in an image (spatial resolution), the frequency with which the same part of the Earth's surface is sensed (temporal resolution), and the wavelengths and width of the wavebands (spectral resolution) of the detectors. Sensors are mounted on platforms such as aircraft and satellites and the data are recorded in either analogue (e.g. aerial photographs) or digital form. The latter is the most commonly used today with the information recorded as a series of grid cells (pixels), each coded with a value (known as a DN digital number) from 0 to 255 representing the radiation detected by the sensor for a particular area of the Earth's surface (see Lillesand and Kiefer 2000 for more details).

Most data gathered in this manner are collected by government agencies, although in recent years commercial organizations have begun to develop their own acquisition systems. Some archival data are available at no charge, but most supply channels charge for datasets. These will need various preprocessing techniques to be applied before the data are usable in any studies. Commercially available image processing software such as ERDAS Imagine, ERMAPPER and IDRISI are usually used for this and subsequent data manipulation and analysis.

Early processing involves the removal of errors introduced by sensor and platform movement and irregularities. Once these have been addressed, techniques such as filtering and contrast stretching may be used to enable the user to determine more detail from a scene. Classification techniques may also be applied which group neighbouring cells with similar digital number values together to form larger map units. The resulting classified image could, for example, be of different land-use types such as urban, arable, grassland, deciduous forest, coniferous forest, heathland and water.

Remotely sensed data are now routinely used in studies ranging from coastal process monitoring to hurricane tracking and mineral exploration. The numerous platforms and sensors give comprehensive coverage at a number of scales at a relatively low cost. Drawbacks to use result from atmospheric conditions such as cloud and dust, as well as our still limited understanding of Earth surface and radiation interactions.

Studies using GIS and/or remotely sensed data can be exciting and innovative. For student projects one of the big drawbacks is the time needed to familiarize yourself with the software and the development of the database. Both image processing and GIS are sophisticated software with numerous user manuals; however, with an open mind and dedication, fascinating insights into geographical phenomena may be gained through them.

References

Burrough, P.A. and McDonnell, R.A. 1998: *Principles of Geographical Information Systems*. Oxford: Oxford University Press.

DeMers, M. 1999: *Fundamentals of Geographic Information Systems*. New York: Wiley.

Heywood, I., Cornelius, S. and Carver, S. 1998: *An Introduction to Geographical Information Systems*. London: Prentice Hall.

Lillesand, T.M. and Kiefer, R.W. 2000: *Remote Sensing and Image Interpretation*. New York: Wiley.

Longley, P.A., Goodchild, M.F., Maguire, D.J. and Rhind, D.W. 2001: *Geographic Information Systems and Science*. Chichester: Wiley.

Internet Resources

Websites

- USGS GISystems tutorial: http://www.usgs.gov/research/gis/title.html
- US National Center for Geographic Information and Analysis core curriculum in GIScience: http://www.ncgia.ucsb.edu.giscc
- Guide to GIS resources on the Internet, including training for specific software and data resources: http://sunsite.Berkeley.edu/GIS/gisnet.html

- Another guide to GIS resources with a more European focus: http://www.geo.ed.ac.uk/home/gishome.html
- Online GIS glossary: http://www.geo.ed.ac.uk/root/agidict/html/welcome.html
- ESRI glossary of GIS terms with a focus on those used with the ArcGIS software: http://www.esri.com/library/glossary/glossary.html

Software

- IDRISI GIS website: http://www.clarklabs.org
- ArcGIS software site: http://www.esri.com
- Integraph software site: http://www.integraph.com/dynamicdefault.asp
- Core curriculum remote sensing tutorial: http://umbc7.umbc.edu/~tbenja1/santabar/rscc.html

- Canada Center for Remote Sensing tutorial: http://www.ccrs.nrcan.gc.ca/ccrs/eduref/tutorial/tutore.html
- NASA Goddard Space Center and NASA and US Airforce Academy tutorial: http://www.fas.org/irp/imint/dosc/rst
- ERDAS Imagine image processing software: http://www.erdas.com
- ERMAPPER image processing software: http://www.ermapper.com

Data

- Ordnance Survey of UK website: http://www.ordnancesurvey.co.uk
- Worldwide warehouse for digital spatial data: http://www.gisdatadepot.com

31

Getting the Best Out of Lectures and Classes

David B. Knight

As a student enrolled in university and attending lectures (with large numbers of students) and classes (with far fewer students), you can develop skills to ensure that you get the best out of different types of instruction. For all of your studies, *writing and thinking go together*. Participating in a class or simply listening to a lecture is not enough. To achieve your best you must attend to your writing skills as a way of enhancing your thinking skills. This chapter identifies common strategies that apply to the *learning process* in general and lists several points that pertain to small-class situations specifically.

Who is in Charge?

You are in charge. Although many courses are prescribed by the curriculum, you are responsible for your own learning. What you put into the learning process will have a direct bearing on what you get out of it. Obviously others help you by offering their insights, factual knowledge, conceptual understanding, methodological skills, and to give encouragement and feedback as you undertake your studies, but ultimately you are the integrator and, thus, learner. Think of a three-part teaching–learning system:

- The instructor *offers* (readings, lectures, seminars, lab and fieldwork assignments,

etc.), *responds* (to questions) and *evaluates* (assignments and exams).
- You *listen* (to lectures), *participate* (in class interaction), *read* (assigned and additional readings) and *do* (lab and fieldwork assignments).
- You, the learner, *take in, think about, mull over, question, analyse, integrate* and *'report back'* (via essays, reports and examinations).

Before focusing on skills you can apply, a word or two about your instructors may be helpful.

Your Lecturers

You will know your instructors or lecturers as geographers, but remember they are also individuals. Accordingly, they will differ in their teaching methods. Consider two extremes. Instructor A walks into the lecture hall, stands behind a podium, reads lecture notes without paying attention to who is in the class or how his words are being received, and then abruptly leaves at the end of the allotted time. Instructor B regularly breaks away from notes to look at those present and 'read' faces as a way of gauging how well what is being said is being received. She poses questions to see if the students understand the material, and welcomes questions and answers them before continuing to lecture. She may spend time after class chatting with students about the lecture material or course assignments. Your instructors may approximate to or differ from instructors A and B.

To appreciate and benefit from contrasting teaching styles, be open and versatile so you can adapt to whatever comes. Hopefully, you will never encounter an instructor like the one identified by the poet W.H. Auden: 'A professor is one who talks in someone else's sleep.' Even if an instructor seems muddled there will still be a goal for the day's lecture, so your task will be to listen carefully to what may be a rambling discussion in order to identify that goal, otherwise the lecture may not be clear. Also, listen carefully for gems of insight. This point is made simply as a reminder that what you get from a learning situation is up to you, so even when a person is seemingly having difficulty communicating with the class, you can nevertheless benefit by being especially attentive to what is being said and by taking good notes. Ignore any instructor's personal mannerisms and be receptive to what he or she says; what is said is more important than how it is said. Note taking is a key skill for obtaining meaning from what you hear.

Why Take Notes?

Studies show that taking notes facilitates learning; recall is vastly improved for those who write good notes. Indeed, you are seven times more likely to recall information one week later if it is recorded in your notes. Even holding a pencil can trigger our minds to listen better to a lecture. You will decide on the degree of detail to record, but more is generally better than less. Do not try to copy down everything you hear. Above all, avoid verbatim recording; use a combination of the instructor's words and your summations.

Getting Prepared for Each Lecture and Class

Your class never starts at 2:00 p.m. or whatever time it is scheduled to meet. It starts some time earlier, when you decide to prepare for it. What should you do?

- Keep all reading and lecture/class notes, handouts and review comments together in a loose-leaf three-ring or pressure-spring binder (not a spiral-bound notebook).
- Read assigned and recommended material in advance of the class, making good reading notes. If questions arise during your reading, jot them down on a page in your binder for later consideration.
- Review your notes from the previous class.
- Start a fresh page for each day's notes. Record the date at the top of the first page and number each page. Note the lecture's title from the course outline or when it is given in class.
- Create an abbreviation system with which you are comfortable and record the abbreviations in a safe place, with a copy in the binder. Table 31.1 includes examples of abbreviations you might use. These are suggestions. If you make up your own abbreviations, keep the list in front of you until you are able to work from memory. Rehearse using your system before the start of term, perhaps taking notes of radio shows or family conversations. Once courses begin, continue to develop abbreviations as more terms become easily recognizable to you.
- Organize your notes effectively. This is as important as actually writing them. Two ways to divide your notepaper and use it are identified here. Figure 31.1a is derived from the Cornell system of note taking. There are three sections, the largest of which is the note-taking space or record column for notes recorded during class. After class, review these notes, highlighting major points by writing key

Table 31.1   Examples of abbreviations that can be used in note taking

cf = compare; in comparison; in relation to	w/ (or) c = with
CL = climatology	wh/ = which
EP = environmental perception	w/o = without
Esp = especially	↑ = increasing
G = geography	↓ = decreasing
FG = fluvial geomorphology	⊀ = see (as in a note to self,
PP = political process	'I see what the instructor
Spa = spatial	means' versus '?')
T = territory	
wh = when	* = most importantly

questions or pithy comments in the recall clue column. Then reduce your notes (within 24 hours of taking them) by critically reading the first two columns and identifying key words and phrases, and recording your own reflections, ideas and relevant questions in the section for 'summary' or 'reflections, ideas, questions'. Thus, the three completed sections will together provide a study guide.

The second scheme (figure 31.1b) is similar, although it provides less room for notes by having the 'reflections, ideas, questions' column at the right side of the page. One of these page formats may be good for you, but know that research has demonstrated that students perform better when they are allowed 'to encode in the way they prefer', so you may want to develop your own method. Whatever you

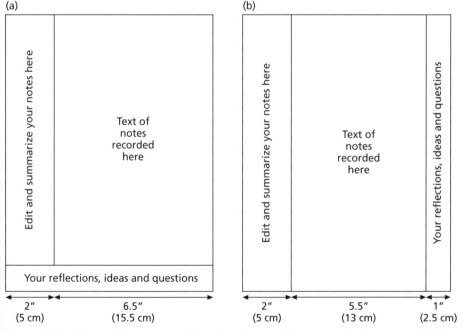

(a)

Edit and summarize your notes here

Text of notes recorded here

Your reflections, ideas and questions

| 2" (5 cm) | 6.5" (15.5 cm) |

(b)

Edit and summarize your notes here

Text of notes recorded here

Your reflections, ideas and questions

| 2" (5 cm) | 5.5" (13 cm) | 1" (2.5 cm) |

Figure 31.1   Note-taking systems. The Cornell note-taking system is shown on the left. Based on Walter Park, *How to Study in College*, 5th edition (Boston: Houghton Mifflin, 1993).

do, *be systematic* when taking and reading notes, preparing for class, writing lecture/class notes and completing follow-up reviews (see below).

How Can I Listen and Concentrate?

Skilled listening is essential. You need to be a skilled listener whatever the style of instruction. How can you improve this skill, if you think such is needed? Since a lack of interest and concentration are dual enemies of good listening, work to develop a mindset that will help you be open to and concentrate on whatever it is you are expected to listen to and study, no matter how hard the subject matter may be. If you listen poorly, you may not get a good grade. Blaming an instructor for being boring and uninteresting or moaning about how useless the course is will not get you either a better grade or sympathy. To get the most out of a class or lecture, you must listen with care. So, be both ready and attentive. A couple of tips that will help you are, first, be prepared to go to class. Read the assigned material. Take good reading notes, using point form to record your summation of the material read and also any questions you have. If you write down a quotation, use quotation marks (and include publication information, including page references, so you can find the appropriate source again). Memorize definitions and disciplinary jargon. Second, be prepared to listen. Your mindset is important. Before the event, try telling yourself something like: 'I am about to attend a lecture on X and I am keen to know what the instructor will say.' Even better, try not to fake it! Or say, 'I am going to listen', and follow through by identifying why you want to listen. Remember that listening is intentional and purposeful.

What the Lecturer Will Probably Do

Lecturers' teaching styles vary. However, the following points generally apply. Few lecturers or instructors summarize assigned readings; it is assumed you have done that for yourself. Students may be asked about the readings, so be prepared. In class, instructors will draw from a wider literature than the assigned readings, from fieldwork, or a set of lab experiments. Lectures often include new thoughts, some not yet published, so be attentive if something you hear is not in the assigned readings (which you will have already read). In terms of organization, instructors may: compare and contrast the findings of assigned material with new material; outline a concept or principle; identify and work through a problem; present and discuss a hypothesis; explore a case study; discuss an assignment; work through a technique; or make a major point and then give examples.

Given the variety of possibilities, you need to be attentive to what and why something is being said. Most instructors identify the goals and organization for the day. You may be given an outline so it will be easy for you to take notes. Paying attention to a well-organized instructor's style will sharpen your appreciation for both the flow of a lecture and its content.

A good lecturer will have an *introduction* (which may identify the overarching context), *thesis* (or statement of purpose for that day's class), *repetition* (but perhaps in a subtly different way from the first statement), *elaboration* (with information or data to support or expand upon the basic issue), *rephrasing* (of the thesis and the elaboration, perhaps from a different perspective rather than simply summarizing the elaboration), and *summary* or *conclusion* (whereby the lecture is brought to a logical end).

Active Listening

If you have prepared well, you will be *an active listener*. How can you improve your chances of getting the best out of a lecture or a class so that, in turn, you can give of your best at examination time?

- Attend class/lecture! Do not rely on a friend to take notes for you. You should be there to listen, write and think about what is said. However, consider having a 'buddy' to compare notes with afterwards, in case either of you has missed something.
- Help yourself. Avoid distractions outside the room: sit away from windows. Sit away from students who like to squirm, talk or otherwise distract. Sit where you can see (the instructor, the screen) and hear. Studies reveal that sitting near the front–middle is best for most people, whereas the back and sides are worst.
- Pay attention to the organization of the lecture. Ask yourself questions about overarching organizational issues. Has the instructor set today's material into the overall context of the course? Has an outline been provided? If so, can I use it as my organizational structure for when taking notes? What is the object statement? What background information is provided? Supporting evidence? Is this deductive or inductive reasoning?
- Stretch. Occasionally move your body, especially your legs and feet, and take a few deep breaths every now and then, but without interrupting the instructor or other students.
- Concentrate on what is being said: focus on the instructor's words. Try to ignore an instructor's boring monotone or any other distinctive characteristic. Be careful not to slip into thinking about next weekend's game or about a friend. If you find yourself drifting into dreamland, coach yourself to say 'Pay attention!' 'Listen to what is being said.' 'What is happening now?' By 'speaking'

to yourself (silently, of course), you will improve your concentration. Your goal is to be an active listener. Also, writing notes will help you concentrate.

Some Additional Points for Taking Notes

- If English is your second language but your studies are in English, use English when note taking. Studies indicate that retention of knowledge is increased by using the language of instruction.
- Do not write the whole time. Stop note taking from time to time to see and not just hear the instructor. Such 'breaks' help your overall concentration.
- Do not write everything verbatim. You might omit items if the instructor speaks faster than you write.
- Copy into your notes anything, including sketch maps, the instructor thinks is important enough to write on the board or identify with an overhead.
- Write down key words and phrases, examples, names, places, dates, equations, numbers and references. Use abbreviations.
- Record definitions offered by the instructor. These may differ from those in the assigned readings. You can reflect on differences later.
- If a question pops into your mind, write it down within brackets [ ] to indicate that it is your thought.
- Be especially attentive towards the end of the class period to the instructor's summary points and any course information. Do not pack up too soon!
- If there is a question period at the end of the lecture/class, listen carefully, and jot down pertinent questions and answers. Do not hesitate to speak up, for even if you think your question is trivial, it may be the question everyone else wanted to ask.
- Include all course handouts at the appropriate locations in your binder. If this

178 *David B. Knight*

is not possible, put a notation in your notes about the handouts and keep all of that course's handouts together.

Listen for Clues

Be attentive to the instructor's use of leading phrases and key questions, such as, 'The argument I want you to consider is as follows'; 'Professor Johnston's opinion is worth considering. He has written . . .'; 'Why should geographers research HIV and AIDS?'; 'What process could have led to this result?'

Be attentive to linking expressions used by the instructor to signal a change of direction in delivery. These include words and phrases for:

• Cause and effect – accordingly, because, therefore, this logically leads to.
• Concession – even though, given that, in the light of, of course, perhaps I'm wrong, could it be that there is another interpretation?
• Contrast – conversely, despite, given this then, however, on the other hand.
• Elaboration – for example (or e.g.), in other words, that is (or i.e.), let me put it this way, to take this point further, this example will make the point clear.
• Emphasis – did you hear that?, especially, now this is important, let me repeat, most importantly, specifically, let me emphasize this point.
• Numbers (as for lists) – first, finally, my last point, then, thirdly, ultimately.
• Repetition – also, even more, in addition, in other words, too, to repeat, since you were probably thinking about lunch I'll repeat that point.
• Summary or conclusion – in conclusion, finally, for these reasons we can conclude, let me summarize, my final point, to wrap up, what I have been saying today can be summarized as follows.

Be aware of the instructor's tone of voice. For example, the last point in a list may be delivered with a tone of finality, or perhaps sarcasm is being used to see if you are in fact listening.

Listen for pauses. The instructor may be mulling over a point before speaking again, signalling that a new direction in the discussion is about to occur, looking at the class in anticipation of posing a question, or checking to see if the students are showing puzzlement. Maybe he or she has swallowed a fly, as one of my geography instructors once did – in mid-sentence! In other words, do not just listen to the words but be aware of how the words are delivered, why, and ask what they mean.

Three Post-class/lecture Reviews

The learning process does not end when you leave the lecture hall or classroom. There are three follow-up reviews you can do. Studies have demonstrated that reviewing notes clearly results in superior recall.

Immediate post-lecture/class review

• Before you leave the lecture hall or classroom, quickly scan your notes to catch if you missed part of a definition or perhaps some data. If so, ask for them from a student colleague.
• Do not waste time rewriting your notes. Instead, review.
• Review your notes while they are still fresh in your mind (within 12 to 24 hours). As suggested above, write summary comments, reflections and questions in either the right-hand column or bottom space (if you use one of the formats in figure 31.1).
• Recall what you learned from the pre-class readings by rereading your reading notes and integrating them with what you have just learned. Points raised from the readings may help you answer questions you posed for yourself during the class or during your review.

Periodic review

Immediate post-class or lecture reviews of your notes can be invaluable. However, a second type of review is also important. The periodic review considers a cluster of related course materials. At the start of term, establish times during the term for reviews that encompass a series of topics. Since you will be taking several courses, stagger the periodic reviews so that you will have only one or two per week, starting, perhaps, in the second or third week of term. If the course material has logical breaks, use them as your review markers, otherwise establish your own. The periodic review permits you to integrate readings, lectures, labs, fieldwork, and your observations and questions. By doing both post-class and periodic reviews, you will develop a deeper and broader appreciation for the material being offered to you.

Full-course review

The full-course review is the third type, to be completed prior to final examinations. This review will be easy if you have done the other reviews.

Out-of-class Contacts

Your instructor will probably have posted 'office hours' during which he or she will gladly respond to questions you may have. Do not overuse such times for the instructor has numerous students. However, if you have a need for contact to deal with specific issues that you feel are not being addressed, either during an end-of-class question period or during the open office times, then request a specific time for a meeting.

Prepare for the meeting. Think carefully about why you want to speak with the instructor and make a list of any items you want to identify. Order them: time may slip by so quickly you will not be able to get to minor points. As a courtesy, give the instructor a copy of your questions when you enter

the office. Additional sources for help and feedback may include a graduate teaching assistant or perhaps a tutor to whom you are assigned. Use them!

Conclusion

Getting the best out of classes and lectures has to be done by you, using the skills identified in this chapter. By preparing adequately, having a positive attitude, actively listening to what is being said, taking good reading and lecture notes, and reviewing in the several ways noted, you will be able to integrate and understand any course's material. In other words, a class or a lecture is not an isolated event. It is a part of the *learning process*, a process that starts at the beginning of the term and continues until the course is completed. This stated, however, always remember that material from each course you take provides a building block upon which you can add other (course) blocks, so do not think that just because you have 'done that' you can forget what you have learned!

The best way to hurt yourself is to miss a lecture or class and not to do the expected work in the order assigned or by the timetable expected. By keeping yourself 'on track', you will find that the learning process can be both enjoyable and rewarding. Above all, remember that listening, writing and thinking go together. With so much riding on your note-taking ability, it is wise to take time to perfect the integration of these three skills. Finally, then, doing a good job of note taking right from the start will save you hours of agony.

Further Reading

Northey, M. and Knight, D.B. 2001: *Making Sense in Geography and Environmental Sciences: A Student's Guide to Research and Writing*, 2nd edition. Toronto and Oxford: Oxford University Press.

32
Writing Essays and Related Assignments

Rachel Pain

Those who are profound strive for clarity; those who are not strive for obscurity.
Friedrich Nietzsche (1844–1900)

Why Write Essays?

There are lots of arguments against geography students writing essays. The most common is that they have no relevance for the sorts of things geography graduates do in the 'real' world. A development agency officer analysing a community consultation, a civil servant writing a ministerial brief or an environmental campaigner working on publicity material is unlikely to begin, 'In this essay I will consider the question by …'. Still, we continue to set essays and related written assignments because the skills they involve are transferable to these vocations and more – they remain a good way for you to demonstrate your ability to synthesize and evaluate a wide range of material, construct a coherent argument and express yourself clearly.

Swimming, Sinking and Learning to Write

Traditionally, not only were essays the dominant type of assessment, but a mystique existed around the task of writing them. Lecturers assumed that very capable students had an innate understanding of this, and that those who didn't would pick it up by osmosis during their time at university. This is slightly facetious, but you can see the danger of the 'sink-or-swim' model – students who cotton on early or have been coached to write essays at school tend to do well, while those who don't may sink. They may also be good at geography, but never use the expected (yet unspoken) techniques. Today this model of assessment is widely viewed as unfair, and is uncommon, even in the oldest universities. Most lecturers would recognize that it is only fair to share the 'rules' of essay writing, tell you how your work will be assessed, and give feedback which helps you to make improvements as well as show you where you went wrong. In other words, while many students are much more concerned about what to write than how to write it, writing is a skill you can learn and improve on; one of the most valuable you'll develop during and beyond your degree. All this is not to say that imagination and spontaneity don't count – good technique, and creativity and flair, work together. As I hope to demonstrate in the advice that follows, writing and thinking are joined at the hip: mastering one will help you do the other as well as you can.

Answering the Question

This is the golden rule. It seems obvious, but is regularly overlooked in exam answers and coursework essays. However interesting and intelligent your essay is, if it isn't wholly pertinent to the question set, it has to be penalized. So find out precisely what the question is asking, and ensure you understand what each word means within the context of the essay. Think about what the content words ('glaciers', 'Margate' and 'actor network theory') and the command words ('outline', 'discuss' and 'compare') entail, and in what overall direction the question is pointing you. Think about permutations, as there are often valid alternative ways of understanding and answering each question.

Researching and Planning

Try to read widely. Look for sources other than those on the reading reference list given. Use plenty of journal articles, which are often more up to date and easier to get hold of. Before you start researching it's a good idea to jot down your own thoughts to clarify what you think (even if you've never thought about it before) and to give you a starting point. As you read, your ideas should begin to develop, and you will compile factual material, ideas, arguments and evidence that need to be included somewhere in the essay. Gradually you can begin ordering and sorting your material into a provisional plan for writing. This is an iterative process and does not end until you read through your finished essay for the last time. Although the same information may justifiably be arranged in different ways, most essays have an introduction, a systematic body of analysis and a conclusion.

Introducing Your Essay

A concise and clear introduction makes an impression and puts the reader in a favour-able frame of mind from the start. There are no hard and fast rules about what should go into it, and nor should it be too long, but the following suggestions may help. First, you may want to set the issue in context – for example, consider its relevance to geography or the wider world. Second, you should indicate somewhere what you understand by the question and the key words and issues; define terms if necessary, and show how you are interpreting the question. Third, briefly outline your essay. You may want to rewrite your introduction after you have finished the rest of the essay, so that it relates more closely to what you have actually written – again, this iterative development is a good sign.

Structuring Your Essay

The lack of a clear structure makes assessing the ideas and arguments in the essay a difficult task, and seems to indicate a confused and disorganized writer, so use your plan to organize the information, arguments and examples you are going to use in a logical and comprehensible way. Bear your ultimate goal (your main argument) in mind throughout, and work towards it logically. You can signpost each stage of the essay to achieve this – start each section by making the general point the section covers and connecting it with the wider argument, then go on to illustrate or elaborate it. Once you are good at structuring essays you might want to think about using structure creatively to achieve an effect on your reader – arguments can be made more persuasive through careful timing of key points or important evidence.

Staying Relevant

Stick to the question – this is all that is required of you. You will get no marks for irrelevant material, and lots for concentrating on and developing the key issues involved.

So if a question asks you to 'account for the spatial location of salt marshes on the Lancashire coast', don't spend several pages explaining how salt marshes developed elsewhere – only mention this where it has an important role in your answer, and then be sure to indicate why you are including this material. If you often get feedback suggesting you go off on tangents, ask yourself as you write each paragraph if what you are writing is still relevant. If necessary, spell out its connection to the question and how it fits in with what you have said so far.

Making and Supporting Arguments

Most essay titles will demand you to present contrary evidence or arguments and evaluate them in a careful and reasoned way. At the same time, you should have confidence in your own judgement and opinions. Tragic assessment moments are created by well-referenced and structured essays which are worthy but dull, because we can't see the student's own thinking, interpretation, opinions or ideas. Human geography in particular is inherently political, so take up a position and fight a corner. Some lecturers are renowned for having strong opinions – these are often the teachers you will remember all your life. But bear in mind that the nature of the academic endeavour is that difference and debate should be embraced, so you don't have to agree, either with them or with any particular texts that you read. Provide evidence and substantiation for any line you take, making good use of examples and case studies, and don't be surprised if you change your mind as your work develops.

Avoiding Bias and Stereotyping

Most geography departments have policies that discourage sexist, racist and other dis-criminatory forms of writing. Even if yours does not, you are likely to be penalized for it. When we fall into writing that reproduces stereotypes – which we all do from time to time – not only does our writing reflect and compound prejudice, but the clarity and precision of our work are compromised. Inclusive writing involves being critical and reflective of what you write. It is obviously important if you are dealing with countries, cultures, ethnic, gender or social groups different to your own, but sometimes we are even less sensitized to clangers about our own cultures. Here are some examples from student work:

'Older people are frail with limited mobility and social interaction' – beware of generalizing (this is only true of a minority of older people).

'Man's impact on the environment' – beware of the universal man (unless you really do mean males only) – try 'the human impact on the environment'.

'If Africans would embrace global capitalism like Europeans have, Africa could develop normally' – this is riven with assumptions about the superiority of Europeans and capitalism (again, poor style compounds the limited content and thought).

Developing Your Own Style

Although some students have already developed a natural and mature writing style by the time they get to university, many have not. This is something that comes with practice, and is well worth working on. Be critical of the geographers you read: does the way they write help to get their message over clearly and succinctly? Where you are impressed, emulate their style. Try to keep your writing clear and concise, and avoid the use of lengthy sentences with complex subordinate clauses. Certain lecturers and tasks ask that you avoid writing in the first person, although in human geography it is becoming more common to personalize writing (if in doubt, ask). Finally, never underestimate the impact which poor spelling (and poor spell-checking), bad grammar and slang have upon the overall feel of your essay, and never waffle in order to fill up space. Lecturers know all about waffle, and detect it in others with relish. Quality is more important than size.

Acknowledging Your Sources

Some of what you write may be truly inspired, but most will draw on other people's material and ideas. Referencing can be a tiresome mechanical task, but providing detailed and precise information on all the sources you have consulted for a piece of work is essential to differentiate between the following:

- material (ideas, arguments, statistics, etc.) you have reproduced directly from a book, article or website (you should use quotation marks, and give author's surname, date and page numbers);
- material you have taken, interpreted and drawn into your essay in your own words (use author's surname and date);
- your own ideas, arguments and interpretations (no referencing required).

The rationale behind systematic referencing is not simply to check for plagiarism – 'has she downloaded this from Essayz-4-U.com?' – as subject experts who have read thousands of student essays, your lecturers have a well-developed nose. Rather, good referencing encourages effective synthesis of your own and others' views, adds weight to your arguments and counter-arguments, indicates to your reader which sources you have consulted, and allows other researchers to refer back to the original material. References are cited and listed in a standard form, usually 'the Harvard system', where references are given in the text as author-date, and the full citation is given in a list at the end of the essay. Many geographical journals, such as *Area* or *Transactions of the Institute of British Geographers*, provide clear examples of this format. Every source cited in the essay should be listed at the end. Your department should supply you with a guide explaining the technicalities.

Concluding

A good conclusion is extremely important – an effective final paragraph can even partially compensate for a poor essay. Like the introduction, you should keep it fairly concise and steer clear of introducing new ideas, material or examples. It's always important to summarize the main points you've made, and be sure to express very clearly your verdict on the question – answer it! You may then want to look outwards again to the bigger picture, or make one or two predictive comments about future trends. Your last sentence can make an impact, so think about it carefully.

Box 32.1 Checklist for writing essays

When you've written your essay, have a read through and ask yourself the following questions:
- Have I answered the question, the whole question and nothing but the question?
- Have I written in a clear and fluent style?
- Have I included a clear introduction and conclusion?
- Have I structured and signposted the essay so that it's easy to follow?
- Have I read a range of sources and referenced them consistently?
- Have I made my own position clear and argued a case?
- Have I avoided sexism, racism and other prejudices in writing?
- Have I used spelling and grammar correctly?

Finally . . .

Remember that the ability to finish an essay minutes before it is due in is a worthy skill, but will probably lose you marks. A short time spent checking and amending your essay can pay off. Box 32.1 on p. 183 offers a checklist for the final read through.

The tips and techniques in this chapter are not intended to be prescriptive, as assignments are set in different ways and lecturers have varied expectations of your work.

Some are more open to alternative ways of writing and structuring essays than others. So long as they make this transparent, and you try to meet their expectations, you can be creative in expressing your geographical knowledge and ideas clearly and effectively in new ways.

Further Reading

Kneale, P.E. 1999: *Study Skills for Geography Students: A Practical Guide*. London: Arnold.

33

Making a Presentation

Chris Young

Talk in Front of the Class – Never!

'I hate giving presentations' is a comment occasionally heard being muttered by students in university geography departments. This is understandable since students generally worry about speaking in front of their peers, despite the fact that no such worries exist when they are in the student union bar. The main concern is usually nerves resulting from a feeling of being exposed. Despite this, most recognize their importance and those same students can later be heard saying, 'That wasn't so bad'. While you may believe that reading for a degree in geography is about improving your academic performance and increasing your knowledge of your favourite subject, these days it is also about enhancing your career prospects. With this in mind your geography degree will encourage you to develop and improve a large number of skills that employers want graduates to have. One of these skills is an ability to communicate orally in an effective and fluent manner.

Presentations may take several forms, some formal, some less formal. Usually there is an expectation that you develop your ability as you move through your degree. This may involve an increase in the amount of time that you speak for, or it could involve your having greater freedom with the material. Whether you are asked to give a formal lecture, or a presentation on a topic within a seminar, or report on projects, fieldwork or other work you have done, there are a number of tips that will make the job easier.

Where Do I Start?

As with most pieces of work, one of the first questions you should ask is: what am I expected to do? There are two aspects that are important in presentations: the skill demonstrated in making the presentation and the academic content. Unfortunately, different tutors or lecturers will almost certainly be looking for different abilities when assessing presentations, and no rules can be given that will apply in all situations. If your tutor provides a list of assessment criteria, use them to guide your preparation and think critically about the presentation you are preparing.

Even if the presentation skills are the main focus, before you can give a presentation the geographical content clearly needs to be determined. This is essentially the same as for any other piece of academic work – you should identify the right material and ask the right questions. It is highly likely that you will need to demonstrate an analytical approach to the information and show clear evidence of applying your geographical knowledge to the task in hand. To ensure that you can do this, start working on the topic early – don't leave everything until the last minute.

One of the controls on what can be said, and the detail you can provide, is the time available to you. Presentations may be short, possibly five or ten minutes, or they may last up to 20 or 30 minutes, allowing you the time for increased detail and greater depth. However long you have, plan carefully to avoid over- or under-running. Both show poor planning and even the less formal presentations should be as planned as possible. One of the most common problems that plagues student presentations is the attempt to use all the information that you have – even if it is not fully relevant. It is unlikely that you will have the time to impart everything you know about the topic – so don't try. Draw out the key points and decide which information you can leave out. You can then select a few examples for illustration.

Once the relevant material is selected it needs to be carefully organized and structured. One way to do this, and help you keep the audience interested, is to 'tell a story'. It can be useful to start by identifying the broad context. You could introduce the appropriate geographical literature or clearly identify an area of contention that is going to be examined throughout the presentation. Try to organize the information and develop your arguments in a clear and logical order, ensuring that the main issues are identified at each stage.

Pitch your discussion at the right level for the audience – is it non-specialist or specialist? You do not want to confuse or be patronizing. Keep jargon to a minimum, although you need to ensure that you use appropriate geographical language. Some technical or specialist terms may be unavoidable, in which case it may be necessary to provide clear definitions – a handout can help to do this.

In these preparation stages it is really important to organize your work discipline. This is especially true if you have been asked to work as part of a team, since this will involve regular consultation. You and the team will need to prioritize tasks and ensure that they are completed by a deadline. This can be achieved by using a planning sheet. The important requirement is that you have plenty of time before the presentation to practise and check the timing and flow of what you have prepared. This helps you to be confident and relaxed when you make the presentation. As with any type of work, the more you prepare the better the end result will be, since familiarity breeds confidence. If you feel confident in what you have done, you will feel more confident in presenting it to others.

A number of tips can help here. If you are working as part of a team, have a good working knowledge of each other's topics. Make sure that you understand what you are talking about and know the information thoroughly yourself. Make sure that the work is precise, accurate and of a high standard. This especially applies to your visual materials (see below).

It can also help to use other students as a trial audience. They can help you to identify where improvements can be made. If you can identify problems in advance, then you are most of the way to solving them. However, be careful – practising on your own will always be faster than in the final presentation, so if your timing is right under practice, then you may need to cut something. Once you have the material clear in your head, then you are ready to make the presentation.

Making the Presentation

What makes a good presentation? We have all sat and listened to both good and bad speakers and we all appreciate different aspects of presentation style. There are, no doubt, tutors or instructors whose style you like and those you don't – what is it that you like, or dislike, about what they do? Can you model your style on the best, and what should you try to avoid?

You will probably be most nervous about making your presentation immediately prior to starting. This is a little unfortunate since your introduction is the key to setting the stage. If you can start well, however, nerves are usually forgotten and you can start to enjoy yourself and think more about the material and where it is going.

The best way to start is to outline your story and set out your objectives clearly. Since you need to relax, a good ploy is to grab the audience's attention with something controversial or amusing such as a cartoon or a quote that can be put up as a visual. On your part you need to show enthusiasm, and one way to ensure this is to smile. Some movement can also show enthusiasm and help generate interest by providing emphasis. However, don't move about too much because it can be distracting. For example, try to avoid waving your hands about wildly. Equally, avoid standing still with your hands in your pocket – this shows boredom and lack of interest.

Although the material may be written out partially or in full, try to avoid reading since this can suggest poor preparation. While reading is inevitable, in places you should, as far as possible, look up and maintain eye contact with your audience. If you watch the audience you can see if people are interested and taking notes (if appropriate), if a point was missed or not understood, or whether they are bored silly – you can then take appropriate action. It also helps to give the audience the feeling that they are involved. While in some cases your tutor may prevent you from using notes, if you do use notes keep them brief – it can help to use the visuals as your source. The key is practice, so that you are confident with the material without having to read it.

When talking, make your voice clear and loud enough so that those at the back can hear everything you say. To maintain interest and show your enthusiasm, vary the speed and alter the intonation. If the audience is expected to take notes, then speaking slowly is important. One of the common faults occurs when you read, since this usually results in material being delivered too

fast. To help the audience you need to make explanations clear, and it can help to repeat important points in several ways to ensure that you get them across. However, repeating the same words will not necessarily make anything clearer (a potential problem if you read). Try to restate the information from another angle, for example with a case study.

Another common problem can occur with the pronunciation of technical or specialist terminology. Check before you start how to pronounce key terms or names, especially if they are geographical in origin. Don't worry if you make a mistake or if you leave something out – we all do. No one will probably know, but if you know that a mistake has been made it is best to correct it so that you do not lose credibility. It is highly unlikely that you will be penalized for owning up!

Visual Aids

The most useful aid available to you when making a presentation will be your visual material. The most common form of visual material is the overhead transparency. However, you may also wish to use slides or extracts from videos, or other more technologically sophisticated media forms such as PowerPoint. Visual materials should be carefully prepared. You must make sure that you know exactly why you are using them. For example, PowerPoint is a powerful tool which allows text, graphics, pictures, video and slides to be used together, but it can also be gimmicky and you can waste a lot of time producing gimmicks rather than concentrating on appropriate and effective material.

Using visual materials can help reduce your notes and help to prevent reading. They can reduce confusion, increase audience participation, reinforce key points *and* allow you to face your audience. However, there are a number of common pitfalls that are best avoided.

Check that the projector or any other facility is working correctly and that you know how to use it properly. Also check that the screen can be seen from anywhere in the room. Stand to one side of the projector, not in front of it. Whatever form of visual aid used (words, tables, diagrams, maps or graphics), make it *large* and clear when projected (a font as big as 24pt may be necessary in larger rooms since normal typescript usually means that anyone at the back cannot read it). Maps or tables from books may need to be redrawn and simplified, highlighted or enlarged to give the maximum impact. A useful rule is no more than six *large* words on one line. This has a number of advantages, not least that it keeps the message short and succinct. The use of colour can be beneficial but some colours, primarily yellow and orange, do not show up easily and are best avoided. Too many or overfull and complex visuals can swamp the audience, so select the information carefully. Leave visuals up long enough for the audience to take in. If the material is complex, it may be wise to provide handouts.

How Do I Stop?

It is important that you don't stop suddenly without some form of conclusion. It is always useful to summarize your main points. If you find you are running out of time, then you need to have the confidence to cut something. However, try not to cut the easiest bit – the conclusion – because this is where you identify your main points succinctly.

One way to signal the end of the presentation is to invite questions. To do this successfully you must be confident with the material because when handling questions you need to provide thoughtful answers. In some cases the audience will ask questions to confirm they have understood. This may occur either during and/or after the presentation and is normal. You should not worry if questions are asked – at least it means that the audience is paying attention! Always try to answer such questions, but always be prepared to own up if you do not know the answer.

An alternative way to finish is to indicate where or how the work could be developed further. This might broaden or focus the work more and could leave open questions for the audience to think about. You will not have covered everything, so give people something to go away with. Do not let them be totally passive.

Lastly, make sure that you finish on time. This helps to show good organization and preparation.

Box 33.1 Checklist of key points to bear in mind when making a presentation

Preparation
- Find out what the requirements are – and follow them.
- Start preparation early and determine the content, keeping in mind the time available.
- Plan and structure the material in an organized way. There are three key aspects to this:
  - asking the right questions;
  - applying your geographical knowledge to develop your argument;
  - using an analytical approach.
- Know your information (and that of others if working in a team) and practise to check your timing.

Making the presentation
- State your objectives clearly and outline where you are going.
- Keep your notes brief and avoid reading.
- Keep in eye contact with the audience.
- Speak slowly, loudly and clearly and vary your intonation.
- Repeat important points.
- Make visuals large, clear and succinct.
- Watch the clock and keep to time.
- Finish with a conclusion which summarizes your main points.
- Lastly – smile, enthuse, relax and enjoy.

Conclusion

Making a presentation as part of your geography degree is a learning experience which can be enjoyed. Presentations will develop your confidence in a number of ways – not least in your ability to communicate information, ideas and arguments effectively to both specialist and non-specialist audiences. You are presenting the results of your own research in a way that can show your enthusiasm for your subject far more easily than any written report. There are a lot of things to think about, but if you prepare fully and are organized this will allow you to relax and enjoy yourself.

Finally – think about your appearance.

This can be an integral part of setting the right atmosphere – audiences react strongly to every aspect of your appearance. A dragging shirt-tail or an old pair of jeans with the knees hanging out can be very distracting. You may not need to be too formal, but making sure that you are presentable can help to give the right impression.

Further Reading

Kneale, P.E. 1999: *Study Skills for Geography Students: A Practical Guide*. London: Arnold.

Young, C. 1998: Giving oral presentations, *Journal of Geography in Higher Education*, 22(2), 263–8.

34

Coping With Exams: Dealing With the Cruel and Unusual

Iain Hay

If you feel intimidated by university exams, you are not alone. But while a little fear may be good for your exam performance, too much can be disabling. This chapter should help reduce your levels of fear by providing guidance on ways you can prepare to complete exams to the best of your ability. After providing a little background information on the reasons exams are inflicted on you and the various forms this unusual punishment takes, the chapter focuses on two key components of good exam technique: preparation and undertaking the exam. The suggestions made in the following pages are not a precise formula that will apply equally well to every single person. In my experience, however, this guidance goes a long way to improving exam results and reducing stress levels for very many students from a wide range of backgrounds. Nevertheless, as you become increasingly accustomed to university exams, you might wish to modify, emphasize or disregard some aspects of this advice to suit your own habits, temperament and lifestyle.

Why Have Exams?

Believe it or not, exams serve three important educational purposes. These are to test your factual knowledge, your ability to synthesize course material, and your ability to explain and justify your informed opinion on some specific topic. Some exams might require that you satisfy only one of these objectives (e.g. short multiple choice exams may only examine your ability to recall information), while others may demand all (for example, oral examination of a thesis).

Types of Exam

Exams take three forms (table 34.1). Because the closed-book type is most common, the discussion that follows will focus on it. A good deal of the material covered applies equally well to the other types of exam. Supplementary guidance on other exam forms is set out at the end of the chapter.

Students who do well in exams have usually mastered the skills of exam preparation and know some of the secrets of successfully negotiating different forms of exam. Let us look at these matters in turn.

Preparing for an Exam

Months to go . . .

Good preparation is central to exam success and it needs to begin from the first day of term. Difficult as it is, you must review as the teaching term progresses. Rewrite lec-

Table 34.1 Types of exams and their characteristics

	Closed-book
Characteristics	Requires that you answer questions on the strength of your wits and ability to recall information. Consulting any material in the exam room other than that provided by the examiner for the purposes of the test is not permitted.
Sub-types	● Multiple choice ● Short answer ● Essay answer
	Open-book
Characteristics	You may consult reference materials such as lecture notes, textbooks and journals during the exam. Sometimes the sources you are allowed to consult will be limited by your examiner.
Sub-types	● Exam room ● Take-home
	Oral exam (viva voce)
Characteristics	Used most commonly as a supplement to written exams or to explore issues emerging from an honours, Master's or Ph.D. thesis. You may have to give a brief presentation before participating in a critical but congenial discussion with examiners about your written work.

ture notes and keep up to date with assigned readings. This will make it easier to understand lecture material throughout term and help you remember it at exam time.

As soon as possible, find out about the exam. Ask teaching staff what types of question may be asked, the time allowed, the materials needed, and so forth. Ask if you will be examined on material *not* covered in lectures and tutorials. Seek direction about how to focus your supplementary reading. On these early foundations you can build solid exam preparation.

Weeks to go . . .

Find a comfortable, quiet and well-lit place where you can study undisturbed. Prepare a study calendar. Allocate specific days to the revision of each topic. An example of an exam study calendar is shown in table 34.2. The fortunate student in this example has only two exams, each of which contributes the same proportion to the final grade. The student is performing equally well in both subjects. Accordingly, roughly equal time is dedicated to each subject. If the exams were weighted differently from one another or if the student was performing better in one subject than another, it would be wise to reapportion time appropriately. Be sure to stick to your revision schedule.

Do not procrastinate. Study is hard work, but the longer you delay, the more difficult the task becomes. Be positive. Have faith in your ability to plan, manage and produce your own success. Make the focus of your revision comprehension, not memorization. Be sure you understand what the course was about. When you have a grasp of course objectives, you will be better placed to make sense of content and to answer exam questions. If they are available, review past exams to get a sense of format and likely content. Be cautious, though – exams may change from year to year! Although it is difficult, try answering past questions under exam conditions. Ask your lecturer to confirm that your trial answers are on the right track. Speak to your lecturer too about any

Table 34.2   Example of an exam study calendar

Date (June)	Morning activity	Afternoon activity	Evening
10	End of classes	Relax, buy groceries...	Arrange notes and reading material
11	Arrange notes and reading material	Physical Geography	Physical Geography (sport/exercise)
12	Physical Geography	Physical Geography	Physical Geography
13	Human Geography	Human Geography	Relax (sport/exercise)
14	Human Geography	Human Geography	Human Geography
15	Physical Geography (sport/exercise)	Physical Geography	Physical Geography
16	Physical Geography	Physical Geography (sport/exercise)	Physical Geography
17	Physical Geography	Physical Geography exam	Relax
18	Human Geography	Human Geography	Human Geography (sport/exercise)
19	Human Geography	Human Geography (sport/exercise)	Human Geography
20	Human Geography exam	Celebrate	Continue celebrating

subject material you do not understand. If you are having other difficulties of a non-academic nature that are affecting your study, speak to a university counsellor. You will not be the only person facing these problems.

Exam preparation can be a stressful enough time without making radical changes to your lifestyle that may impair your exam performance. Maintain your regular diet, sleep and exercise patterns. If you are in the habit of exercising regularly, keep doing that. Most people find that exercise perks them up, makes learning easier and enhances exam performance. Don't overdo it, though! Exam time is no time to shock your body with a conversion from couch potato to Olympian.

Hours to go . . .

On the day of your exam eat or drink something to give you sustained energy – not just a temporary 'buzz'! If you do not think you can eat, try drinking a flavoured milk, fruit juice or something similar. Do not face an exam on an empty stomach. Before you leave

for your exam, make sure you have your student identification card (in most universities you are required to present your ID in order to sit the exam), pencils/pens, ruler, paper, eraser, watch, lucky charms and a calculator (if required). Exam booklets and scribbling paper will usually be provided. Be sure to get to the right exam in the right place at the right time. Allow for the possibility of traffic delays, late buses and bad weather. If you do miss the exam for some reason, *see your lecturer immediately* to explain the situation.

Zero hour: techniques for sitting and passing written exams

Before an exam almost everyone feels tense. However, if you have prepared effectively, heightened anxiety will probably help you perform at a higher level than if you were quite calm about the whole affair. Breathe deeply and stride into the exam room with a sense of purpose. You know your subject. Here is your opportunity to prove it.

When you are allowed to view the exam paper ensure you have all pages, questions and answer sheets. Check the reverse side of every page for extra questions that may be hiding there. On rare occasions, printing or instructional errors may occur. If this seems to be the case, let the invigilator know immediately. Next, read the instructions carefully. How much time do you have? Which questions need to be answered? What is the mark value of each question? *Repeat* this process after you have answered the first question to confirm that you are doing things correctly. Lecturers find it disheartening to mark an improperly completed paper by a good student. It is even more upsetting to be that student.

If you have not been able to do it before the exam, work out a timetable. On the basis of marks allocated to them, calculate the amount of time you should devote to each question (see table 34.3). This time budget shown as table 34.3 could be modified usefully by allowing about 10 minutes at the end of the exam to proofread answers.

Table 34.3 Example of a three-hour exam writing schedule

Question 1	5 marks	9 minutes
Question 2	10 marks	18 minutes
Question 3	15 marks	27 minutes
Question 4	20 marks	36 minutes
Question 5	50 marks	90 minutes
Total	100 marks	180 minutes

Take time at the start to read each question carefully and to make notes on scrap paper. Carefully choose the questions you will answer and think critically about their meaning. Jot down ideas as you look over the questions. Use your preliminary notes as the basis of an essay plan for each answer. A plan might consist of main points arranged in logical order to give a coherent structure to your answers. If you happen to run out of time, the marker may refer to your plan to get an impression of the case you intended to make. Take care to distinguish essay plans from final answers in your an-

swer booklet (for example, a pen stroke through the plan).

Now, at last, you can actually begin writing your answers. Begin with the questions you can answer best. It is often helpful to complete the easy questions first to build up your confidence and momentum. Moreover, if you find yourself short of time, you will have shown your best work.

A common mistake is failing to answer the question asked. Some people opt instead to write a prepared answer on a related topic. Markers want to know what you think and what you have learned about a *specified* topic. The right answer to the wrong question will not get you very far! Examiners will assess your level of understanding of particular subjects. Do not try to trick them; do not try using the 'shotgun technique', by which you tell all that you know about the topic irrespective of its relevance to the question; do not try to write lots of pages in the hope that you might fool someone into believing that you know more than you do. Concentrate instead on producing focused, well-structured answers. *That* will impress an examiner.

Grab the marker's attention. Examiners usually have many scripts to assess. They do not want to see the question rephrased as the introduction to your answer. They do not want to read rambling introductions. Instead, they want you to capture their attention with clear, concise and coherent answers. Emphasize key points. Consider underlining words, using headings and including phrases which draw attention to important matters (for example, 'The most significant issue is . . .', 'A leading cause of . . .'). Bullet/numbered lists and correctly labelled diagrams can also be helpful. Whatever you do, spare the padding. Get to your point.

It is very difficult to mark the exam script of someone whose writing verges on the illegible. If you have problems writing legibly, write on alternate lines or print to ensure that your work can be read. Keep your English expression clear. You will overcome many grammatical and punctuation problems by using short sentences. These tend to have greater impact than long sentences too.

Attempt all required questions. It is usually easier to get the first one-third to half of the marks for any written question than it is to get the last half or third. It is foolish therefore to leave required questions unanswered. If time is a problem, write an introduction, outline your argument in note form and write a conclusion. This will provide the marker with some sense of your ideas and may see you rewarded appropriately.

When you have finished all your answers, proofread. Check for grammatical errors, spelling mistakes and other problems. Add important material you missed in the first attempt. Finally, keep calm. If you begin to panic or go 'blank', stop writing, breathe deeply and relax for a minute or two. Do not give in to frustration and storm out of the exam room. Why risk solving a problem *after* you have left the exam but while it is still on? And no matter how desperate things may seem, do not cheat! Exams are carefully supervised. Copying and other forms of academic dishonesty rarely go unnoticed.

Sensible exam technique is critical to success. If you follow the advice outlined above, you will have taken some major steps towards a distinguished exam performance.

Multiple Choice Exams

Aside from 'knowing your stuff', good results in multiple choice exams depend on understanding peculiarities of this form of exam. There is much more to success than simply selecting (a), (b) or (c). Let's look at some of these matters.

Go quickly through the exam answering all those questions you can complete easily. Leave difficult questions. Return to them later if you have time. Read *all* the answer options, deleting those which are clearly incorrect. This will help you narrow down your choices. Avoid answers which include abso-

lutes. In the world in which we live, 'never', 'always' and 'no one' are rarely true. Do not be distracted by a pattern of answers. If, for example, every answer appears to have been (b), that does not mean that the next answer 'ought' to be a (b). Nor does it mean that the next answer is not (b), or that some of your previous answers are wrong. If there appears to be no correct answer, choose that option which seems closest to correct. Unless you have been advised that penalties are imposed for giving incorrect answers, answer every question and, if you must, guess! Finally, do not 'overanalyse' questions. Be cautious about revising initial responses when you are proofreading. If you find yourself hesitating about making a change, you might do well to leave things alone. You probably got it right the first time.

Oral Exams

Formal oral exams – or viva voces – are sometimes used as part of upper-level undergraduate or postgraduate assessment. They can be quite intimidating. However, if you think of an oral exam as a formal version of informal discussions you might have had with colleagues about your work, it should not be too daunting. Examiners will probably want you to elaborate on material you may not have had the opportunity to include in a written paper or in your thesis. They may also encourage you to think about alternative ways of approaching your topic. Look forward to a viva voce as an opportunity to explore a subject about which *you* may be the best informed.

In the way of specific advice for preparing and completing an oral exam, consider the following points. Present yourself in a way which is both comfortable and which suits the formality of the occasion. As a rule of thumb, dress slightly more formally than the examiners. During the exam maintain eye contact with everyone, sit comfortably in an alert position, and do not fidget. Speak clearly and give brief but lucid answers. It is

Table 34.4   Common questions in oral exams

- Why did you select your research question?
- How does your work connect with existing studies? Does it say anything different? Does it confirm other work?
- Why did you select your methodology? Are there any weaknesses in the approach you adopted?
- What practical problems did you encounter? How did you overcome them?
- Are any of the findings unexpected?
- What avenues for future research does your work suggest?
- What have you learned from your research experience? What would you do differently if you had your time again?

useful to have thought about questions you might be asked before the exam begins (see table 34.4). If you do not understand a question, ask to have it rephrased. Take time to think about your answers. You should also feel free to challenge the examiners' arguments and logic, but be prepared to give consideration to their views.

Open-book Exams

Open-book exams are quite deceptive. To deal effectively and quickly with them you must know your subject thoroughly. The format simply allows you access to specific examples, references and other material that might support *your* answers to questions. It is still *you* who must produce the intellectual skeleton upon which your answer is placed. For this reason you must study as you would for a closed-book exam. In addition, you should prepare ready reference notes that you can understand and access quickly. You must also make yourself familiar with the texts you plan to use. If appropriate, mark sections of texts in a way that will allow you to identify them easily (for example, highlighter pen, 'post-it' notes). Do not mark other people's books!

Take-home Exams

Take-home exams appear to offer ample time to ponder questions and formulate good answers. This may be true, but like other exams, preparation remains a key to success. Be sure you have available appropriate reference material with which you are familiar. You should have read through, taken notes from and highlighted sufficient resources to allow you to complete the exam satisfactorily. Arrange these materials in a way that allows you to find specific items quickly (for example, alphabetically by author, under subject headings or by date of publication).

Keep to a timetable such as that shown in table 34.5. The student in this example has used an eight-hour working day (say, 8:00 a.m.–noon, 1:00 p.m.–5:00 p.m.) to calculate the amount of time available for each exam question. Evenings might be spent proofreading answers.

Table 34.5   Example of a two-day take-home exam timetable (i.e., about 16 working hours)

Question 1	20 marks	@ 3.5 hours
Question 2	30 marks	@ 4.5 hours
Question 3	50 marks	@ 8 hours
Total	**100 marks**	**16 hours**

If you are given 48 hours to complete an exam, you do not have to work on the questions that long. Your aim should be to produce concise, carefully considered, well-argued answers, supported with examples where appropriate. You are not meant to be writing for the entire time! Instead, think carefully and focus clearly on the questions asked.

Finally, a practical point: if you are preparing answers on a computer, be sure you have sufficient paper, ink cartridges or printer ribbons to allow you to print everything when you have finished late at night and every shop you know is closed and all your friends are out of town.

Not So Cruel After All . . .

With adequate preparation, exams need not be objects of fear and loathing. Instead, they can be just the opportunity you need to demonstrate to your sceptical lecturer that you know the subject, that you can synthesize material covered during term, and that you have developed carefully thought-out opinions on specific topics. Good luck!

Further Reading

Barass, R. 1984: *Study! A Guide to Effective Study, Revision and Examination Techniques*. London: Chapman and Hall.

Hay, I. 2002: *Communicating in Geography and the Environmental Sciences*, 2nd edition. Melbourne: Oxford University Press.

35

Research Design for Dissertations and Projects

Brian Hoskin, Wendy Gill and Sue Burkill

In most geography honours degree courses students have to write a dissertation in the final year. This is a substantial piece of work, which is significant in helping you to achieve a good result. If you are at the stage where you are about to begin a dissertation, you may be feeling unsure of how to get started. The good news is that *you* are in control. You are able to choose a topic or geographical issue that you are really interested in (see box 35.1) and pace the work to suit yourself; the most challenging aspect is probably the emphasis on working independently.

This chapter outlines what doing undergraduate 'research' means and helps you to focus your research proposals. The various boxes provide practical tips to help you plan your ideas from the beginning, encouraging you to build on your abilities and enthusiasm. Finally, a research action plan is presented to help you manage your time realistically and effectively throughout the dissertation process.

Box 35.1 Early ideas

- Which geographical issues/topics interest you?
- 'Where' (the location) might you like to carry out your research?

What is Undergraduate Research and Why is it Important?

The emphasis in a dissertation is on you doing some original research. What does research in this context mean? What is meant by original? In this chapter we shall assume that research means 'a process of systematically seeking answers to questions' (Lindsay 1997: 5). It is a problem-solving activity and if you cannot see what the issue or problem is then the research may not be worth doing. You answer questions and solve problems in many contexts throughout your degree; however, it is the need for *originality* that sets the dissertation apart. Your work must involve something new. This does not mean that you cannot consider anything that has been looked at before – most scholarship builds in some way upon the work of others – but some part of your work must be significantly original. This may be the method or approach you adopt or the context you are exploring (Parsons and Knight 1995: 6). You should be 'adding benefit' for other researchers and you will be contributing to the 'total research knowledge' of the geography community (Kneale 1999).

In addition, undergraduate research should be a showcase for your skills and intellect. The final report should give the reader the impression that you know and understand something about the nature of geography and its techniques and methods (Clark and Wareham 2000). Finally, research involves communicating your ideas, results and conclusions to others in a well-written and carefully presented document.

It is clear that a dissertation is hard work, but there are lots of good reasons why it is so important. For example, the theories, methods and knowledge acquired will enhance your ability to do well in other coursework/exams; it may be a useful component to add to your CV and it may provide your tutors with further information about your research potential.

Box 35.2 Why the research is important

- It will deepen your understanding of aspects of geography.
- It may open up career opportunities.
- It will improve your 'life skills'.
- It will give you tremendous personal satisfaction.

Getting Started: The Research Proposal

The first stage in your research is one of the most difficult – you need to have a 'good idea', one that can be managed in the time available, uses your particular skills and abili-

---

Box 35.3 Where might your inspiration come from?

- A module, course or fieldwork that has excited you.
- A lecturer's or tutor's enthusiasm for a research topic.
- Ideas presented in a previous student's dissertation.
- An article in a journal/newspaper.
- A local or national controversy.
- A career interest or hobby.
- For further ideas see Parts I and II of this book.

---

ties and is of interest to you (see box 35.3).

This idea will have to be carefully developed into a *research proposal* that explores conceptual and practical aspects of your dissertation. You will need to establish what will be studied, why it is important, and where and how you will study it.

Selecting a Topic

Deciding on the topic is the likely starting point for most dissertations. Doing a dissertation offers you the chance to pursue an area of personal interest and to present arguments on a subject or issue you believe is important. It is vital to choose a topic that will hold your interest through the highs and lows of up to 12 months of study. You should be aware by now that there are many different opinions on exactly what constitutes geography and what, therefore, constitutes a 'valid' dissertation project. For example, personal enthusiasms do not always translate too well into appropriate geographical investigations and it might be useful at an early stage to discuss your ideas with your dissertation tutor.

As always, there are a number of constraints on the work you can do, although these are more often practical than theoretical (see box 35.4). You will probably have about a year to take your research from conception and design to final presentation. This might seem like a long time at the outset, but it can practically rule out certain types of study – you cannot, for instance, study the impact of winter sports for a dissertation due for submission in the following spring!

Choosing the Location and Scale of Study

The selection of the site for your study might well be a primary consideration in your selection of a topic. For practical purposes you may have to focus your research on a small

Box 35.4 Practical considerations/ questions to ask yourself about your topic

- Is there sufficient relevant background information available to you?
- Just how relevant is your proposed study (in relation to particular branches of geography)?
- Do you have the funds and resources available to allow you to pursue the topic?
- Can you complete your study in the time available?
- For further ideas see Parts IV and V of this book.

scale in a local area, although it might still be possible to contextualize your work against work done at a larger scale. You will need to select a location that you have easy access to as your data collection might take several weeks, and it is worth considering somewhere that you can easily visit during your summer vacation. Many students consider undertaking projects based upon overseas fieldwork; such projects might seem exciting, but they are accompanied by particular problems ranging from language barriers to the financial cost (see box 35.5).

It is being increasingly recognized in geography that research is not something that has to be conducted 'out there' or 'in the

field' to qualify as a legitimate study, and you might consider that the 'site' for your research could be a filmic representation, some music, a novel, a computer-generated landscape or a place on the Internet (although it is best to check the precise requirements of your institution).

Selecting Methods and Techniques

Your methods and techniques would conventionally be chosen on the basis of their effectiveness or expediency *after* selecting a topic, issue or problem and location. For example, you would be ill advised to decide that you wanted to undertake participant observation or a vegetation survey and then go looking for an issue to apply these methods to. However, an enthusiasm for humanistic geography or computer-based mapping, for instance, might serve to point you in a particular direction *before* you have chosen a topic.

The research methodology may be difficult to define in the initial stages of your dissertation, but it will help you decide whether or not your proposal is feasible (see box 35.6).

Box 35.5 Location/scale considerations

- Is your study site readily accessible?
- How much data might you need?
- Can you revisit your study site frequently (if necessary)?
- Do you need to obtain permission to work in your chosen locality?
- Have you carried out a risk assessment?
- Overseas fieldwork can be exciting, but make sure you have a 'plan B' location (for example, close to home or university) just in case things don't work out.
- Overall – be prepared to be flexible.
- For further ideas see Part III of this book.

Box 35.6 Identifying your methods/ techniques

- Background reading will help identify a range of relevant methods/techniques for your research (see the short bibliography at the end of this chapter).
- Are the chosen methods appropriate for the topic?
- Do the methods suit your capabilities and interests?
- Do you have sufficient time, money and expertise to employ the methods selected (or is there help available)?
- Is there appropriate equipment available for your use?
- For further ideas see Part III of this book.

Box 35.7 Your research action plan

Action	Date to be completed proposed	actual
Decide on your topic	....	....
1st meeting with your tutor/supervisor	....	....
Identify initial readings	....	....
Decide on an appropriate methodology	....	....
Set up indexing system for all sources used	....	....
Identify and make contact with relevant people	....	....
Visit study site where appropriate (review risk assessment)	....	....
Conduct pilot study (if necessary)	....	....
Review methodology (if necessary)	....	....
Conduct main study	....	....
Present results	....	....
Interpret results	....	....
Summarize your findings	....	....
Produce first draft	....	....
standardize margins, font sizes etc.		
Identify any gaps	....	....
Work on edits	....	....
Produce final version	....	....
Submit your dissertation	....	....

It is important to keep your methods and techniques manageable; many promising dissertations have come unstuck through overambitious or inappropriate research methodologies. It is helpful to carry out a pilot project to establish whether the research you are planning is realistic. This will give you the confidence to go ahead (or reject) your ideas at an early stage; it also helps you to design an effective sampling method if you are planning a quantitative survey. You must be prepared to be flexible in your approach and be able to cope as new questions arise and obstacles appear (and they almost certainly will do!).

The Research Action Plan

Having identified your topic, location and research methodology(ies), you are well on the way to completing your research proposal. But remember that you may need to review/revise your plans at each stage of the process outlined above. The next stage is a plan of action to allow you to carry out the work effectively (see box 35.7). Many people put a lot of energy into their data collection phase and then run out of steam. You need to keep your enthusiasm high through the presentation and interpretation stages too. This means you will need to practise and develop your time management skills. So produce a realistic timetable for each stage, allowing some time for slippage (identified through the proposed and actual completion dates in box 35.7).

Some Practical Considerations

Careful planning and an appropriate topic are key ingredients for a good piece of research. The eventual quality of your work will benefit hugely if you take on board a few basic principles right from the start (box 35.8).

Box 35.8 Some basic principles

Writing

- Start writing sooner rather than later – there are usually sections or chapters that can be written before the research has taken place, e.g. a description of the topic/issue, or information about the location of the study.
- An appropriate language is essential and you should aim to adopt a clear, fluent style.
- Be prepared to make changes to your early writing since your style will develop as you become more involved with your research.
- Get started on your references/bibliography as soon as possible (this is a very time-consuming task, particularly if you leave it until the last minute).

Organizing

- You may find it useful to have a series of folders for the various sections of your work, which can be added to as and when you find information.

Structuring

- Be selective about the information you include – it's all too tempting to include everything and to adopt an 'I've found out about this and I want you [the reader] to see all my efforts!' approach (this is not a good move if you want to produce a quality piece of work).
- Make sure that there are clear links between the sections within your chapters, and indeed between chapters – try to plan these into the work from the beginning.

Box 35.9 What makes a good dissertation?

You can achieve this if:

- You have a good topic.
- You have placed the work in its relevant geographical context.
- You have referred to the appropriate literature.
- You have used and justified an appropriate methodology.
- You have collected/used an adequate amount of data (but remember, quantity is no substitute for quality).
- You have presented your results clearly.
- You have interpreted/discussed your findings (and referred back to previous relevant literature).
- You have offered a balanced view.
- You have presented the work well (and in accordance with your institution's regulations).

So What *Does* Make a Good Dissertation?

A good dissertation should be interesting to read, well researched, well written and well presented. You will know that your dissertation is developing well if you can confidently say that you are addressing all the suggestions in box 35.9.

Is There Life After Your Dissertation/Research?

Without a doubt! The dissertation/research will help you to develop your own 'specialism' in geography – something you may wish to pursue to a higher level (for example, studying this further on a Master's course or even working towards a doctorate) – the opportunities are endless! It will certainly help you in your future career as you will have had the opportunity to identify your own skills, your strengths and also your weaknesses. Do not underestimate the significance of this piece of work, but more importantly, enjoy it!

Further Reading

Burkill, S. and Burley, J. 1996: Getting started on a geography dissertation, *Journal of Geography in Higher Education*, 20(3), 431–8.

Clark, G. and Wareham, T. 2000: *Geography@University: Making the Most of Your Geography Degree and Course*. Cheltenham: Geography Discipline Network.

Flowerdew, R. and Martin, D. (eds) 1997: *Methods in Human Geography: A Guide for Students Doing a Research Project*. Harlow: Longman.

Kneale, P.E. 1999: *Study Skills for Geography Students: A Practical Guide*. London: Arnold.

Lindsay, J.M. 1997: *Techniques in Human Geography*. London: Routledge.

Parsons, A.J. and Knight, P. 1995: *How To Do Your Dissertation in Geography and Related Disciplines*. London: Chapman and Hall.

Robinson, G.M. 1998: *Methods and Techniques in Human Geography*. Chichester: Wiley.

36

Analysing Data

Allan Pentecost

What are Data?

Data are 'known facts used for inference or reckoning', or 'quantities or characters operated on by a computer'. The term 'data' therefore has two definitions. Data input into computers are usually numbers, but might also include a series of letters. Also, 'known facts' include all kinds of information, not just numbers. 'Analysis' was traditionally used by chemists to explain processes used to determine the composition of a substance. The results of a chemical analysis *are* the data, so analysis, when applied to *data*, must mean something else. Data analysis provides us with two kinds of information – a summary of information, and a means by which our data can be compared with data obtained on other occasions, at other places, or by other people. Data analysis provides a succinct description of information. Datasets can be collected in several ways. For example, numbers of nesting birds may have been counted in a defined area and season but in different years. Or bird numbers may be compared with numbers obtained from a model of bird populations. Most of this chapter will be devoted to analysing numerical data, but it is worth noting that series of letters or other characters can be analysed using similar methods.

Types of Data

Numerical data may be classified into four basic types and it is important to recognize these types before starting any analysis. The first type to be described, which in many ways is the simplest, is explained with an example. A biogeographer has investigated the number of bird species found nesting in trees on a large island and the results are presented in table 36.1.

Table 36.1

Nesting species	Nesting numbers
Macaw	12
Robin	6
Oilbird	2
Common Potoo	4
Sappho Comet	1
Streamertail	3

The data consist of counts of individual birds, so all measurements are whole numbers. There can be no negative numbers or fractions in this dataset. Six species were encountered, the most frequent being the macaw. Such data are termed *categorical* and arise when we are sorting data into categories. To provide a description of such data we can use two quantities: the most frequent category or 'class', which is called the *mode*, and the *frequency range*, which gives the range

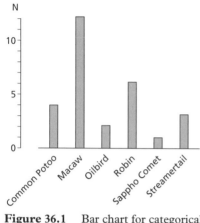

Figure 36.1   Bar chart for categorical data. The chart shows the frequency of nesting bird numbers. The order is immaterial but is presented alphabetically for convenience.

Table 36.2

Scale	Circled by villagers
1	0
2	1
3	1
4	0
5	1
6	0
7	2
8	4
9	7
10	1

of numbers of individuals found in each species. For the above example, the range is 1–12. These two quantities provide us with a summary of our results and the entire dataset can be illustrated as a *bar chart* (figure 36.1), where bird species are listed alphabetically along the horizontal axis.

Another unrelated study is undertaken to gauge residents' attitudes to a new shopping complex to be built next to a village. A questionnaire is distributed and villagers are asked to respond to the statement, 'The new shopping complex will provide employment opportunities in a village where unemployment is high'. They are asked to answer by circling one of the numbers 1–10 laid out below.

1 2 3 4 5 6 7 8 9 10

strongly   strongly
agree   disagree

The scale is used to assess how strongly respondents feel about the statement. A value of 5 would indicate a 'neutral response', while 7 indicates mild disagreement with the statement. Suppose 17 villagers complete this part of the questionnaire, then the data might look like table 36.2.

The data consist again of whole numbers

with no negatives. However, this time, the dataset shows *order*. Numbers in the right-hand column are presented as frequencies of response which move from 'strongly agree' to 'strongly disagree'. They can only be set out sensibly in this order. For categorical data, order is not important and is often given alphabetically for convenience. For the questionnaire data, the order *is* important and such measurements are made at the *ordinal level*.

With ordered data, a description can be provided using two useful statistics measuring the 'central tendency' and the 'spread'. For ordinal data the best measure of central tendency is the *median*, or 'middle value'. The frequency histogram in figure 36.2a presents some ordinal data on a scale 1–11 and the median divides the histogram into two equal areas. It lies in the class representing the ordinal value 4 in the histogram.

To find a measure of 'spread', the histogram is divided into four equal areas and the two adjacent to the median define the *interquartile range* (IQR). Within the range we have 50 per cent of our measurements, centred around the median. In figure 36.2a, the interquartile range is given by the difference between the third and first quartiles, spanning ordinal values 3–6 on the histogram.

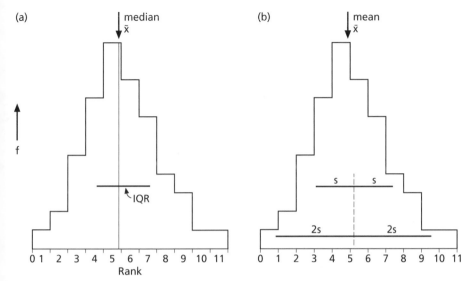

Figure 36.2 Two frequency histograms indicating some common descriptive statistics.
(a) Frequency distribution of an ordinal dataset indicating the position of the median and
the interquartile range (IQR).
(b) Frequency distribution of an interval/ratio dataset showing the position of the mean
and the standard deviation from the mean (s). One standard deviation includes about 70
per cent of these observations and two deviations (2s) include about 95 per cent. The
two histograms are identical in outline to show comparisons between the statistics. The
median and IQR can also be applied to interval/ratio data, but the mean and standard
deviation are not appropriate for ordinal data. These frequency distributions only approximate the normal curve as the measurements show positive skewness (see text).

At the highest levels of measurement, data
are measured with more precision, with the
numbers bearing well-defined relationships
with each other. Examples are field areas
(measured in hectares, for example) and air
temperature (°K). Such numbers have a true
zero point and are said to be at *ratio level*.
Related to them are *interval-level* measurements with no true zero point, such as temperature measured in °C where negative
numbers appear below the freezing point of
water.

For interval- and ratio-level measurements, the usual measure of location is the
average or *sample mean* (often shortened to
mean). A measure of spread is the *sample
standard deviation*, given the symbol *s*. The
standard deviation is calculated using a formula involving 'sums of squares' which can
be found on scientific calculators, usually the

s_{n-1} key. If one standard deviation is plotted
either side of the mean (i.e., ± 1 s), it is frequently found that about 70 per cent of the
total observations are enclosed within the
range, and it therefore includes more data
than the interquartile range. In addition, the
standard deviation spans an equal distance
either side of the mean, but the interquartile
range rarely spans an equal distance either
side of the median.

At a distance of two standard deviations
from the mean about 95 per cent of observations are frequently included but the precise number depends on the actual shape of
the frequency curve. The mean, standard
deviation, median and interquartile range are
examples of *descriptive statistics*. The standard deviation's value lies mainly in its relationship with an important 'model' curve
much used by statisticians. This is the 'nor-

mal curve' (figure 36.3) which has a complicated equation to describe it but a simple form. For this curve the mean is located at the apex of the curve and is equal to the median and the mode. The curve has some interesting properties. For example, it is symmetric about the mean and one standard deviation either side of the mean encloses 68.3 per cent of the total area under the curve. In figure 36.2b, the mean is found to be slightly larger than the median, so this curve only approximates a normal distribution. When the mean exceeds the median, this often indicates a *positively skewed* frequency distribution, where the right-hand tail of the distribution spreads out further than the lower, left-hand tail of the curve. Such curves are not symmetric about the mean.

The descriptive statistics outlined above are used in data analysis more often than any other statistic as they provide useful summaries of datasets and tell us something of the frequency distribution. However, much of our interest in data analysis involves comparing one set of data with another. For example, I could hypothesize that there has been a decrease in wetland in England between 1850 and 2000 because agricultural practices have increased over that period, leading to loss of wetlands to drainage. To support the hypothesis I needed to collect some data. I selected random areas of equal area within the country and measured wetland area using a planimeter with the aid of 1850 and 2000 maps. It would have been impractical to sample the entire area of land, so a series of *sample areas* were taken randomly under the assumption that they would be representative of the whole. This saves time, and in most investigations we are forced to take samples since measuring the entire area (or entire population in the case of animals or plants) is impracticable. It is then possible using *inferential statistics* to perform a statistical test on these samples to decide if my hypothesis is acceptable. Something can be learned from the descriptive statistics alone, such as the mean or median,

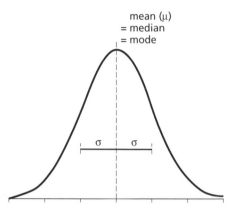

Figure 36.3   A normal curve showing equality of the median, mean and mode. The mode is the most frequent class/observation. Note that the mean and standard deviation are given special symbols on the normal curve (μ and σ respectively).

which could demonstrate that a change in area has occurred, but this becomes unsatisfactory if the difference in the statistics is small. For example, if the mean area of wetland in my samples had fallen from 6.1 per cent to 5.7 per cent between 1850 and 2000, it could be argued that the difference merely reflects uncertainties in the areas determined in the original surveys or, more likely, the areas which were randomly sampled. More progress can be made by conducting a statistical test.

To proceed further I need to be more specific about hypotheses because statistical testing is a formal process requiring clear definitions. They are framed around descriptive statistics such as the mean and standard deviation. In statistical testing we state two complementary hypotheses, the *null* and the *alternative* hypothesis. An appropriate statistical test for this example would be a z-test with the null hypothesis 'mean wetland in 1850 and 2000 is the same', and the alternative hypothesis 'mean wetland area has decreased between 1850 and 2000'. The *null hypothesis* implies no change and on completion of the test only one of the hypotheses will be accepted. To complete, a sample value of the test statistic z must be obtained. It can either be calculated by hand or by using a statistical software package such as Minitab. The test statistic z, a simple function of the mean and standard deviation, is compared with a *critical value* of z. The latter is obtained from the normal curve and found in *critical statistical tables*. If our test statistic is less than our critical z, then the null hypothesis is accepted. If not, it is rejected, which would imply that the wetland area *has probably decreased* between the two dates. For our result we would also state a 'level of significance'. This is a percentage probability found from the critical values (for a hand calculation) or from the printout from the computer. For example, if a *5 per cent level of significance* had been obtained, then there is a 5 per cent (or less) chance that the differences we see could have arisen by chance.

All statistical tests involve the setting up of hypotheses, but tests do not give 'black-and-white' results. For example, an analysis of national lottery figures demonstrates that the likelihood of winning a big prize on one ticket is less than one in a million. An appropriate statistical test would lead to acceptance of the null (no-win) hypothesis, yet of course, people do occasionally win. Because statistical analysis provides probabilities, they will only demonstrate how likely an outcome will be in the long run.

There are many types of statistical test to suit different situations and levels of measurement. There are rules to be followed and in fact the z-test, given in the example above, would have required some manipulation of the data before it was correctly applied.

Which Test Do I Use?

This is probably the most frequently asked question on data analysis courses. Tests are used either to compare two or more sets of sample data or to compare sample data with theoretical 'model' data. The latter 'goodness of fit tests' lie outside the scope of this

chapter, but most statistics texts provide information on the topic. When comparing two or more datasets, several tests can often be applied, providing effectively the same results. This will be apparent from table 36.3, which gives a (rather simplistic) guide to the common kinds of tests available.

To begin the decision-making process it is important to distinguish between two basic types of comparison. Do you want to: (a) compare one set of measurements with another to determine whether their measure of location or dispersion is significantly different (e.g. the wetland example above), *or* (b) determine how one set of measurements changes with another (e.g. how temperature changes with altitude or pollutant concentration with time)? This is an important distinction. As a rule, (a) involves just one variable, such as area, while (b) involves two variables, such as temperature and altitude, but there are exceptions. Once this distinction has been made, you need to determine the level of measurement of your data, which will be either nominal, ordinal or interval/ratio, as explained above. When you have done this, refer to table 36.3.

If your data are at the nominal level the choice is quite limited, but a range of chi-square and G-tests is available and these are capable of analysing quite complex and large sets of nominal data. Because these data are categorical, the numbers analysed will be integers.

For ordinal data the choice is greater, though ordinal datasets are in fact quite uncommon. However, in the social sciences, ranking of effects and responses is frequently encountered and in this field ordinal methods will often be found useful. Ordinal statistical tests are most conveniently undertaken with a statistics package as, unlike analyses on other datasets, they require ranking procedures which can be tedious and prone to error when performed by hand.

The most frequently used measurements are at interval/ratio level where the choice of tests is greater still. They are classified into *parametric* and *non-parametric tests*. The

Table 36.3

(a) Compare one set of measurements with another

	Type of test	Parametric (P) or non-parametric (N) test	Conditions of use	Examples and comments
Categorical data	G-test and chi-square tests	P	Data must be categorized. Avoid categories containing small numbers or zeros if possible.	Data must be obtained randomly. **Ex.** Compare numbers of birds' nests in two different types of shrub.
Ordinal data	Mann-Whitney test	N	Used for independent samples. Frequency distribution of the two samples being compared must be approximately of the same form.	**Ex.** Compare social status (via questionnaire or census) of individuals in two villages.
	Kruskal-Wallis test	N	Used to compare three or more samples. Frequency distribution of the samples should be approximately of the same form.	**Ex.** Compare attitudes of 5 groups of people to new housing estate.
Interval/ratio level	t/z-test of independent samples	P	The two samples should be independent and for small samples (<30) should approximate the normal distribution. Sample variances should be homogeneous.	If some of the conditions are not fulfilled a transformation may be applied or a Mann-Whitney test used. **Ex.** Compare rock-weathering rates at two different altitudes or aspects.
	t/z-test of dependent samples	P	The two samples should approximate the normal distribution and be related to each other.	A Wilcoxon matched-pairs test may be used as an alternative. **Ex.** Compare contamination levels at several fixed sites before and after a pollution event.
	Single-factor analysis of variance	P	The two/many samples should be approximately normally distributed and the sample variances should be homogeneous.	If some of the conditions are not fulfilled a transformation may be attempted or a Kruskal-Wallis test applied. **Ex.** Determine how wheat yield varies with different fertilizer applications in 5 plots.

(b) Determine how one variable (sample) changes with another

Ordinal data	Spearman rank correlation	N		**Ex.** Compare attitude towards new housing development with age of respondent in questionnaire.
			The two variables to be compared should be approximately bivariately normally distributed.	If the condition is not fulfilled a transformation may be applied or a Spearman test used instead. **Ex.** Find how the Mn concentration varies with Fe concentration in soil.
Interval/ratio level	Product-moment correlation coefficient	P		
	Model 1 regression analysis	P	A change in one of the variables (called the dependent variable) must depend upon a change in the other. The dependent variable must be approximately normally distributed. The independent variable should be measurable with precision. The relationship between the variables should be linear.	If some of the criteria are not fulfilled a transformation can be applied. **Ex.** Find the straight-line equation relating the temperature of springwater (dependent variable) to the altitude (independent variable). Use equation to predict temperatures of water at different altitudes.

former rely on properties of the normal curve and to be used correctly, the data must possess a certain form. These forms vary with the tests undertaken but often include a normal distribution for the raw data and homogeneity of the variances. Separate tests are available to determine normality and variance homogeneity and they are available in most statistics packages such as Minitab and SPSS. If the data fail these tests, alternative methods of analysis can be used, or a *mathematical transformation* can be applied. For example, if you are analysing percentages, it would be most unusual for the data to follow a normal distribution and an *arcsine* transformation should be undertaken. When these transformed data are retested, they are often found to provide a good fit to the normal distribution. Alternative *non-parametric* (also called *distribution-free*) tests are often used when the data are non-normal. Some examples are given in table 36.3. Their main disadvantage is that for more complex analyses suitable non-parametric tests are sometimes unavailable.

Reference is also made to the independence of the datasets in table 36.3. This applies to most tests and needs to be examined. In the wetland example, if 500 random 1 km² areas had been selected from the English maps of 1850 and a different 500 random samples taken from the 2000 map, then the two series of samples should be independent (showing no influence over each other). If instead the *same* 500 sites had been chosen for both studies, then the samples are not independent and a different statistical test would be needed. Such 'paired' comparisons are often found to be preferable as they limit the overall variability, providing more satisfactory outcomes.

Where the aim is to determine how one set of data changes with another, a different scenario is presented (table 36.3). Here, the data will often describe two varying quantities in the environment such as air temperature and altitude. Ideally, both sets of data would be obtained at the same time at a

range of altitudes and a test could be undertaken to see if the sets were associated. If the only interest is to determine the degree of association, then a correlation coefficient can be calculated. Coefficients range from −1 to +1 and values close to unity indicate strong associations. Often, however, a mathematical relationship is required between the datasets. In this case, regression analysis can be used to generate a simple equation relating the two variables. Again, a range of conditions applies for parametric tests, transformations can be attempted and non-parametric tests are also available.

Finally, it should be noted that all of the above tests are capable of accepting one of two hypotheses concerning the data, and either a 'difference' or a 'change' will be observed or not. The tests provide no information about the *amount* of change or difference between two or more samples. For these, *confidence intervals* need to be calculated. Confidence intervals can also be applied to individual datasets and provide further useful information on the variability of the sample.

Summary

1 Data comprise 'known facts' and consist of numbers or non-numerical characters.

2 Numerical data can be classified as nominal (categorical), ordinal and interval/ratio.

3 Useful summaries of data are provided by the descriptive statistics: median and IQR (ordinal data); mean and standard deviation (interval/ratio data).

4 Datasets can be compared using descriptive statistics, but in general a better approach involves the statistical testing of null and alternative hypotheses.

5 Statistical tests vary according to: (a) the level of measurement; (b) whether data are to be compared with respect to location/dispersion or whether a change in

one set varies with a change in another; (c) whether datasets are dependent or independent.

6 Parametric statistical tests are based on properties of the normal distribution and the data to be analysed will often need to be screened prior to testing.

7 Non-parametric (distribution-free) statistical tests can be used on a wide range of data with minimal screening.

Further Reading

Chase, W. and Bown, F. 1997: *General Statistics*, 3rd edition. New York: Wiley.

Pentecost, A. 1999: *Analysing Environmental Data*. Harlow: Pearson Education.

Rogerson, P.A. 2001: *Statistical Methods for Geography*. London: Sage.

Walford, N. 1995: *Geographical Data Analysis*. Chichester: Wiley.

Wheater, C.P. and Cook, P.A. 2000: *Using Statistics to Understand the Environment*. London: Routledge.

37

Approaches to Physical Geography Fieldwork

David L. Higgitt

Fieldwork in Practice

Down in the Yangtze Three Gorges, China, limestone cliffs squeeze the great river into a succession of boils and whirlpools. Unlike the scenic gaze of passengers on tourist ships, our mission is fieldwork. Somewhere high above us on the valley sides should be the records of past floods. Summer flows, swollen by the monsoon and compressed between the canyon walls, can be 40 m higher than in the dry season. When an exceptional flood occurs the waters are forced even higher, depositing sediments that provide clues for assessing flood risk. But first these sediments have to be found and sampled: back to the fieldwork. Jumping from the boat, the gear is unloaded and we begin a long climb up the boulder-covered slope. We move carefully above the annual flood level into scrub vegetation. There is a familiarity to the native shrub whose sharp thorns are employed in western gardens to deter intruders but now challenge geomorphologists in search of flood sediments. We push through, accepting scratches. The temperature has risen to 35°C. The air is limp and humid. Perspiration floods our brows and stings our eyes. My companion, a few paces ahead, emits a yelp. Snake! All these factors are listed in our risk assessment but now every step is slow and deliberate, cautious of the threats of instability and invertebrates. Eventually, the sediments are located, sampled and surveyed. Weary bodies, scratched and bruised, return contented in the acquisition of a couple more data points for the project. And some people claim that numbers don't bleed!

What does this tell us about physical geography fieldwork? It can be uncomfortable, exhausting, physically demanding, frustrating and repetitive. But the exhilaration and enjoyment that follow a fieldwork campaign live on long after the scratches have healed and the frustrations are forgotten. My basic message is that enthusiasm is the key ingredient for physical geography fieldwork and there is little that can simulate the buzz from conducting a field investigation well. But writing about eagerness does not provide much practical guidance. Here are a few tips for getting more out of physical geography fieldwork.

Preparation

Preparation is the key to successful physical geography fieldwork. In my experience students often find it more difficult to plan a physical geography project or dissertation than a human geography subject but, conversely, more straightforward to execute and write up. Careful planning is, therefore, crucial and concerns experimental design, logistics and safety.

Experimental design

A good project will have clear and interesting objectives, requiring the selection of a topic worthy of investigation. There is a huge range of worthwhile topics, each constrained by the availability of time, equipment, experience and guidance. It is not surprising that many dissertations reflect the interests of staff or are developed from subjects covered in earlier modules. Having a reasonable grasp of research issues in a particular subject area provides a good foundation for project formulation. It is also important to recognize the limitations of a dissertation format and not attempt something too general, unwieldy or complicated. In practice, it is likely that observations and impressions formed during the fieldwork will help to refine the nature of the problem and suggest new ideas. For this reason a pilot study is especially useful to ensure that the right sort of research questions have been asked.

Logistics

A pilot study helps not only to decide on the suitability of the research design but also to evaluate its feasibility. Can the fieldwork be conducted in the length of time available? Are the field locations reasonably accessible and have the necessary permissions to work there been obtained? Is the right equipment available at the right time? Make sure you identify what equipment is needed at an early stage, ensure that it is available and undertake the necessary training to use it competently. Unfortunately, if you identify that your project requires access to a very expensive piece of kit or sample analysis that is not readily available in your own department, it is likely that you will have to think again. Some departments may pose limits on the number of analyses. It is not uncommon for a student to arrive back at the laboratories with a truckload of soil samples, only to have the time or resources to analyse a small part of the collection. There are no bonus marks given for number of samples taken – this represents poor planning and

wasted effort in the field.

Many field measurements are difficult to accomplish individually and in any case local safety rules may preclude lone working. Reliable field assistance can be essential to the smooth running of any project. Many parents are persuaded to experience the joys of fieldwork, but be warned that their enthusiasm for arduous and repetitive work, especially in adverse weather conditions, may be slightly less than yours. An attractive option is to link a dissertation to an ongoing research project or expedition, which can improve the logistical support in terms of assistance and resources and provide a strong context for the study. Some departments will encourage this, while others may specify an independent study. Many students ponder the feasibility of undertaking physical geography abroad (see chapter 38 by Katie Willis in this volume and Nash 2000). There are some clear advantages in terms of glamour and excitement, but equally many pitfalls. It is more difficult to fully evaluate the logistics of a project at an overseas location and it is unlikely that a pilot study or supplementary fieldwork can be undertaken. The one-shot nature of an overseas dissertation makes it a risky option. On the other hand, the prospect of raising sponsorship by forming a student expedition means that a dissertation could form the excuse to undertake an educational visit that might otherwise be impossible. Most departments will have assessment criteria that are location-neutral and going overseas will not yield additional marks for bravery.

Safety issues

Working in the outdoors has safety implications and in most countries there is some legal requirement for a formal assessment of risk. For independent project work, such as a dissertation, the emphasis may be on the student to identify and evaluate potential hazards. In this context, a hazard is anything related to the environment or activity that might cause harm, and risk is the chance

Table 37.1   The five steps of risk assessment

Step	Procedure	Detail
1	Identify the hazards: what factors have potential to cause harm?	• Physical • Biological • Chemical • Anthropogenic
2	Who might be harmed?	• Self • Assistants • General public • Livestock • Environmental impact
3	Evaluate the risks: what is the likelihood and severity of identified hazards?	What measures are needed to reduce risks (e.g. protective gear, modified sampling plan)?
4	Record the findings	This is the crunch for demonstrating that the assessment has been carried out
5	Review the assessment periodically	Have there been any changes to the fieldsite since the assessment was produced?

of someone being harmed by that hazard. In many departments dissertations cannot begin until a risk assessment has been filed, and in others it is awarded some of the marks. Completing a risk assessment is not difficult, as illustrated in table 37.1 (more detail in Higgitt and Bullard 1999).

Risk assessment helps to identify the feasibility of working in particular terrains. Consideration should also be given not to overestimate how much can be done in a day. While lecturers endeavour to design inclusive exercises on field classes that are not too physically demanding, it is evident that a reasonable level of personal fitness can be a prerequisite to enjoying the experience. Plan to work within your own (and your assistants') capabilities. The checklist in table 37.2 may be of some use.

Data Collection

Though physical geography has evolved

from the qualitative description of landscape, few projects will get far without the acquisition and analysis of quantitative data. There are few things in life more frustrating than expending large amounts of effort in compiling a dataset which is inadequate for testing the key ideas. Care is needed in deciding the quantity, distribution and accuracy of data necessary. What are the appropriate spatial and temporal sampling frameworks? If you are planning to do laboratory work on samples collected in the field, then you should consider the number, size and nature of samples required (see chapter 39 by Heather Viles in this volume on laboratory work). Whatever your fieldwork entails, it is vital to keep good records – and a field notebook is an essential purchase.

Examining the spatial variability of a phenomenon is feasible. Examples might include particle size variations along a beach, heavy metal pollution across a floodplain or vegetation communities on a salt marsh transect. Examining temporal variability is

Table 37.2   A personal checklist for assessing project logistics and safety

Query	Indicate arrangements made
Access arranged?	
Contact address for any collaborative organization	
Research assistance available?	
Personal fitness adequate?	
Training for use of equipment?	
Health and first aid requirements?	
• Immunizations required?	
• Medication available?	
• Protective gear required?	
• First aid kit carried?	
Personal safety and communication?	
Emergency procedure?	
Insurance adequate?	

interesting but requires some caution because the length of time available for fieldwork is often confined. As most dissertation fieldwork is undertaken during the summer vacation, a phenomenon that displays marked variation at this time (e.g. diurnal variations in proglacial stream discharge) is suitable, but one which is suppressed (e.g. nitrate pollution – largely a winter phenomenon) is less logical. Examining storm response can provide a fascinating study, but what happens if there are no suitable events during the fieldwork period? There are ways around this. Most regulations will place some emphasis on primary data collection – the demonstration that you have the skill and competence to collect and analyse field data – but the use of previously collected (secondary) data can augment a project. For example, longer-term records of climatic variables, river discharge, suspended sediment or pollutant concentrations can provide the context for a more detailed short-term study. Similarly, historical information derived from maps, air photographs, postcards, paintings and documentary sources enables studies of longer-term morphological change of coastlines, river channel patterns, land use and so on. Practical guidance for extracting information from historical maps is provided by Hooke and Kain (1982).

Proxy records such as pollen preserved in sediment sequences permit studies of environmental change, while glacial deposits such as hummocky moraine can be used to examine the history of deglaciation. Projects might examine the recovery of the landscape following a recent disturbance such as a landslide, flood or land-use change. Using paired catchments to compare different land uses can be a productive approach and one which might be scaled down to examining small plots of different vegetation covers or management techniques. Whatever the approach adopted, it should be designed to generate representative data which are useful towards the project objectives. This is also dependent on the availability of suitable measurement techniques, and these are discussed next.

Appropriate Techniques

It is impossible to provide a thorough overview of suitable techniques for physical geography fieldwork in a book of this length, let alone a single chapter. Fortunately there are many volumes available that are able to provide more details (Goudie 1990; Parsons and Knight 1995; Jones et al. 2000; Southwood and Henderson 2000).

Mapping and surveying

Mapping and surveying once lay at the very heart of geography and a student will need to know at least the basics. Needless to say, competence in compass and map reading is a safety requirement in remote locations. A good base map will be essential for marking the locations of samples and landscape features. Geomorphological mapping has a range of standard symbols to depict landform and slope features (Cooke and Doornkamp 1990). Laminating a suitably enlarged base map with clear plastic can be very useful, enabling features to be marked on with OHP pens. Mapping directly onto an aerial photograph is also efficient. Be aware that the acquisition of decent maps in some overseas locations is difficult and that there are copyright controls over the reproduction of base maps and aerial photographs in the final report.

Simple surveying procedures using an Abney level or clinometer are often sufficiently accurate for the purpose, but more precise elevation may require a surveying level or electronic distance measurer (EDM) and care needed to tie into benchmarks. The advent of the global positioning system (GPS) raises some new research opportunities (Higgitt and Warburton 1999), but the user should be aware of capabilities and the level of detail required for a particular application. Hand-held GPS receivers available at relatively low prices are good enough to identify the location of sample sites, but mapping and surveying applications require differential GPS using more sophisticated equipment.

Geomorphology

Compilation of measurements of landform dimensions such as width, depth, height, length, shape, gradient and aspect can generate relatively large datasets in a short time which are useful for testing ideas about landform development or comparing similar landforms in different settings. Such measurements are bracketed under the general term geomorphometry. Examples in fluvial geomorphology include channel cross-section dimensions, meander wavelengths or stream networks, and in karst geomorphology include dimensions of closed depressions or joint sets on limestone pavements. GIS techniques might be employed to analyse spatial relationships within geomorphometric data.

Many geomorphological projects examine material properties, many of which can be conducted in the field. The rock-mass strength index (Selby 1980) is a good example. It scores points for six properties, five of which require nothing more complicated than a tape measure. The key property of rock hardness is measured using a Schmidt hammer. Similarly, indices of rock weathering can be derived by measuring thickness of weathering rinds, depth of cracks and microtopography of surfaces, using callipers, rulers and profile gauges, respectively. More specific measurements of material properties require transport of samples to the laboratory. It should be noted that some rock exposures are protected by conservation regulations and it is better, on environmental grounds, to sample material that is already detached, unless there are particular reasons why fresh material must be collected. Wear safety glasses at all times when rock sampling. Fluvial gravels or other coarse deposits are cumbersome to transport and can be split into size categories using a sampling frame. Alternatively, use callipers or a ruler to measure the axes of a sample of clasts.

Soils, sediments and vegetation

Soil or sediment sampling begins with adequate description of the fieldsite. Standard procedures are available for describing soil properties, logging the stratigraphy or examining the fabric of a sediment profile. Digging a small pit is useful for examining soils, while sediments are often exposed in eroded stream banks or obtained through coring. Quaternary studies frequently use deep cores from wetlands. The sediments can be described and sectioned on site or returned to the laboratory. Examination of pollen, diatoms, forams and other microfossils is appropriate but takes considerable time to master. Once sampled, pits should be refilled immediately to prevent injury to livestock.

Many soil studies are linked to variations in vegetation. Species identification can be tricky without some prior experience, but there are many pocket guides that assist (e.g. Fitter et al. 1996). Vegetation data can be analysed in several ways to derive measures of species composition or habitat classification. In some locations detailed maps of vegetation cover characteristics may have been published.

Water and weather

Hydrological and meteorological measurements are relatively straightforward to take using inexpensive equipment. Many projects examining these processes attempt to demonstrate temporal variability so care must be taken at the experimental design stage to ensure that significant short-term variations will be observed in the measurement period. Relating short-term variations in air pollution to weather patterns, or in water quality to rainfall, may yield interesting results. Monitoring rainfall, temperature, humidity, wind speed and direction uses standard equipment. Traps can be designed to collect rainfall or dust samples from particular directions. Specialist equipment is needed to monitor air pollutants. There are many aspects of water quality that can be monitored in the field such as temperature, pH, conductivity and dissolved oxygen content. Kits are available for some chemical analysis but may not be sufficiently precise for meaningful analysis and samples must be returned to the laboratory, which in turn requires consideration about sample collection and storage. River discharge measurements are most often undertaken by salt-dilution gauging or by velocity-area calculations using a current meter. A large number of velocity readings are needed across the width of the stream to ensure accuracy.

Studies of water and weather that seek to examine variations at one or more points over time need to consider the representativeness of sample points and the required frequency of sampling. I was greatly impressed during a field trip by a group who wished to monitor rock surface temperature in a desert wadi throughout the night – until I realized that I had to accompany them for safety reasons! In hindsight, a data logger could have saved some hassle.

Into the Field . . .

Assessors of fieldwork projects and dissertations are looking for a combination of originality, sound experimental design, competent analysis and logical conclusions. There are a number of feasible and worthwhile topics where careful planning and enthusiastic fieldwork will not only be rewarded by the examiners, but should provide a memorable experience.

References

Cooke, R.U. and Doornkamp, J.C. 1990: *Geomorphology in Environmental Management: A New Introduction,* 2nd edition. Oxford: Clarendon.

Fitter, R., Fitter, A. and Blamey, M. 1996: *Wild Flowers of Britain and Europe,* 5th edition. London: HarperCollins.

Goudie, A.S. (ed.) 1990: *Geomorphological Techniques*, 2nd edition. London: Unwin Hyman.

Higgitt, D.L. and Bullard, J. 1999: Assessing fieldwork risk for undergraduate projects, *Journal of Geography in Higher Education*, 23, 441–9.

Higgitt, D.L. and Warburton, J. 1999: Applications of differential GPS in upland fluvial geomorphology, *Geomorphology*, 29, 121–34.

Hooke, J.M. and Kain, R.J.P. 1982: *Historical Change in the Physical Environment*. London: Butterworth Scientific.

Jones, A., Duck, R.W., Reed, R. and Weyers, J. 2000: *Practical Skills in Environmental Science*. London: Pearson Education.

Nash, D. 2000: Doing independent overseas fieldwork 1: practicalities and pitfalls, *Journal of Geography in Higher Education*, 24, 139–49.

Parsons, A.J. and Knight, P. 1995: *How To Do Your Dissertation in Geography and Related Disciplines*. London: Chapman and Hall.

Selby, M.J. 1980: A rock-mass strength classification for geomorphic purposes: with tests from Antarctica and New Zealand, *Zeitschrift für Geomorphologie*, 24, 31–51.

Southwood, T.R.E. and Henderson, P.A. 2000: *Ecological Methods*, 3rd edition. Oxford: Blackwell.

38

Fieldwork Abroad

Katie Willis

It was only when we were coming in to land at Mexico City airport in July 1990 that I realized I had missed out a very important part of my preparations for my first piece of overseas fieldwork. I was due to spend two months in the southern Mexican city of Oaxaca studying women's employment as part of my Master's degree. I had spent months refining the questionnaire, learning Spanish, reading up on the theory, and collecting names of contacts in Oaxaca. What I had not done was think about my accommodation in Mexico City, where I would be staying for two days before catching a bus down south.

It was late at night, I had been travelling for nearly 20 hours, it was my first time in one of the world's mega-cities, and I didn't know where to stay. I ended up approaching a couple of Dutch backpackers who were on the same flight and asking them where they were going to stay. We shared a taxi to the hotel, which made me feel much more secure, and I was able to help them as they couldn't speak Spanish. The hotel itself was a real fleapit, but at least I knew my newfound friends were in the room next door.

This has certainly taught me a lesson, and since then I have always made arrangements to be met at the airport, or to have accommodation organized for my arrival. Of course, this was not the only problem I have had with fieldwork, but it exemplifies the need to consider the practical dimensions as well as the academic aspects of fieldwork abroad.

Preparation

Planning should start as early as possible. While preparation for all fieldwork is important, when you are travelling abroad there is often a greater need to 'get it right first time' because you are less likely to return. The topics covered in this section are discussed in greater detail in Robson and Willis (1997).

Research preparation

Some students feel that there is no point in making preparations before they go because they will only be able to sort things out once they see what conditions are like 'on the ground'. Nothing could be further from the truth. While you are bound to have to make adjustments once you arrive, it is better to have ironed out the main theoretical problems and checked on the feasibility before you go, rather than having to make major changes in the field.

When you are devising a research project, bear in mind both the theoretical context of your research and the research practicalities. Table 38.1 summarizes three projects conducted by Liverpool University students during fieldwork in Santa Cruz, California. As you will see, preparation before departure was key. Theory and methods may be appropriate in an abstract sense, but you need to check feasibility on the ground. For example, do you have permission to take sediment samples from a particular lake? Is

Table 38.1   Outlines of three undergraduate projects in Santa Cruz, California

	Project 1	*Project 2*	*Project 3*
Theme of project	Bilingual education	Ethnic labour segregation	Waterfront redevelopment
Suitable for location?	Yes – Santa Cruz has a large non-English-speaking population.	Yes – Santa Cruz has a diversity of ethnic groups.	No – This process is not going on in Santa Cruz.
Proposed methods	Interviews with Education Department and case study of schools.	Questionnaire survey and analysis of census.	Interviews with planners, developers etc.
Feasibility of project?	Interviews and case study schools were set up before departure. Contacted by email.	Census material available from library. Questionnaire survey limited because no Spanish speakers in the student group.	A project on redevelopment would be possible e.g. in city centre, but waterfront focus is unsuitable.
Outcome	Excellent project.	Project altered so looking at employers' perspective rather than employees'.	Weak project because group did not refocus their project.

the archive going to be open during your visit? Will government officials be willing to talk to you? If the feasibility of the project is questioned, make sure you have some form of back-up plan, or reformulate your research question accordingly. A useful overview of issues involved in research in developing countries is Deveraux and Hoddinott (1992).

Health and safety

Safety is key for all fieldwork, but overseas you have to remember that you may be living as well as working in an unfamiliar environment. Make sure that you have all the appropriate vaccinations and take malaria tablets where necessary. It is important that your doctor is aware of what you will be doing overseas as the normal 'tourist' advice may be inadequate. You should always take a basic first aid kit with you. Proper insurance is vital, both for your health and your possessions. Do not even think of travelling without it.

You should ensure that the area where you will be working is not a high-risk area for civil unrest. Not all local conflicts overseas are reported by the media, so finding out more is advised. The Foreign and Commonwealth Office in the UK (website address in the references) is an excellent source.

Bureaucracy

As an undergraduate, you probably won't be spending more than about a month undertaking fieldwork, so will probably be able to travel on a tourist visa (where appropriate). Liverpool University students have conducted research in countries as diverse as the United States, South Africa, Malaysia, India, Uganda and Trinidad and Tobago while on a tourist visa. They didn't experience any problems with officials, partly because of the length of time they were staying and also because the kind of research they conducted was not regarded as sensitive. You may want to check with the embassy/high commission of the country where you will be working, but do not say that you will be 'working' or 'studying' as these terms mean very specific things to immigration officials and could cause a range of unnecessary difficulties.

Language

Trying to become fluent in a foreign language for an undergraduate dissertation is probably a bit of a tall order. However, you will need to consider how you are going to deal with language issues if your research involves interviews or questionnaires. Will you need a translator? If so, who will this be? Relying on a translator can be problematic and you need to be aware of how this extra 'filter' between yourself and the interviewee can affect your results.

Even if you do not have to interview people in a foreign language, it is worth trying to pick up some useful phrases. As a physical geographer you may not have to speak Swahili for your research in Kenya, for example, but a few phrases will help you get around and may help you 'fit in' in your fieldsite.

What to take

As always, try to travel light. Take advice from tutors, travel guidebooks and local contacts about what you will need and what is available at your fieldsite. If you are part of an expedition to a remote area, your requirements will obviously be very different from others who will be based in large urban areas. In addition, in some places items may be available but at a vastly inflated cost, for example contact lens solution, tampons and sun cream.

If you are taking electrical goods, for instance a laptop computer, then make sure you know the voltage and socket details of your destination.

In terms of clothing, again take advice. You will need to consider both the climatic conditions at your destination and also social norms about appropriate dress. If you

are conducting research on tourism then dressing like a tourist could be acceptable, but even if the temperature is in the 30s, dressing in shorts is unlikely to make you welcome in a government office, or in certain cultural settings. For my Mexican fieldwork, I had a collection of cotton dresses with bright floral patterns. While these have never seen the light of day in Britain, they were ideal in Mexico for reasons of both comfort and appropriate garb.

As with all overseas travel, try not to take all your money in cash. Find out if your cashpoint card will work in your destination, or if travellers' cheques are useful. If you decide to take travellers' cheques, then make sure you get them in the correct currency (don't use sterling in the United States, for example). If you don't have a credit card, then consider applying for one in case of emergencies. Always make sure that you have the numbers of any cards or travellers' cheques written down and packed in various places in your luggage, along with the emergency numbers in case of loss. You should also take photocopies of your passport, insurance details and your lists of contacts.

Before heading off to the field it is also worth setting up an email account that can be accessed anywhere in the world (hotmail or Yahoo, for example). This is good for keeping in touch with people at home, and also for those emergency queries to get to your supervisor.

In the Field

Research

Although you have a limited time in the field, make sure that you don't launch into your fieldwork the minute you get off the plane. In some cases you will need to adapt to a different time zone or extremes of heat or cold, but in all cases taking some time to assess your fieldsite is important. Taking a step back and assessing your planned research could save you a great deal of heart-

Table 38.2   Overseas fieldwork preparation

1   Research
Is your research topic feasible in your chosen location?
Are your methods feasible?
Do you have permission to work on your chosen fieldsite where appropriate?
Do you have a draft questionnaire/interview schedule where appropriate?
Do you have a translator where appropriate?
Do you have a back-up plan if your chosen project falls through?
Do you know how to get hold of your tutor if you have research problems?

2   Health and safety
Have you had all the appropriate vaccinations?
Do you have your malaria tablets where appropriate?
Have you got insurance?
Are you sure your fieldsite is free of local conflicts?
Do you have a first aid kit that is appropriate for your destination?
Do you have someone you can contact in your country of destination?

3   Bureaucracy
Do you have a valid passport that has at least 6 months left to run?
Do you have all the appropriate visas?
Do you have photocopies of all documentation?
Do you have a letter from your department, preferably in the local language, outlining what you are
doing and confirming that you are a student and a trustworthy person?

4   Packing
Do you have all the equipment you will need?
Are any electrical items going to work out there?
Do you have appropriate clothing?
Do you have a way of getting money in an emergency?

5   Other
Have you organized accommodation at all appropriate sites?
Have you got an internationally accessible email account (if you are going to be near an Internet café
or equivalent)?

ache later on. If you are conducting a questionnaire, for example, make sure that someone familiar with the local context has read it through and you have piloted it properly. Physical geographers may want to examine alternative sites to ensure that their preselected fieldsite is the most appropriate.

If, on your arrival, your well-prepared project is obviously not going to work, do not panic! The archive may be unexpectedly closed, the rainy season may have come early, or accessing individuals for interviews may prove to be impossible. As you are in the field it is worth trying to refocus your project. Your back-up project may be appropriate, or other measures may have to be taken. Talking to your local contacts and emailing your tutor may be useful courses of action.

When interviewing or conducting questionnaires, data 'accuracy' may be an issue. This may be because of language problems, or it may be because your interviewees are keen to give you, as a foreign researcher, the 'right' answer, or feel that presenting a particular aspect of their lives may be beneficial to them. Recognizing this possibility and trying to have alternative sources of information may help (see Casley and Lury 1987; Howard 1997: 20–4).

Make sure that you keep fieldnotes throughout your stay. For human geographers these should include general observations as well as the specific 'findings' of your research. You may not be planning an ethnographic study, but the passing comments of a taxi driver or a brief article in a newspaper may trigger a very useful line of enquiry.

It is crucial that you constantly review your work as you go along. Again this is good fieldwork practice, but its importance is heightened by the limited time you have in the field. It is a good idea to begin to code questionnaires in the field, to outline basic trends in qualitative data, or to plot physical geography measurements. This gives you a headstart on your analysis, but more importantly helps ensure you are on the right track.

Safety

No sample or interview is worth putting yourself at risk for. As at home, take appropriate precautions, whether you are in an isolated location or in the middle of a bustling city. In some locations a knowledge of poisonous plants, insects and snakes is crucial. Dressing and behaving appropriately may help you, particularly women, avoid unwanted attention. Remember that in most places the main hazard is road traffic (Elmhirst 1996; Seymore 1996).

Ethics

Any research has an ethical dimension, but the particular power relations when researchers from economically richer countries are working in economically poorer countries have been highlighted (see, for example, Patai 1991; Sidaway 1992; Wilson 1992; Madge 1993). The process has sometimes been termed 'data mining' and has been equated with forms of neo-colonialism. For some people this power inequality is such that they do not feel comfortable doing research in this context. For others, however, the way forward is to do the research but in as harm-free and collaborative a way as possible. While you, as an undergraduate, have very few resources, time and influence to make a difference, it is crucial that any work you do is characterized by respect and an awareness of context. You should also try to feed back any key findings to the people that helped you.

While most students researching abroad are in an unfamiliar environment, for some undergraduates they will be returning home to conduct research. This can have obvious benefits in terms of local knowledge, but it does not automatically mean that you will be an insider. Your position in class, gender and ethnicity terms may affect the ways in which people interact with you (see Amadiume 1993).

Bringing samples/data home

Having completed your fieldwork, getting your data or samples back home safely is key. Questionnaire coding sheets can be photocopied and/or saved on disc and posted home, as can fieldnotes or other qualitative material. You should then keep the originals with you in hand luggage – this may make you unpopular with cabin crew, but it is worth it. If you are bringing back sediment samples, be very careful about what you say on any customs forms. Anything that could be 'organic matter' is often viewed with suspicion, so 'geological specimens' is sometimes a useful generic term to employ.

Dissertations Based on Fieldwork Abroad

Most geography undergraduates choose not to travel overseas for their dissertation research, partly because of cost, but also because of the competing demands from paid work and/or family commitments. However, overseas fieldwork is possible in some cases and table 38.3 gives you some idea of recent dissertations, including some from Liverpool University. The human geography bias is a reflection of the logistical (particularly safety) issues involved in overseas physi-

Table 38.3   A selection of dissertations based on fieldwork abroad

1   The provision of housing for female-headed households in Cape Town, South Africa
2   Informal sector entry and operation in Port of Spain, Trinidad
3   An investigation into the incorporation of women into development schemes in Madhya Pradesh, India
4   Ecotourism? The examples of Xcaret, Xpu-Ha and the Sian Ka'an Biosphere Reserve, Mexico
5   The role and performance of non-governmental organizations in development: case study of Vision TERUDO, Kumi District, Uganda
6   Hydraulic geometry of a supraglacial stream on Fallsjokull, Iceland
7   Who are the travellers? An examination of alternative tourism on Ko Phangan, Thailand
8   Flexible specialization in the province of Florence, Italy

cal geography research if you are not involved in a larger expedition.

For those of you able to travel abroad to conduct research, make the most of this opportunity. It can be an excellent basis for a dissertation, but it can also represent an important personal experience that will have a lasting impact. Make sure that you consider both academic and logistical issues before you go and you will be able to enjoy this experience to the full.

References

Amadiume, I. 1993: The mouth that spoke falsehood will later speak the truth: going home to the field in Eastern Nigeria. In D. Bell et al. (eds), *Gendered Fields: Women, Men and Ethnography*, London: Routledge, 182–98.

Casley, D.J. and Lury, D.A. 1987: *Data Collection in Developing Countries*, 2nd edition. Oxford: Oxford University Press.

Deveraux, S. and Hoddinott, J. (eds) 1992: *Fieldwork in Developing Countries*. London: Harvester Wheatsheaf.

Elmhirst, B. 1996: The newest demon, *New Scientist*, 151, 36.

Foreign and Commonwealth Office website: www.fco.gov.uk.

Howard, S. 1997: Methodological issues in overseas fieldwork: experiences from Nicaragua's Northern Atlantic Coast. In E. Robson and K. Willis (eds), *Postgraduate Fieldwork in Developing Areas: A Rough Guide*, 2nd edition, London: DARG Monograph 9, RGS (with IBG), 19–37.

Madge, C. 1993: Boundary disputes: comments on Sidaway (1992), *Area*, 25, 294–9.

Patai, D. 1991: US academics and Third World women: is ethical research possible? In S. Berger Gluck and D. Patai (eds), *Women's Words: The Feminist Practice of Oral History*, London: Routledge, 137–54.

Robson, E. and Willis, K. (eds) 1997: *Postgraduate Fieldwork in Developing Areas: A Rough Guide*, 2nd edition. London: DARG Monograph 9, RGS (with IBG).

Seymore, J. 1996: Trafficking in death, *New Scientist*, 151, 34–7.

Sidaway, J. 1992: In other worlds: on the politics of research by 'First World' geographers in the 'Third World', *Area*, 24, 403–8.

Wilson, K. 1992: Thinking about the ethics of fieldwork. In S. Deveraux and J. Hoddinott (eds), *Fieldwork in Developing Countries*, London: Harvester Wheatsheaf, 179–99.

39

Laboratory Work

Heather A. Viles

Although many geography students (and probably a good few geography lecturers) are unaware of the fact, most university geography departments have good laboratory facilities. Like computing suites and libraries, laboratories help geographers answer research questions. In order to make best use of laboratories you need to know what questions to ask and to have some idea of the sorts of questions that can be answered using the equipment and facilities available in your university. Many geography departments offer laboratory practicals as part of particular courses, which will give you a taste of how laboratory science works, but for many students dissertations or other projects provide an ideal opportunity to develop laboratory skills.

Two main types of laboratory work are carried out by physical geographers. First, there is *analysis* of samples collected in the field. In this case, the researcher is answering questions such as 'what is my sample made of?' and 'how old is it?' This type of analytical work usually amplifies and builds upon field observations. For example, you may wish to compare two river terrace deposits. Field survey and observations will provide basic information about the height, size and morphology of the deposits, and subsequent analysis on samples collected in the field will permit more detailed description of sediment characteristics and, perhaps, dates of deposition. The second type of laboratory work is *experimentation* and,

again, for most geographical projects this complements field studies. In experimentation the researcher is asking questions such as 'how does my sample behave under certain conditions?' Experimentation can be very useful in process studies in physical geography, such as investigations of weathering. You may want to study the development of cavernous weathering features. Field observations of the nature and distribution of such features in an arid area could then be used to design a realistic experiment to test how they might develop.

Geographers work on several types of samples, usually collected from the natural or built environment. Soils, sediments (such as samples of dune sands or marine sediments), rocks, water samples and various organic materials (such as leaves, tree cores, animal dung) are all commonly examined by geographers. Types of samples used range from cores extracted from bogs, lakes, mires or ice by heavy coring equipment, through small bags of sediment and soil, to lumps of rock hammered from outcrops and sealed bottles of water.

Low- and High-tech Approaches

For both analysis and experimentation there are two main approaches that can be followed. First, *low-tech* methods can be used, which provide a cheap and simple (but often rather time-consuming) way of doing things. An example of a low-tech method is the determination of calcium contents in a water sample using titration, which can be done using very basic laboratory equipment and is straightforward to do, but requires patience and can be extremely boring! Second, there is a whole suite of *high-tech* methods using expensive equipment which are usually quick, can analyse large numbers of samples and provide highly accurate results. An alternative method of measuring calcium concentrations in water, for example, is to use an atomic absorption spectrophotometer (AAS), which provides quick analyses once it has been properly set up and calibrated.

Before you get very far with designing a project involving laboratory work, it is important to find out what methods are available to you and how they work. Most analysis techniques available in geography laboratories provide information on the physical, chemical and biological nature of samples. Commonly available techniques include methods to determine the following:

- Grain-size distributions of sediment.
- Chemical contents of water samples (or solid samples after dissolution) – often in the form of anions (e.g. sulphate, nitrate and chloride) and cations (e.g. calcium, iron and lead).

Most geography laboratories also have a range of microscopes for use in sample preparation, identification of microfossils and determination of grain shape, among other things. Some laboratories also possess equipment to carry out specialist analyses such as:

- Pollen, diatom and other microfossil analyses.
- Dendrochronological studies (which involve the use of tree rings to investigate climate, pollution or geomorphic change).

- Dating – such as radiocarbon and luminescence.
- Analysis of organic chemicals or isotopes.
- Rock mechanics.

You will get a feel for what's on offer by talking to lecturers and researchers in your department, who will almost certainly have a good supply of project ideas involving any specialist equipment which they will let you use. Dedicated experimental facilities are less widely available in many laboratories, but cover such things as environmental cabinets, flumes, laboratory-based rainfall simulators and wave tanks. However, many useful experiments can be done using very simple equipment – given a good degree of imagination and resourcefulness.

Most laboratories will have protocol or instruction sheets for a wide range of techniques (often available on departmental websites). However, there is much more to laboratory work than simply using one or more methods, and as the rest of this chapter discusses, you need to consider the following:

- Safety and good housekeeping.
- What questions you are asking and how best to answer them.
- Effective sample preparation.
- Good recording, analysis and presentation of results.

Safety and Good Housekeeping

Safety in the laboratory is of paramount importance and involves taking care of yourselves and other people, equipment and samples. All laboratories have safety rules and regulations which must be read before any work is started. For all laboratory work undertaken by students, a formal risk assessment needs to be undertaken and recorded on COSHH forms. Most safety guidelines are based on common sense. Some general rules are:

- Always follow the instructions of the laboratory technician.
- Always ask if you are unclear on anything.
- Follow all instructions carefully.
- Own up immediately if anything goes wrong, or if you break anything.

Good housekeeping is also a key aspect of effective laboratory work. You should regard the laboratory like any shared space, and not leave mess around. When you have finished an analysis you should ensure you clear everything away (or follow your particular laboratory technician's instructions). Never leave samples lying around – they are likely to become contaminated, lost or thrown away.

Questions to be Asked

Well before you start any laboratory work as part of a project it is vital to consider what questions you are asking and why. If, for example, you are studying beach sediments, there are many different analyses you could do to provide a description of the physical, chemical and biological properties of the sediments. You have to decide which analyses are going to tell you something interesting and useful – and that, of course, depends on what questions you are asking. Some techniques take a long time and do not tell you very much; others are quick and will provide highly pertinent information. It is also vital to ensure that you collect meaningful and representative samples in the field (see chapter 37 by David Higgitt). To return to the beach sediment study, if you have only collected one sample from one beach, no amount of laboratory analysis will tell you anything very meaningful. Sampling in the field needs also to take account of what laboratory techniques are going to be used on the samples – as some analyses or experiments require large amounts of sample, whereas others need only trace quantities. In general, grain-size analysis of sandy beach

material requires a few tens of grammes of sample, whereas microscopic analyses of shape and surface textures require only a few hundred individual grains. Sampling also needs to take account of what statistical tests require in order to make meaningful comparisons between datasets, or to investigate trends in data. Simple statistical tests (see chapter 36 by Allan Pentecost) will allow you to compare data from different sets of samples.

Effective Sample Preparation

Each technique that you can use in the laboratory requires samples to be prepared, and there are often many possible preparation methods. For example, solid samples may need to be digested in acid to produce a liquid sample, sediments may need to be dried and organic components removed, and rock specimens may need to be cleaned and polished. Choosing the correct preparation technique is vital to obtaining good results, and you must ensure that you get it right. The two words to consider here are 'appropriateness' and 'accuracy'. Although there are many ways of preparing samples for grain-size analysis, for example, only some will be appropriate to particular research questions. If you are investigating beach sands you may or may not want to wash them prior to grain-size analysis to remove salts. You need to decide what is the best strategy to follow given the research questions you are addressing. Accuracy in preparation is vital for many analytical techniques – and if not done properly, the whole analysis will be wasted as you will get meaningless results. In order to carry out chemical analysis, for example, dilution may be necessary and this must be done with great care, otherwise you may not know exactly how the concentration of elements in your prepared sample relates to the concentration in the original sample.

Recording

Once you have decided what you want to do and how to do it, and collected your samples, you are ready to begin the most important phase of laboratory work. At this stage it is vital to adopt good working practices that will ensure you get meaningful results. The first step is to buy a decent notebook and ensure you keep all laboratory protocols, notes and results in it. The second step is to ensure that you keep a clear record of everything you do, and to make sure that you are sufficiently accurate and careful in the key preparation stages before analyses or observations. Many analyses require you to know the initial weight or volume of your sample to a certain degree of accuracy, in order to calculate concentrations of various components. If you don't take such measurements accurately at the start, you might as well not bother doing the rest of the analysis as your results will not mean anything. Another example would be the analysis of pollen or other microfossils from core samples. Careless cutting up of a core or shoddily prepared pollen slides will mean that you will not be able to produce good results however hard you peer down the microscope and however much time you have invested in learning to identify different pollen grains.

When you start the analysis or experiment it is highly important to record the results effectively and keep them safe. This might seem obvious, but I have the dubious honour of having accidentally thrown away half my hard-collected grain-size analyses on some particularly smelly salt marsh muds that I studied for my dissertation. Hours of work somehow disappeared into a wastepaper basket somewhere. Many analyses have standard reporting sheets (many of which are now available as Excel spreadsheets – or could easily be produced in such a format); hand-held computers can be used to record many results to save writing time, and dataloggers are often available to store frequently collected data (such as tempera-

ture readings inside an environmental cabinet programmed to mimic a diurnal cycle in a desert). Many high-tech pieces of equipment record results to disc, which can then be downloaded and manipulated.

Finally, having collected your laboratory results you will have to analyse them using statistical techniques and present them graphically or in table format. This is where the laboratory work really starts to come to life, as this is the stage at which you are finally using the results to answer the questions you originally posed. Some techniques, such as grain-size analysis, have standard methods of data analysis and presentation for summarizing the data from one sample. For others you will have to develop your own methods. If this all sounds rather depressing, don't panic. There is an easy solution, which is a pilot study.

Doing a Pilot Study

A pilot study will help answer all the issues raised in the previous sections. Spend a few hours at a nearby fieldsite, similar to the type of environment you wish to work on, collect some samples and try some analyses in the laboratory. In many cases, where your proposed site is easily accessible, this pilot study can form part of the finished project if it is successful. If it is not successful, then you have saved yourself a large amount of wasted effort in both laboratory and field. If you are considering doing a project on leaf litter in the tropical rainforest, clearly it is going to be difficult to fly out to Amazonia, or wherever, for a day or two to do a pilot study. Even in this case, a pilot study in a nearby woodland would at least get you thinking about the issues involved (albeit probably at a smaller scale!) and allow you to practise laboratory techniques on similar types of samples. Pilot studies enable you to resolve issues of field sampling methods, numbers and size of samples required, the difficulty and time required for different laboratory techniques, and more fundamen-

Table 39.1 Examples of successful dissertation projects involving laboratory work

Project aims	Fieldwork	Laboratory work
To compare the nature of a set of natural and nearby artificial beaches, in order to assess the success of the artificial beaches.	• Survey beach profiles. • Discuss artificial beach design with engineers. • Collect representative samples of natural and artificial beach material.	• Grain-size analysis • Mineralogy and shape analyses
Disturbance and vegetation change in a Scottish catchment.	• Collect core from suitable lake/bog.	• Pollen analysis on different levels of the core
How far does heavy metal pollution from a mine spread down an ephemeral river valley?	• Survey of river deposits down valley. • Collection of sediment samples.	• Heavy metal contents of sediment samples • Grain-size analysis
Does water in pools on a coastal limestone platform cause dissolution of the limestone?	• Map pools. • Monitor pool-water chemistry and volumes over tidal cycles. • Collect water samples.	• Analysis of calcium contents of water samples

tally, whether you actually enjoy this sort of research.

Suggested Projects

To finish, table 39.1 contains examples of the sorts of dissertation projects which utilize laboratory techniques to help answer geographical questions. As you can see, in some cases only one laboratory technique is used, in others a whole suite of methods is needed. In some projects laboratory work is a small adjunct to fieldwork, in others the laboratory programme forms the entire project.

Further Reading

Allen, J.R.L. 1985: *Experiments in Physical Sedimentology*. London: Allen and Unwin. Some good suggestions for experimental studies of interest to physical geographers.

Goudie, A.S. (ed.) 1994: *Geomorphological Techniques*, 2nd edition. Oxford: Blackwell. Information on a whole host of laboratory techniques and methods for many different types of geomorphological investigation.

Watts, S. and Halliwell, L. (eds) 1996: *Essential Environmental Science: Methods and Techniques*. London: Routledge. Chapter 1 on the good scientist and chapter 5 on general laboratory equipment and techniques are particularly useful as general introductions.

40

Questionnaire Surveys

Gary Bridge

'I Know! I'll Do a Questionnaire!'

It's close to midnight and you're in a sweat. Tomorrow you have to hand in the proposal for your undergraduate dissertation. You know that you want to do something on the increasing socio-economic status of many inner London neighbourhoods (a process known as gentrification), but you're not sure how to do it. You must prove to the examiner that you have done some actual fieldwork. Then it comes to you in a flash of inspiration (or perhaps because you can't be bothered thinking about it any more) – 'Of course! I'll do a questionnaire!' You collapse into bed for a contented night's sleep.

So far you have made two crucial errors. First of all, you have only defined an area of research and not a specific research problem. What is it that you want to know about gentrification? What is your research question? For example, you might ask why gentrification occurs where it does, or who is doing the gentrifying and why, or what are the feelings of working-class residents about the social change taking place in a neighbourhood. Each of these questions would require a different research method and different sources of information.

The second crucial error, and this often applies even where the research problem is well defined, is the assumption that a questionnaire is the best method available. That is because it has come to be associated with

so-called 'hard' social science involving large surveys and statistical analysis. All too often the questionnaire is seen as the cure-all for the problem of doing fieldwork. However, a questionnaire is only effective when it is the *most appropriate method* of providing the information needed to address a *well-defined research problem*. Let us say, for example, that you are interested in finding out whether gentrification is a back-to-the-city movement of the suburbanized middle class or a within-city movement of middle-class residents who are choosing not to suburbanize. You will have to know where gentrifiers have come from (i.e., their previous addresses). A questionnaire survey of residents in the Docklands in London might seem like the obvious way of getting the information. However, other sources of information may be available. For example, you may be allowed access to local estate agents' records which will list purchasers' previous addresses. This would be a simpler and less costly way of getting the required information.

Even when it has been established that a questionnaire survey *is* the most appropriate method of gaining the information to address the specific research question, success is not guaranteed. The success or failure of a questionnaire survey is determined by three things: (1) the sampling theory, i.e., are you asking the right people; (2) questionnaire design (wording of the questions, layout of the questionnaire, etc.); and (3) analysis and interpretation of the results.

'A survey is a method of collecting information directly from people about their feelings, motivations, plans, beliefs, and personal, educational, and financial background' (Fink and Kosecoff 1985). Surveys can take the form of questionnaires or interviews. Questionnaires are distinguished from interviews by the fact that they are either self-administered (i.e., filled in by the respondents themselves, as in a postal questionnaire) or filled in by a researcher (in person or on the telephone) with no prompting or interaction with the respondent other than to ask the questions themselves. Interviews, in contrast, whether formal or informal, involve more dialogue between interviewer and respondent. Interviews tend to delve more deeply into people's attitudes, beliefs and feelings (see chapter 41 by Jacquelin Burgess in this volume). They usually involve qualitative analysis of the information gained (often in the form of case studies), whereas the information gained from questionnaires is usually subjected to quantitative analysis involving statistics.

Sampling

A questionnaire survey starts with the definition of the population of interest and procedures for contacting a sample of that population, and ends with the analysis of the data from the questionnaires.

When conducting a questionnaire survey it is seldom possible to question all the members of the population of interest. In the example above it would be too expensive to question all the residents of the Docklands. A sample of the population must be taken. To do this you must have a clear notion of the population of interest. Unless you know that, the sample is meaningless. Ask yourself, 'Who should be asked and how do I contact them?' What is the sampling frame that is appropriate to the population you are interested in (e.g. electoral rolls, telephone directories, trade directories, tax registers)?

Does the sampling frame fairly represent the study population? For example, in Britain the electoral register is the traditional way of sampling residents in a neighbourhood, but it only records those residents who are over the age of 18 who have registered to vote. The most appropriate way of selecting households or individuals from your sampling frame is determined by sampling theory. The aim of sampling theory is to avoid bias and ensure that your sample is as representative of the total population of interest as possible. There are a number of sampling methods, and by consulting the textbooks you will be able to decide which is the most appropriate for your study. The size of the sample is also important. It is necessary to allow for non-responses, especially in postal questionnaires. The general rule on sample size is 'the bigger the better', and the upper limit is likely to be set by practicality, e.g. how much postage you can afford, how many streets you are willing to walk. For most statistical tests the minimum sample size is 30.

All these questions can be resolved by applying common sense to your particular research question and by consulting the textbooks (especially Dixon and Leach 1978, for an introductory guide; Fink and Kosecoff 1985, for a straightforward account; Moser and Kalton 1971).

Finally, if you are doing a residential questionnaire, which member of the household is to answer the questions? If you are doing such a questionnaire in person, unannounced, pay a visit to the local police station before you start, so that they know you are in the area, and always carry identification.

Analysis and Interpretation

You will have to follow the procedures laid out in the textbooks (especially Wrigley 1985; Clark and Hosking 1986) to help you process, analyse and interpret results of the survey. You might have a fantastic questionnaire, with a high response rate, and then ruin your study with poor analysis. Do not necessarily leap for the most sophisticated software. Statistics must be used critically. Within limits there is probably a statistical test that will do anything you can think of. The problem is knowing what it is you are looking for rather than knowing about the statistics. When you have got the problem straight, then you can look up the appropriate statistics in a textbook.

Questionnaire Design: The Difficult Middle Bit

As I have argued, sampling and analysis are dependent on getting the research problem straight and then referring to the appropriate textbooks. The most difficult part of a questionnaire survey is stage (2), collecting the information, and so this stage will occupy the rest of the discussion.

The form of your questionnaire will differ according to whether it is postal or interview. Deciding between the two will probably be determined by the nature of your study and practical limitations. For example, if potential respondents are scattered all over the country, then a postal questionnaire is the only practical method. If you are in a position to choose between postal and interview questionnaires bear in mind the following pros and cons of each method.

Postal questionnaire: pros and cons

PROS

1   It cuts down on travelling time and legwork.
2   There is no interviewer bias.
3   It gives the respondent time to answer difficult questions.
4   It is good for personal and embarrassing questions.

CONS

1   The questions must be easily understandable and unambiguous – this requires a lot of work in the design of the questionnaire (see below).
2   The answers cannot be rechecked with the respondent.
3   There is no respondent spontaneity.
4   The respondents can see all the questions before answering, which gives them an insight into the line of your questioning, and they may therefore tailor their responses to fit your reasoning, so biasing the responses.
5   Who is answering? Even if you are specific in your instructions about who should answer, you can never be sure they have been followed.
6   Supplementary observational data are not available. If you are actually there you can see what the respondents look like, how they respond to the questions and what their environment is like. This is useful contextual information for the survey.
7   There is a low response rate and waste of resources. Response rates for postal questionnaires tend to be low. A response rate of 30–40 per cent from a survey of residents in an ordinary neighbourhood is considered good. Rates may be higher if you are surveying a particular interest group (e.g. other geography students) or if there is something in it for the respondents (e.g. you are using the information to promote their grievances or offering gifts for responding). This last possibility is an unlikely one for a geography student.

To achieve the absolute minimum requirement for statistical analysis, a sample size of 30, you would need to send out at least 100 questionnaires, given average response rates. That means at least 200 postage stamps – two for each respondent, one for the outgoing questionnaire and another on the self-addressed envelope enclosed with the questionnaire so that the respondent can send the completed questionnaire back to you. You may also want to send out reminders after a couple of weeks or so. It is easy to see how the costs mount up. If you can hand deliver or collect all or some of the questionnaires this will help reduce costs but of course will be a drain on your time. It is important to give people sufficient time to fill in the questionnaire before sending reminders. You should allow them at least one weekend.

As well as the self-addressed envelope the postal questionnaire should be accompanied by a covering letter from your academic institution. This letter should explain who you are, the purpose of the survey, how they have been selected for the survey and the reason they are being approached. A guarantee of confidentiality is also essential. Don't be officious but, equally, don't be apologetic. An example of a covering letter is given in figure 40.1. It is not a formula to be followed rigidly, but the tone is important.

Good layout of the questionnaire and the ordering of questions is essential, especially for a postal questionnaire, and this will be discussed later.

Questionnaires administered by the researcher: pros and cons

PROS

1   There is a higher response rate than for postal questionnaires.
2   Fewer resources are needed, provided that travel is minimal.
3   You may get positive feedback on the design of the questionnaire, so that you can adapt the design as you go.
4   It provides the chance to clarify the questions.

CONS

1   It is time-consuming.
2   It may be inconvenient for the respondent.

3   You may invade a person's privacy.
4   There is the problem of you, the interviewer: commercial polling organizations rarely employ students since they are the last people likely to get a sympathetic response. Don't let this last point put you off. Good interviewing depends on the personality of the interviewer as much as age or sex. It also depends on a well-constructed questionnaire which, as already mentioned, is crucial for postal questionnaires too. So we now turn to the issue of questionnaire design.

Questionnaire Design: Committing Yourself to Paper

There are no hard-and-fast rules for questionnaire design. It will vary according to the nature of the topic and the people who are to be canvassed. Thus a well-crafted questionnaire is a product of a clearly defined research objective, a sensitivity to the potential respondents, trial and error (using a pilot survey or, if that is not possible, by passing the questionnaire around friends, family and colleagues to make sure that they understand the questions in the same way that you do) and, the most valuable commodity of all, *common sense*.

Although there are no golden rules for questionnaire design, there are some handy hints based on the past experiences (and mistakes) of other researchers.

Asking the right questions

Questions of content are made much easier if you have thought about your objectives and about the final method of analysis. One of the biggest mistakes made in dissertations is the reliance upon meaningless questions. How do the terms in the question relate to the abstract categories of your analysis? Have you included all the necessary questions and are all the questions necessary?

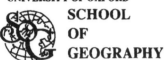

UNIVERSITY OF OXFORD

SCHOOL
OF
GEOGRAPHY

School of Geography
Mansfield Road
Oxford OX1 3TB
England

Tel: (0865) 271919
Telex: 83147 VIA.OR.G
Fax: (0865) 270708
(attn: School of Geography)

Direct line:

                                    30th March, 1990

        Dear Resident,

                I am a Geography student at Oxford University
        and am currently doing research on the social changes
        occurring in the Sands End area.

                The views and experiences of local residents,
        such as yourself, are a crucial part of the research. Your
        address is one of a number that have been chosen on a chance
        basis. It would be of great help if any <u>one</u> member of your
        household, aged 18 or over, could spend a few moments
        filling in the brief questionnaire enclosed.

                <u>All the information you give will remain
        anonymous and confidential. It will be covered by the Data
        Protection Act</u>.

                A stamped/addressed envelope is provided for
        you to return the completed questionnaire by post.

                I hope you can find the time to help me with my
        research.

                Yours faithfully,

JANET address: GEOGMAIL @ UK.AC.OXFORD.VAX

Figure 40.1   Specimen letter of introduction

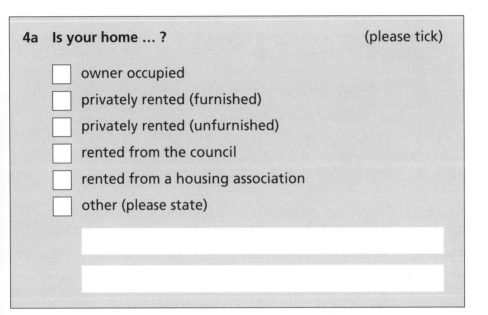

Figure 40.2 An example of a closed question.

Asking the right questions in the right way

There are four main considerations here: (1) the type of answer required, (2) the words themselves, (3) bias and (4) ambiguity.

(1) *Is the type of answer required fact, opinion or attitude?* Different question formats will be appropriate for the different types of answer needed. Closed questions (like multiple choice questions in an exam) are often suitable for factual questions (see figure 40.2). Closed questions have the advantages that they are quicker for the respondent to fill in, are more precise and are easier to analyse. They are also easier to code. Coding means giving a number to each of the possible responses so that the answers can be fed into the computer. Instructions on coding can be found in any survey textbook.

Attitudes can sometimes be recorded using a closed-question format in the form of a rating scale. Rating scales come in various guises (e.g. nominal, ordinal, interval,

graphic, comparative, additive) depending on the sophistication of the information required. The tenure question (figure 40.3) is an example of a nominal or categorical rating scale. Again, this is textbook stuff. It is important to note that the form of the answers, whether categorical (yes/no), continuous (age) and scaled (using diagrams), has relevance for the type of statistics that can be used. An example of a rating scale to capture attitudinal information is given in figure 40.3.

Sometimes questions have no obvious answers or you may want the respondents to answer in their own words. In this case open-ended questions must be used. These can always be coded afterwards. An example of an open-ended question is found in figure 40.4.

(2) *The words themselves.* The second element of good questioning is using the right words. Keep them simple. Use everyday words that have immediate meaning to people and avoid jargon and

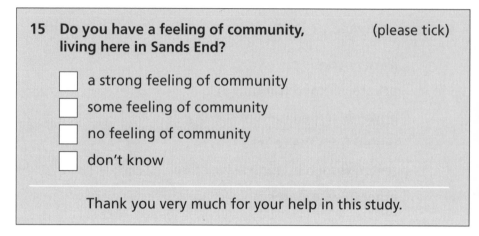

**15   Do you have a feeling of community,   (please tick)
living here in Sands End?**

☐ a strong feeling of community

☐ some feeling of community

☐ no feeling of community

☐ don't know

Thank you very much for your help in this study.

Figure 40.3   An example of a rating scale to capture attitudinal information.

specialized words (unless you are surveying a specialized group of people where those words have specific and acknowledged meanings). Even apparently simple terms like 'friend' and 'community' may mean quite different things to different people. In general, do not use a complicated word where a simple one will do; for example, 'live' is better than 'reside'.

(3) *Bias.*   Certain names, places or phrases are emotionally charged and they can unfairly influence questionnaire responses. For example, a questionnaire of geography students might ask:

1   Would you attend a lecture given by Dr Spock?
2   Would you attend an 8:00 a.m. lecture given by Dr Spock?
3   Would you attend an 8:00 a.m. lecture given by Dr Spock, the expert on soil profiles?

Options 2 and 3 add more information, but they may also bias the answers.

Another source of bias is when you as researcher are unaware of your own position on a topic. You need to check for this by showing the questionnaire to your friends and family, and people you know less well, to get a range of reactions to the questions. A fairly blatant example of researcher bias in the gentrification study might be the question, 'in what ways do you think yuppies have ruined the neighbourhood?'

Bias may also be introduced by asking questions that are too personal. Asking the respondents 'How much do you earn?' may bias answers upwards, or, at worst, put respondents off altogether. Often alternative formats can be used to cope with questions that are too personal. In this case listing a number of income brackets (£000–£9,999; £10,000–£19,000) for the respondent to indicate which band he or she falls into would be a more sensitive way of asking the question.

(4) *Ambiguity.*   Ambiguous questions are usually ones that contain more than one thought. For example, 'do you think that the local government should cut its education or sanitation programmes?' This question is vague as well as ambiguous. It contains two thoughts (cutting education and sanitation). It is also not clear whether it is asking whether education in general or sanitation in particular should be cut. The easiest way to avoid ambiguous questions is to follow

This last section asks about your opinions
of Sands End as a place to live.

**13   What are the advantages and disadvantages
of Sands End as a place to live?**

Please make your answers as full as possible.

Advantages

Disadvantages

Turn over

Figure 40.4   An example of an open-ended question.

the rule of 'one thought per question'. More than one thought requires additional questions.

Asking the right questions, in the right way, and in the right order

Ordering of questions is important, especially for a postal questionnaire in which you

need to grab potential respondents' attention and keep it without putting them off. Easy factual questions usually come first. This helps respondents relax into the questionnaire and, often, helps you establish who they are. More complicated material should come later. Sensitive questions should be put towards, but not at, the end. Questions should follow a logical order. Where you

Q21d How often do you see this person? (please circle)

daily / weekly / monthly / yearly / rarely

Q21e Where do you usually meet?

Q22a Think of your favourite evening's entertainment. Who (other than your wife/husband or partner) would you most like to be with you on such an evening?

Q22b Which of the following terms best describes this person? (please circle)

relative / co-worker / neighbour / friend / acquaintance / member of same organization

Q22c Where does this person live? (please circle)

Sands End / elsewhere in Fulham / elsewhere in London / elsewhere in UK / abroad

Q22d How often do you meet this person? (please circle)

daily / weekly / monthly / yearly / rarely

Q22e Where do you usually meet this person?

Q23 What are the advantages and disadvantages of Sands End as a place to live? Please make your answers as full as possible. Advantages

Figure 40.5 An example of a poorly laid-out questionnaire

8e **Where does s/he work?**

work establishment   location (street / district / country)

9a **Do you have any children?**

☐ yes – continue Q9   ☐ no – go to Q10

9b **How old is/are your child/children?**

10a **Do any members of your family, friends or relatives live in Sands End, apart from those who live with you?**

☐ yes – continue Q10   ☐ no – go to Q11

10b **For those people connected to you who live in Sands End but not in your home, please state the type of the relationship in each case (e.g. brother, cousin, friend)**

11 **Which of the following 2 statements**   (please tick)
below comes closest to your opinion?

☐ The most important job for the government is to
make certain every person has a decent steady job
and standard of living.

☐ The most important job for the government is to
make certain that there are good opportunities
for each person to get ahead on their own.

Turn over

Figure 40.6   An example of a well-laid-out questionnaire

have to change a line of questioning or where you are asking background questions that respondents might feel are unrelated to the central topic, explain briefly why you are doing this. In general proceed from the most familiar to the least.

Questionnaire layout: you've got the look

Good layout is essential for a postal questionnaire. It must be clear which questions are to be answered and how they are to be answered. Do not try to save paper. Again, common sense is the best guide. Questionnaires that are cramped, with poorly defined sections, are likely to be binned. One such poorly laid-out questionnaire (figure 40.5) and one with a better layout (figure 40.6) are given as examples. Questions are in bold type and the level of detail required in each answer is directly specified. Rarely will all questions apply to each respondent. Filters must be used. Filters are used in figure 40.6. Arrows are also useful in guiding respondents through filters. However, they can only operate successfully if the questions they are directing the respondent to are actually on the same page.

Underlining sections, or even single questions, gives a compartmentalized visual image that looks tidy. It also gives the respondents a greater sense of accomplishment as they complete each section. Don't forget to ask respondents to turn over the page. There is nothing more frustrating than receiving a carefully completed questionnaire with the back page blank because the respondent did not realize that he/she had not finished. Number each page for the same reason.

Length. Up to a certain point the length of a postal questionnaire does not seem to be a serious deterrent. It is whether the questionnaire interests the respondent and looks good that counts. It should be as short as reasonably possible, all other things being equal. This requires precise questioning and relevance at all times (with the exception of dummy questions used to soften up respondents for difficult questions). It is possible to be too short, through peremptory questioning or trying to cram questions into too small a space. This will only irritate respondents. Precision and relevance are the watchwords.

Length may appear to be less of a problem for interview questionnaires, but you will probably find that you will often be asked, 'How long will this take?' Don't understate the length. If you feel panicky and are afraid of losing a respondent, say that you can end the questioning whenever he/she wishes. Respondents rarely cut you dead half-way through unless they really have to go or unless, of course, you have offended them or bored them rigid.

And Finally...

Questionnaire surveys should be enjoyable. You are conducting original research and discovering more about people. You will probably be pleasantly surprised by how cooperative they are. And remember, the best questionnaire asks precise questions, of the right people!

References

Clark, W.A.V. and Hosking, P. 1986: *Statistical Methods for Geographers*. New York: Wiley.

Dixon, C.J. and Leach, B. 1978: *Questionnaires and Interviews in Geographic Research*. Concepts and Techniques in Modern Geography 18. Norwich: GeoAbstracts.

Fink, A. and Kosecoff, J. 1985: *How to Conduct Surveys: A Step by Step Guide*. London: Sage.

Moser, C. and Kalton, G. 1971: *Survey Methods in Social Investigation*. London: Heinemann.

Wrigley, N. 1985: *Categorical Data Analysis for Geographers and Environmental Scientists*. London: Longman.

Further Reading

Babbie, E. 1990: *Survey Research.* Belmont, CA: Wadsworth.

Oppenheim, A. 1992: *Questionnaire Design, Interviewing and Attitude Measurement.* London: Pinter.

Peterson, R. 2000: *Constructing Effective Questionnaires.* London: Sage.

Rogerson, P. 2001: *Statistical Methods for Geography.* London: Sage.

41

The Art of Interviewing

Jacquelin Burgess

Street Corner Society, first published in 1943, is one of the classic urban ethnographies. In the Appendix, William Whyte discusses some of the problems he encountered in doing the field research, including difficulties in getting people to talk to him about sensitive issues. His key informant, Doc, gave him some timely advice.

> Go easy on that 'who' 'what' 'why' 'when' 'where' stuff, Bill. You ask those questions and people will clam up on you. If people accept you, you can just hang around, and you'll learn the answers in the long run without ever having to ask those questions. (Whyte 1955: 303)

Doc was right, of course, but the problem facing undergraduates who are required to carry out research for a dissertation or extended project is one of time. You simply do not have enough time to hang around in the field, working on being accepted by the group you want to study. So I write this chapter in the expectation that you have probably already left things just a little late – and we shall indeed consider some of the basic *why, who, what, how* and *then what* questions of interviewing. Research in human geography embraces both quantitative and qualitative methodologies. Interview methods include one-to-one conversations between researcher and informant, and small group interviews or focus groups, where the researcher will facilitate discussions between six to eight people. I shall discuss both ap-

proaches in this chapter but will concentrate more on one-to-one interviews since these represent a basic methodological tool for qualitative research.

Qualitative field research has a long and respectable tradition in sociology and anthropology. Within geography it is most closely associated with cultural and social geography (see Limb and Dwyer 2001). The basis of qualitative research is an emphasis on people as creative human beings who act in the world on the basis of their knowledge and understanding of the society and the institutions which structure their lives and which, in turn, they are able to shape and change. If that sounds a bit of a mouthful, what it means in practice is a commitment to understanding people's experiences through listening to the ways in which they describe and account for aspects of their lives and activities. Not surprisingly, therefore, qualitative methods provide more complex interpretations of feelings and actions than do quantitative studies. The data are usually linguistic rather than statistical; contextual rather than cut out from everyday life; and researchers are engaged with their informants rather than separated from them, as in a questionnaire survey.

What do you need to be a good interviewer? In the jargon of American sociology, 'successful interviewing is not unlike carrying on an unthreatening, self-controlled, supportive, polite and cordial interaction in everyday life' (Lofland 1971: 90). In plain

English, if you like talking to people – you don't go around shouting at them, putting them down, ridiculing them or not listening to what they say – you have the potential to become a good interviewer. The art of interviewing is to be able to conduct a conversation in such a way that the person you are talking to is able to freely express her or his opinions and feelings while, at the same time, enabling you to meet your own research objectives. It is not uncommon to come away from a fascinating conversation with someone about life, the universe and everything, to use Douglas Adams's famous phrase from *The Hitch-Hikers' Guide to the Galaxy*, only to realize that you still do not know whether that individual really does believe that the answer is 42! So the goal is to achieve an end result which satisfies both of you, informant and researcher. What follows is some practical advice on how to achieve that result.

Why Use Interviews as the Basis for Dissertation Research?

In order to reach your decision, you must consider the following points:

- the nature of your research proposal; your objectives;
- alternative methods of achieving your objectives;
- the constraints that will affect the ways in which you can achieve those objectives.

If your research proposal is concerned with aspects of human geography which require that you make interpretations of the feelings, values, motivations and constraints that help shape people's geographical behaviour, or you need to understand the discourses through which different social groups make sense of complex issues, then interviewing may be an appropriate research technique. Whatever the topic, you need to consider

whether interviews would provide the best way of achieving your objectives. Would a questionnaire survey be more appropriate? The answer will be no, if you want to concentrate on how individuals describe and account for their own experiences; if you want to understand the complexities of the problem rather than reducing it to a set of key explanatory variables; if your aim is to undertake a case study rather than a representative sample of a wider population.

Among the major constraints you will face will be those of *access* and *time*. Can you get hold of the people you want to talk to? It is often very difficult, for example, to gain access to members of elite groups, such as the managing directors of multinational companies or the very rich. Time is also important. If it will take a year to win the trust of the community you want to work with, then it may not be possible to continue with the project. The best kind of qualitative research builds over time with the researcher moving backwards and forwards from the field to the interpretation of data and back into the field for more interviews, informed by the experiences of what has gone before. Bear in mind that the transcription and interpretation of interview data take much longer than for statistical data; writing up is a more creative and interactive process than that associated with quantitative analysis. Students often ask how many interviews are 'enough'? It is a balance between resources (time, money) on the one hand, and intellectual content on the other. All qualitative researchers can describe the moment when they stop hearing new stories; in other words, the bounds of the subject have been reached.

What Kind of Interviews Will You Conduct?

Having done the background reading and formulated your research objectives, you will now need to decide:

- who you will interview;

- how you will make contact;
- what kind of interview you will conduct – formal or informal; one-to-one and/or focus group.

The selection of interviewees will be determined theoretically by the nature of your project and practically by the relations you are able to establish in the field. Let us take an example. The question of fox hunting has dogged New Labour since it came to power in 1997, stimulating some of the largest protests about countryside politics witnessed in London for many decades. You are interested in the cultural politics of the issue. On a theoretical level, you know that ideological distinctions between 'city' and 'country' are embedded in English culture, and that hunting is a practice which expresses the material bases of class and power that bind city and country together. Furthermore, the animal rights movement and the hunt saboteurs are engaged in actions which mobilize around ethical issues but which also have class-based politics at their core. Media coverage has been intense – great pictures of 'the country coming to town' as Barbours and Green Wellies march down Oxford Street.

You may decide to study the question from the rural perspective and focus on a locality where hunting is part of the way of life. Or you may be more interested in understanding the nature of protest from the perspective of the animal rights activists. In either case, your aim will be to adopt a strategy of *theoretical sampling*: your choice of informants is made on this basis rather than on random sampling from a whole population. If the topic is a sensitive one, it may be better to interview individuals on a one-to-one basis. On the other hand, if you are most interested in listening to people debate among themselves about the reasons why they believe hunting is right or wrong, then running a focus group with hunt protesters might be better. The number of people you interview or the number of focus groups you run is less important than the quality of in-

Figure 41.1   The country comes to town.
Photograph: Richard Watt Photography.

formation you gain from your interviewees. As you proceed, you may well find that new ideas and issues are emerging from the field research, and that the right to hunt is actually subordinate to a much bigger argument about rights and obligations, especially how individual freedoms – to engage in a sport which has only very recently been socially constructed as 'cruel and unnatural', or to engage in active, political protest without being threatened by police in riot gear and arrested on a 'terrorism' charge – are being renegotiated.

Normally, you should first contact your interviewee by letter, by telephone or through a personal introduction to make an appointment. With ethnographic research generally, the best idea is to make an initial contact with someone who might be able to provide you with other introductions – these people are often members of community-based organizations, for example. This kind of contact technique is often described as *snowballing* – inevitably, you find that you get passed on and the list of 'people you really should talk to' grows at quite an alarming rate. Say who you are and who recommended that you talk to the interviewee. Say what you are doing; try to make it interesting to the other person – why should someone want to give up their time to talk to a geography student? But don't make any rash promises about being able to change the world as a result of your dissertation findings. You will also need to think about problems of confidentiality – people will need to be reassured about what will happen to the information they might give you. If you are interviewing one-to-one, you need to decide where to conduct the interview – in people's homes or offices, or in a public space? If running a focus group, can you find a space such as a room in the local library or community centre where people can come together?

Qualitative researchers normally make distinctions between *formal* and *informal* interviews (Burgess 1984; Limb and Dwyer 2001). Formal interviews most closely resemble questionnaire-based interviews in that the researcher has a clear agenda of issues that he or she wishes to cover. These will usually be written down beforehand, not as set questions but as important topics to be discussed. Occasionally, it is useful to send your interviewee the schedule before you meet to give him or her time to prepare. This also saves time – a useful point if you are dealing with very busy people. Similarly, focus group sessions are designed around an agenda of topics to be discussed by the group which are introduced by the facilitator. Formal interviews differ from questionnaires in one very important respect. The sequence of topics covered in individual and group interviews is determined through the interaction of researcher and informants. The important thing is to go with the flow of the interview, being flexible about the order in which topics are discussed but making sure that, by the end of the meeting, you have met all your objectives. Formal interviews give you the security of knowing that you have covered the full range of issues with all your informants. By contrast, informal interviews much more closely resemble ordinary conversations. The aim is to discover how individuals describe and make associations between different kinds of ideas and experiences. These interviews tend to be much longer and more tangential to the problem. But informal interviews can be full of insights into the life and personality of the person you are talking to.

How to Conduct a Successful Interview

We need to think about three issues here:

- the interpersonal skills you need to conduct the interview;
- different ways of asking questions;
- the recording of information.

Interviewing skills should be learned and practised before you go into the field. They are as much part of the geographical repertoire of research techniques as learning how to use a depth-integrating sampler. The skills needed for individual and group interviews are the same. Taking *interpersonal skills* first, the fundamental goal is to create a rapport between yourself and the interviewee(s). If they like you, find talking to you a pleasant and interesting experience and trust you, then the interview will go well. You can do several things to ease the transition from 'stranger' to 'friendly acquaintance'. Prepare yourself for the interview by being sensitive to the expectations of the person you will be meeting. It is not a very good idea to arrive for an interview with the chief planning officer of a local authority in worn-out jeans and dirty trainers. Neither would it really be appropriate to interview adolescents in the neighbourhood gang in your most formal clothes. Communication covers much more than the language we use in talking to one another. Non-verbal communication through our body language is just as powerful. Think about your posture in the interview. Are you sitting hunched up with arms and legs crossed, clutching your notebook and pencil as if your life depended on them? Your nervousness will communicate itself to your interviewee, who may begin to wonder what is wrong with them – or you. Be relaxed in your posture but continue to convey interest and attention. Make lots of eye contact with the other person – acknowledge that they exist – but also be careful that it does not get out of hand and you end up interrogating them with piercing stares or inviting them to bed! Practise a range of facial expressions in the mirror or on a friend and see how you express interest, pleasure, confusion and uncertainty. See how you take the initiative in asking questions or changing the topic of conversation. In a group context, use your eye contact to engage with all the members of the group, especially those who look as if they need drawing into the discussion.

In the interview itself, you need to think about two related issues: the ways in which you *phrase your questions* and the *pace of the interview*. It is possible to ask questions in different ways that will give rather different kinds of answers. *Closed questions* with the familiar 'what, where, when, how often, how much, who, why' require that the informant give you pieces of information and often leave all the initiative with you. *Open questions* such as 'tell me about…' and 'in what ways do you feel…' are invitations to encourage communication. A good interviewer will use both kinds of phrasings. Perhaps even more importantly, a good interviewer will learn to listen not only to what he or she is being told, but also to how it is being said and what lies underneath the remarks. Listening closely enables you to handle the different kinds of silences which arise in conversations and to pace the interview. When I began my research career, I conducted interviews with local government officers. Playing the tape-recordings afterwards, I was dismayed by the number of times I jumped in with the next question

rather than giving the interviewee time to develop his or her point. The majority of inexperienced interviewers find silences difficult to deal with. It is a reflection of their own anxieties about the interview. Try to identify what kind of silence you are dealing with and then respond accordingly. For example:

- Is it a thoughtful silence? In which case, make the appropriate 'mms' and 'aahs' to ease the interview on.
- Is it a stuck silence? The informant is having difficulty with the question you have asked, perhaps. In which case, rephrase, recapitulate what they have just said or clarify with an example.
- Is it an embarrassed silence? Maybe something has been said or asked which should not have been. If you made the error, say that you hadn't realized that it would cause difficulties, apologize and move on to a safer topic. In general, be sensitive to the feelings of your interviewee and tactfully change the subject if necessary.

The third issue is working out an appropriate method of *recording the answers* to your questions. Clearly, if you are frantically scribbling down everything that is being said to you, you are not going to be able to develop a good rapport with your informant who, in turn, is likely to become more self-conscious and anxious about what you might be writing. Tape-recorders would seem to be the obvious solution, but beware. Many people will simply refuse to be 'on the record' or they will severely censor what they say to you. Others will become acutely embarrassed. Some years ago, our first-year fieldclass was held in north-east England and one project was based on recording the oral histories of retired miners and their families. In setting the project up, I tried to record an interview with an elderly man who had a broad Geordie accent. The session was going very badly, despite the fact that we had previously struck up a very good relationship. I stopped the tape and asked if it was a problem for him. 'Yes, it bloody well is,' he said. 'Your students in London will listen to me and think I'm just an ignorant nobody because I don't talk like them.'

Good practice is to use a reporter's notebook and just jot down key words or phrases, often called *scratch notes*, during the course of the interview. Once the interview has finished, *as soon as possible* find a quiet spot and write down everything you can remember about the course of the interview. Review in your mind precisely how it went, the sequence in which topics came up and what was said. You will find that you can remember very much more than you think you can. Then, once you get back to base, type up a full transcript of the interview. Do not forget to include date, time, name of person, address and telephone number. In a focus group, it is impossible to maintain the flow of discussion, take scratch notes and remember what everyone has said. It is essential to tape-record the meeting, gaining the participants' permission first. The vital point here is to listen to the tape immediately after the session and produce a *running order* of speakers, i.e., identify each person from their voice and the first few words of their contribution. Then, when you come to transcribe the discussion fully, you will have a record of who spoke when. It is extremely difficult to remember people's voices, especially if you do not have any previous acquaintance with them.

In both cases, debrief yourself. Add in other details which will help you later to remember the interview. How did it go? Did you have any problems? What was the person or the group like? Which questions worked well? Was there anything surprising that you hadn't thought about before? This procedure will take a considerable amount of time, but you will gain from it when you come to interpret the data and write your dissertation.

And Then What?

The analysis of qualitative data is not easy and it is the most difficult aspect of the research procedure to describe succinctly. Good discussions of different approaches can be found in Jackson (2001) and Crang (1997). Prepare your transcript by entering sequential line numbers for the whole text – including your questions, and have a very wide margin on the right-hand side of the page to allow you to enter comments and codes that summarize what is being said. Interview analysis proceeds through the development of *coding frames*, which summarize and then fracture the text analytically. Coding frames develop from your original research questions, your understanding of the theoretical concepts you are working with, and, critically, from the interviews themselves. This latter way of coding is often described as *grounded*, i.e., it comes from the data, and contributes to the development of 'grounded theory' (Strauss 1987). Within your interview transcript, you will be able to distinguish factual information from opinions and feelings; stories and anecdotes based on personal experiences from stories based on the media; details of when things happened, who was involved, why certain decisions were made and by whom.

Coding frames develop from reading and rereading the interview texts. At stage 1, you will probably want to code on the basis of factual information and answers to set questions, rather like producing a detailed index of content for each interview. The coding will allow you to trace the substantive elements of the interview and identify when and where themes/items occur across the transcript. Stage 2 will begin to develop more conceptual codes as you focus more closely on the meanings of what has been said. Some researchers will develop a coding frame for key concepts before they apply it to their stage 2 analysis; others allow it to emerge from the data. So, to return to the fox-hunting example, you want to explore the concept of 'rights'. Within the transcript of the fox hunters, there may be strong comments about 'interference from townies', 'the depredation of foxes destroying farmers' livelihoods', the 'importance of maintaining traditional bridleways by hunting across countryside', etc. On the other side, your hunt saboteurs talk about 'how cruelty to foxes is not acceptable', why it is 'wrong that humans can take pleasure from killing animals', about the ways in which new laws about protest 'makes us all criminals'. In developing the stage 2 analysis, the coding frame will identify these different kinds of expressions of 'rights' and begin to make finer connections and distinctions both within, and between, interview transcripts.

The analytical process becomes more refined the deeper you get into it. As you progress, write memos to yourself, memos about the ideas that arise while you are carrying out the interpretation. Use your discoveries to inform subsequent interviews in the field, and to draft pieces of writing that will eventually become your first draft of the empirical material for the dissertation. By the end of the research experience, you will have been able to write an interesting, lively and insightful project that is grounded in the realities of everyday life – an example of genuinely humane geography.

References

Burgess, R.G. 1984: *In the Field: An Introduction to Field Research.* London: Allen and Unwin.

Crang, M. 1997: Analyzing qualitative materials. In R. Flowerdew and D. Martin (eds), *Methods in Human Geography: A Guide for Students Doing a Research Project,* Harlow: Longman, 183–96.

Jackson, P. 2001: Making sense of qualitative data. In M. Limb and C. Dwyer (eds), *Qualitative Methodologies for Geographers,* London: Arnold, 199–214.

Limb, M. and Dwyer, C. (eds) 2001: *Qualitative Methodologies for Geographers.* London: Arnold.

Lofland, J. 1971: *Analyzing Social Settings: A Guide to Qualitative Observation and Analysis.* Belmont, CA: Wadsworth.

Strauss, A. 1987: *Qualitative Analysis for Social Scientists.* Cambridge: Cambridge University Press.

Whyte, W.F. 1955: *Street Corner Society: The Social Structure of an Italian Slum,* 2nd edition. Chicago: University of Chicago Press.

42
Doing Ethnography

Pamela Shurmer-Smith

Ethnography is a term which geographers and others have imported from social anthropology, literally meaning 'writing about people'. These days, it generally means a particular mode of research which involves being in close contact with the people one is studying. However, if one goes back less than 20 years, it generally meant the body of *facts* about the social organization, beliefs and practices of distinctive groups of people. In this meaning, ethnography was descriptive and was contrasted with analysis and interpretation, which were theoretical and interpretive. The change in meaning is an important part of why ethnography is now such a rich and important element in social research.

Ethnography As Was

It seems incredible now, but in the past methodology was not generally taught to social anthropology undergraduate students and even postgraduates were lucky to get more than a few patronizing words of advice before embarking on their fieldwork in foreign societies. As an undergraduate I took a course entitled 'Central African Ethnography'. It involved knowing about tribal societies in that region and had nothing to do with fieldwork or research techniques. This lack of attention to methods and methodology was reflected in the way people wrote; it was considered bad form for anthropolo-

gists, constructed as dispassionate scientists, to write about themselves or use the first person (though sometimes in a map of a village one would find 'the ethnographer's tent' marked). Referring to experiences, emotions or insecurities was regarded as being overly popularist, resulting in nothing more than amateurish travel writing. There was an assumption (pretence?) that any diligent anthropologist would observe the same things in the same situation – of course, this could never be proven since there never were two anthropologists in identical situations. There was a conspiracy of silence about the problems of inserting oneself into a society that was not expecting to be studied, the difficulties of learning the language and appropriate behaviour, let alone the isolation, self-doubt and fear.

Anthropologists have long been identified on the basis of *participant observation*, their preferred research method, whereby the researcher becomes part of the research. This implies simultaneously joining in with the people being studied and being sufficiently detached to observe, take notes about and analyse what is taking place. I am sure that many readers recoiled from this description, thinking that anthropologists must be pretty tacky people, worming their way into other people's lives and then, sneakily, treating them like objects. But this reaction was not always normal.

Like geography, anthropology emerged from the British empire's need to understand the places it colonized, but the processes involved in understanding people are not the same as those of topographic survey and it rapidly became apparent that questions generated in the West can be incomprehensible or irrelevant when asked elsewhere. Anthropologists learned the hard way that they had to discover through local knowledge which were the important questions to ask. They could only do this by living in the societies they wanted to know. Given the assumption of racial supremacy and segregation underwriting imperialism, the participation element of participant observation seemed pretty radical to officials and white settlers. Anthropologists were members of the imperial power who lived for long periods with subject people who were denied a voice of their own; most came to feel that they 'belonged' in some way to both camps. In 1966 all of the lecturers in the Department of Social Anthropology at the University of Rhodesia were deported, partly on the grounds that they propagated knowledge that opposed the racist ideology of the white Rhodesian government, but also because they all had close relationships across racially segregated groups and this was seen as subversive. However, despite this liberalism, the underlying assumptions of colonial participant observation were condescending – it was always one-way, dominant people living with and then writing their own accounts of the lives of those who were dominated.

Sensitive anthropologists realized that ethnographic data were not just lying around 'out there' but were constructed by researchers out of their own participating and observing. They became increasingly uneasy and self-conscious about what was going on, particularly when it was revealed in the 1960s that research in Latin American societies had been part of the CIA's 'Operation Camelot', whereby participant observers were indirectly funded to reveal likely tension points. At this time some people refused to do fieldwork in former colonial settings, turning their attention to their own societies. I was one of these. Whereas sociol-

ogists tried to generalize about western societies, anthropologists brought participant observation back home to particularize and to try to see from different viewpoints. As a young woman with left-wing political sympathies, I spent four years in the 1970s getting inside the lives of elderly right-wing spinsters in a club in southern England. Anthropology 'at home' was, however, always seen as a poor relation of heroic foreign fieldwork.

The New Ethnography

In the late 1980s, what became known as the 'new ethnography' burst upon the academic world. The books edited by Clifford and Marcus (1986) and Marcus and Fischer (1986) were particularly important in arousing an awareness of a new problematic regarding the study of real living human beings. The new ethnography's insights were the culmination of long-standing unease and introspection but, using the language of post-structuralism, iconoclastic young American researchers reached well beyond an anthropological audience to claim that ethnography was less 'fact' than 'fiction' (i.e., made rather than given). They emphasized the processes of making through writing, highlighting the ethnographer as the author of the people written about – writing them into a particular existence. They talked of a poetics and aesthetics of ethnography and they demanded that the convention of the objective observer, writing according to scientific conventions, should be revealed to be a sham.

Suddenly ethnography was no longer mundane and atheoretical; it became central to the theorization of finding out (methodology), the theorization of representation and the morality of both of these. I hope you can see why this meant that it became reasonable to use the term 'ethnography' to apply to what researchers did, instead of the factual information they presented. Participant observation lost its naivety; the new generation of ethnographers were philosophically aware, reflexive and highly self-conscious about what they were doing and how they wrote it up. Acknowledging that objectivity was impossible, they prioritized the revelation of multiple subjectivities. It became important to recognize the right of people at the centre of research to be active partners (not just passive objects of study), including authorship. Whereas notions of scientific objectivity had formerly required ethnographers to hide in the third person, the new ethnography demanded that they come out as 'I' – positioned individuals with age, gender, sexuality, ethnicity, class, personal history and attitudes – so that readers could work out whose point of view they were exposed to.

Ethnography has become fashionable in human geography and other social sciences. As a fervent advocate of the method, I am delighted to see this, but I know that it is fraught with philosophical, practical, moral and personal problems. Like dancing or writing poetry, participant observation is easy to do (you just get on with it), but it is exceedingly difficult to do well. There are few recipes for success. The best way to learn is to read as many ethnographic monographs as possible, to read between the lines in order to work out what the ethnographer was doing, what was going right and wrong and what one would have done differently. Texts are most successful if the author has been open about positionality – their background, status and power – and context (see chapter 24 by Peter Jackson in this volume). When conducting one's research it is essential constantly to reflect back on one's interpretations of what is going on and to chart the way in which one's local understanding seems to evolve.

Participant Observation: Some Advice

One of the greatest problems with participant observation is that, because the method depends on the researcher's socialization into the group being studied, ethnography cannot be rushed. If you cannot spend at least three continuous months with the people you want to write about, you might as well forget about it. Less than this and the best you will come up with is the view the people themselves would like to project, tempered with one's own gut reactions; it will be subjective in the bad sense. It takes a long while to appreciate other people's subjectivity, to realize when they are lying or deliberately trying to impress, when they are joking or winding their ethnographer up (all these things are useful knowledge, but one needs to recognize them for what they are). It takes time to achieve a niche in another social group from which one can genuinely participate. It may be tempting to counter this by doing a study of a social situation one already knows, but here there are very real dangers that one will never be able to transcend one's own original position. Either way, the resulting work is likely to be trite.

Participant observation requires researchers to take risks with their own personalities and to be aware that they will be transformed by the experience – they must separate themselves from their own relationships to build a new set of (temporary) friends and enemies in a quite different context. They will swing between deep engagement and detached, almost clinical, assessment of situations. No matter how sympathetic, an ethnographer always knows that there is exploitation and it is difficult to ignore the inner voice that asks what sort of human being behaves this way. People often assume that confident outward-going people automatically make the best ethnographers, but I have found it otherwise; those who are introspective and worry about other people's reactions are often far more finely tuned to the nuances of social situations and write the most subtle accounts.

The new ethnography increased sensitivity in the doing and writing of research, but it also resulted in what many people regard as an unhealthy self-obsession on the part of researchers. It is best if I do not name names (if you read enough you will find them for yourself), but there are studies which could have the title, 'Me, my problems doing fieldwork and my even greater problems trying to write it all up'. Though it is impossible genuinely to understand the world from someone else's point of view, that should not be an excuse for not trying to reveal the contexts in which other people construct their lives. I believe that one should explain where one is coming from and then represent other people as honestly as possible without too much breast-beating.

Ethnographers need to establish viable roles for themselves, something which is normal within the society (and this rarely includes tape-recording interviews or writing notes in public). Phil Crang (1994) looked at service industry behaviour by taking a job as a waiter, Ben Malbon (1999) exhausted himself through excessive clubbing to find out about consumption, Katy Bennett (2002) learned about patriarchy by becoming a live-in nanny. They were not acting. They assumed genuine ways of being and had to learn how to be. This includes thinking about dress, movement, speech, how to make contact with people. The danger is that one can be so drawn into successful participation that one forgets to be detached enough to observe. There needs to be some means of jotting down records which is not obtrusive (Phil Crang scribbled them on his order pad, but not everyone is lucky enough to have a role that gives them this opportunity). One also gets so exhausted that it is difficult to fight the temptation to put off the absolutely essential task of emptying one's experiences into a daily diary. Here we need to think about the morality of the exercise. It is important that people know

that they are being studied and have given their permission; the question of how often they should be reminded and how much they should see of work in progress is, however, a much greyer area, which you might like to think about (for further ideas, see chapter 45 on ethics by Tim Unwin).

But ethnography still involves writing; there is little point in it unless there is a written representation. Unless one is very insensitive about the feelings of people one has shared experiences with, the task is approached diffidently. I have wasted most of my research by not writing it up for publication – it never seems finished; each time I come close to a representation I think is good enough to give to the people I have been working with, I see another flaw. Every representation seems to be a distortion. This is pretty futile – the final duty of an ethnographer is to write (and to take the consequences if people complain that they do not like what is written about them).

References

Bennett, K. 2002: Participant observation. In P. Shurmer-Smith (ed.), *Doing Cultural Geography*, London: Sage, 139–50.

Clifford, J. and Marcus, G. (eds) 1986: *Writing Culture: The Politics and Poetics of Ethnography*. Berkeley: University of California Press.

Coffey, A. 1999: *The Ethnographic Self: Fieldwork and the Representation of Identity*. London: Sage.

Crang, P. 1994: It's showtime: on the workplace geographies of display in a restaurant in South East England, *Environment and Planning D: Society and Space*, 12, 675–704.

Malbon, B. 1999: *Clubbing: Dancing, Ecstasy and Vitality*. London: Routledge.

Marcus, G. and Fischer, M. 1986: *Anthropology as Cultural Critique: An Experimental Moment in the Social Sciences*. Chicago: University of Chicago Press.

43

Investigating Visual Images

John Morgan

The world we inhabit is filled with visual images. They are central to how we represent, make meaning, and communicate in the world around us.

(Sturken and Cartwright 2001: 1)

In the above quotation, Sturken and Cartwright highlight the idea that the world we live in is more and more dominated by visual images. They point out that our knowledge and understanding of the world come not only from books but from television, films, magazines, advertisements, video games and so on. It will probably not have escaped your notice that geography is a subject that makes great use of visual images. For example, think of the times in your geography education when you viewed slides, watched videos or climbed to the top of a hill to get a view from above. Often these images are treated as though they are a transparent 'window on the world'. The fact that these images are always selected and 'framed' is often ignored.

Increasingly, geographers distinguish between *vision* and *visuality*. Vision refers to what the human eye is physiologically capable of seeing. Visuality refers to the way in which vision is constructed in various ways. This idea that our 'ways of seeing' are socially and culturally constructed has become increasingly important in the social sciences as it has been influenced by the 'cultural turn' (see chapter 24 by Peter Jackson in this volume). Geography as a subject has not been immune from this process, and cultural geographers in particular have studied visual images. Burgess and Gold (1985: 1) express this point well:

The media have been on the periphery of geographical inquiry for too long. The very ordinariness of television, radio, newspapers, fiction, film and pop music perhaps masks their importance as part of people's geography threaded into the fabric of daily life with deep taproots into the well-springs of popular consciousness.

This chapter is written in the hope that you might decide to 'take images seriously' in your own geographical studies. It is written in the belief that the best way to develop an understanding of 'visual culture' is to consume it. That means looking at pictures, adverts, watching television and video, and going to the cinema, and finding ways to think about and understand that experience.

Gillian Rose (2001) provides a useful starting point for thinking about how to approach the study of a visual image. She suggests that an image can be thought about in three ways:

- First, there are questions to be asked about the *production* of the image. For example, who produced it? For what purpose? Who did they produce it for?
- Second, there are questions to ask about the image itself. What is being shown? What are the components of the image? What use is made of colour? Is it an image combined with words?
- Third, there are questions about the

Figure 43.1   Magazine advetisement for Land Rover, 'Been anywhere interesting lately?'
Photograph: Permission Land Rover/Webfoot.

audience for the image. For example, who consumes it? Do they consume it alone or with other people? What do they make of it?

It is useful to bear these questions in mind as you read the following examples.

Been Anywhere Interesting Lately?

Look at figure 43.1. It is an advert for a Land Rover vehicle and appeared in the weekend section of the *Guardian* newspaper. The advert shows a man, living in a fairly typical house and garden, going about the normal weekend business of washing his Land Rover. This is unremarkable. Unremarkable, that is, until you look more carefully at what he is actually washing out of the car. There is a stream of desert red soil running to the

drain, and a scorpion. The caption reads: 'Been anywhere interesting lately?'

When I first saw this advert, it made little impression on me. However, this image, like all images, has two levels of meaning. French theorist Roland Barthes described these two levels with the terms '*denotative*' and '*connotative*' meaning. The denotative meaning of the image refers to its literal, descriptive meaning. Thus, this advert denotes a man washing the debris of his recent travels out of his car. The connotative meaning of the image relies on the cultural and historical context of the image and its viewers' lived knowledge and experience. Thus, this advert connotes the idea that merely to own a car and travel is not enough. The *really* important thing is where you actually go.

Let's look at how the advert 'works' in more detail. We can do this by focusing on the specific elements that make up the advert. First, there are a number of things in the picture that seem to be linked with notions of domesticity. For example, the carefully manicured lawn and the well-tended flowerbeds are symbolic of an ideal of suburban life, as is the act of washing one's car at the weekend. However, the desert soil and the scorpion are 'out of place' in this normal domestic scene. In this case they stand for the 'exotic'. Barthes developed the idea of the '*sign*', which is composed of the '*signifier*', a sound, written word or image, and the '*signified*', which is the concept evoked by that image. In the Land Rover advert, one interpretation is that the scorpion (*the signifier*) stands for 'adventure' and 'excitement' (*the signified*) and that its association with the Land Rover means that the Land Rover comes to stand for the idea of adventure, excitement and travel.

The reader of the advertisement has to be able to interpret the visual codes. This advertisement 'works' because it plays on the reader's understanding of the cultural codes that represent the relationship between people and nature. The advert is a variation on the quite ubiquitous image of men driving 4 × 4 vehicles and pick-up trucks through rough landscapes that again references ideas of rugged male individualism (see figure 43.2).

As you are reading this discussion of the Land Rover advert you may be wondering why any of this matters. After all, most of us seem to be able to negotiate our newspapers and television viewing without spending hours 'deconstructing' images! However, images are an important means through which *ideologies* are produced and onto which ideologies are projected. Ideology is often associated with the idea of 'propaganda' or the attempt to use images to persuade people into holding certain beliefs and values. However, another way to think about ideology is that it is a much more pervasive, mundane process in which we all engage, whether we are aware of it or not:

One could say that ideology is the means by which certain values, such as individual freedom, progress, and the importance of home,

are made to seem like natural, inevitable aspects of everyday life. (Sturken and Cartwright 2001: 21)

In the case of the Land Rover advertisement, our understanding of it relies on a whole set of ideological constructions. The most important of these is the idea of individualism and the idea that such individualism can be realized through escape and travel – or, if you like, adventure. This is a common theme in western consumer culture, and one which is used to sell anything from holidays to chocolate bars. The Land Rover advert draws upon a distinction between the home/ordinary and away/extraordinary. There is also the question of gender relations here, since it is, presumably, the male viewer who is being addressed by this advert (there are other adverts for cars that address female readers, and these tend to be quite different).

This example suggests ways in which print advertisements can draw upon ideas about place and mobility. A number of geographers have explored these themes:

- In what ways do adverts draw upon images of space and place in order to sell the product?
- What types of environments are associated with different products?
- Do adverts such as this promote particular ideologies?
- What meanings do readers make of adverts?

Representing Industrial Change

Cultural geographers are increasingly interested in the ways in which the media are involved in the construction of our ideas about space and place. The second example illustrates some of the issues geographers have considered through the discussion of one popular film.

The 1990s saw the release of a number of

Figure 43.2   Magazine advertisement for Jeep.
Photograph: Permission Chrysler/Pentamark.

Figure 43.2   Scene from *The Full Monty*.
Photograph: Fox Searchlight/Courtesy The Kobal Collection.

films which focused on the problems of unemployment and social exclusion faced by a
growing group often described as the
'underclass'. Films such as *The Full Monty*,
Brassed Off, or *My Name is Joe* map the
changing landscapes that resulted from processes of deindustrialization, mass unemployment and poverty. These themes are
explicit in the opening sequence of the most
popular of these films – *The Full Monty* (figure 43.3). Set in the aftermath of the deindustrialization of Sheffield, the film starts
with footage of a promotional film that pro-
motes Sheffield as 'steel city', a city on the
move, with nightclubs, located at the centre
of Britain's industrial north, with 90,000
men employed in steel production. The film
then cuts to present-day Sheffield, with a
steelyard which is derelict, rusted and abandoned, except for two redundant steelwork-
ers 'liberating' a girder.

The Full Monty explores issues of gender
politics in a post-industrial context, and the
common feature of all the male characters

is that they are experiencing a 'crisis of identity' following their redundancy. This crisis
of masculinity is manifested in a variety of
ways: the redundancy of fathers, the
infantilization of unemployed men and
sexual impotency. The central characters
have lost their work and as such their male
identities are threatened. This is symbolized
in the film by the advance of women into
'men's space'. Thus the working men's club
is appropriated by women for the
Chippendales (male stripping) event, a point
made even worse by the fact that the women
are using the men's toilets.

The film is also a commentary on changing economic activities. The types of work
shown in *The Full Monty* reflect wider patterns of employment in Britain, where nearly
half the workforce is female, often in
casualized and low-paid forms of white-
collar work. In the film, places of active employment are supermarkets, where women
serve and men act as security guards, and
female-dominated factory work, which the

men in the film avoid at all costs. There are many issues that might be analysed in the film. For example, it cleverly makes use of the metaphor of 'stripping'. Thus, the landscape of industrialization is one which has been stripped (of natural resources and now stripped of capital), and in the fact that the craft-skilled men are no longer needed (as one of the characters says, 'like skateboards') and are thus stripped of their identities. In order to regain their pride and identities, the men learn to 'strip', offering themselves for display as commodities (again, the gender reversal is important here, in that this time it is the male strippers who are watched, rather than the traditional female strip show). Indeed, it has been suggested that in order to regain their identities men must repackage themselves as 'commodities', which comes close to the role envisaged for workers in post-Fordist capitalism: workers must be flexible, adaptable, enterprising and skilled at packaging themselves for the demands of the market.

Claire Monk (2000) suggests that the film offers male audiences a 'symbolic, if inevitably problematic, solution' to the problems of male unemployment. She argues that the film expresses the problems of the post-industrial male in a 'feminized' society as characterized by gender, rather than economic, realities. Indeed, she worries about the misogyny of this solution, especially when the working-class community whose passing is being mourned is a community of *men*. The position of women within the film is interesting. While there are some key women characters – such as Gaz's wife Mandy, portrayed negatively as a social climber, expressed geographically in her move to the suburbs and socially in her relationship with a rather smug middle-class man – others are marginal to the narrative. None the less, the film enjoyed great popularity among female viewers, possibly because of the extent to which traditional gender roles are reversed in the film.

This example shows how popular films can be a powerful medium for the construc-

tion of meanings about human and natural environments and in recent years geographers have become increasingly interested in studying them. In many cases, these fictional representations of places have impacts on the actual places. For instance, councillors and local business groups in Sheffield at the same time publicly criticized the image of Sheffield as a city in decline and sought to use the film to attract tourists to the city. They pointed out that Sheffield produces 70 per cent of Britain's engineering and specialist steels, unemployment has fallen below 9 per cent, it has a lively cultural quarter and would soon host the National Museum for Popular Music. They accused the film-makers of replaying the old myths and stereotypes about the north (see Shields 1991).

- Can you find examples of films which have a strong geographical theme? How do they represent places and environments?
- Is there any evidence that images of places can have an impact on those places?
- To what extent do images of places in films affect people's perceptions or 'mental maps' of those places?

Conclusion

The aim of this chapter was to encourage you to think about how some of the visual images found in popular culture reflect geographical themes and ideas. It has focused on print adverts and films, though examples could be drawn from painting, photography, television and the 'new media' such as websites and games. In conclusion, here are some ideas to help you get started in thinking about visual images:

- Start by looking for examples of visual culture which have a geographical theme (that is, they are linked to ideas about space, place and environment).

- Try to develop your own ideas about the visual image(s) you are interested in. Use Gillian Rose's questions to guide you.
- Find out whether other people have written about this image (or others like it). Read as widely as possible, using websites (chatrooms), newspapers and magazines, academic journals and books (you may have to go beyond the geography section of your library).
- Talk to people about your image. You will most likely find that they have different perspectives and ideas.
- Once you have decided upon an image to study, the suggested further reading below will help you get started in thinking about these issues, as well as provide some ideas about how to study visual images in geography.

References

Burgess, J. and Gold, J. (eds) 1985: *Geography, the Media, and Popular Culture*. Beckenham: Croom Helm.

Monk, C. 2000: Underbelly UK: the 1990s underclass film, masculinity and the ideologies of 'new' Britain. In J. Ashby and A. Higson (eds), *British Cinema: Past and Present*, London: Routledge.

Rose, G. 2001: *Visual Methodologies*. London: Sage.

Shields, R. 1991: *Places on the Margin: Alternative Geographies of Modernity*. London: Routledge.

Skelton, T. and Valentine, G. (eds) 1998: *Cool Places: Geographies of Youth Cultures*. London: Routledge.

Sturken, M. and Cartwright, L. 2001: *Practices of Looking: An Introduction to Visual Culture*. Oxford: Oxford University Press.

Further Reading

If you want to read more you should look to John Berger's classic *Ways of Seeing* (London: Penguin, 1972). Crang's *Cultural Geography* (London: Routledge, 1998) is a readable and accessible introduction to cultural geography, and Shurmer-Smith and Hannam's *Worlds of Desire, Realms of Power: A Cultural Geography* (London: Arnold, 1994) is stimulating if slightly more demanding. A good guide to studying visual images is Rose's *Visual Methodologies* (London: Sage, 2001), which gives more information on detailed methodology. Many ideas in cultural geography are linked to other disciplines. An excellent introduction to all types of images is Sturken and Cartwright's *Practices of Looking: An Introduction to Visual Culture* (Oxford: Oxford University Press, 2001). It is beautifully illustrated, and you should find its interpretation of films, adverts, photographs and art extremely thought-provoking.

44
Researching Historical Geography

Robert J. Mayhew

As an A-level student on the brink of going to university to read geography, I was entranced by the film version of Umberto Eco's bestselling novel *The Name of the Rose* (1983). This was a medieval murder mystery, starring Sean Connery as Brother William of Baskerville, a clerical detective on the trail of a lost copy of Aristotle's treatise on laughter, the search for which had left a trail of murders in a monastic library. A few years later, as a budding research student, I found myself rooting around catalogues in the cloistered surroundings of the Bodleian Library in Oxford, in search of a rare eighteenth-century travel book. After hours of searching I finally tracked down the book in question, and on receiving it found the pages were uncut, which meant that it had not been read by a soul in over 200 years. Both *The Name of the Rose* and my travel book conjure up 'the pleasures of the past' (Cannadine 1989), where the diligent enquirer is enmired in the musty mysteries of uncovering truths about past societies, truths long since lost.

Linking History and Geography

Yes, you might say, but what has all this to do with geography? Isn't this just history? On the contrary: from long before the era of Eco's medieval sleuth, since ancient Greece and Rome in fact, geography and history have been inseparable practices (Clarke 1999). In the modern day, geography and history are linked by geographers in two main ways. First, geographers have shown a continued and recently resurgent interest in the *history of geography* (see chapter 46 by David Livingstone). Second, geography has long had a sub-discipline called *historical geography* (see chapter 19 by Nash and Ogborn). The types of questions these two historical rubrics in geography might ask include: what did eighteenth-century Europeans learn in geography classes and how does this differ from their successors three centuries later? What do ibn-Batuta's travel journals tell us about Arabic visions of the medieval world? What was the relationship between geography and the emergence of European empires in the nineteenth century? What was the urban geography of China under the Ming dynasty? What were the economic regions of ancient Rome? How have North American societies polluted their environments over time?

Historical Sources for Geographers

Given that one has a historical question one wishes to pose as a geographer, how does

Figure 44.1   Scene from *The Name of the Rose* featuring the character Sir William Baskerville played by Sean Connery.
Photograph: Neue Constantin/ZDF/Courtesy The Kobal Collection.

one go about answering it? As with any form of enquiry, of course, you first need data, which in history are normally referred to as *primary sources*. Primary sources are any forms of evidence which survive from the era you are interested in. These can simply be divided into *written* and *unwritten* sources. Written sources may come in manuscript or in printed form, and include a wide range of script material from obvious sources such as books and newspapers to advertising slogans and picture-postcard catchphrases. Unwritten sources include paintings and buildings, and also – of especial interest to historical geographers traditionally – landscapes, which inscribe a record of human–environment relationships over time. Three things seem worth noting here about data sources, one general and two specific to historical geography. First, as a general rule, the quantity and variety of primary sources available increase the more recent the era with which you are concerned: the twentieth century gave us the Internet, film, television and radio, not to mention an unprecedented quantity of printed matter, while the ancient Egyptians have left us monuments, hieroglyphics and little else besides. In terms of data availability, in the later period we are liable to be swamped, in the earlier to be stumped! Second, historical geographers frequently deploy literary, artistic and architectural sources, which links their work to that of the humanities as opposed to the social and natural sciences. Finally, historical geography can be practised in the field (more accurately, landscape) as much as in the library: musty books can be traded in for fragrant fields according to taste.

But how do we find the specific data source to answer our particular question? A good starting point is to look at the bibliographies of others who have researched the topic in which you are interested for preliminary pointers to specific primary data sources. But more serious enquiry will demand recourse to records offices and libraries. Fortunately, there is an increasingly impressive range of computerized and CD-ROM catalogues for these institutions, which allow for complex searches for relevant books and other artefacts such as paintings. Further, books and images from the past are becoming ever more accessible thanks to facsimiles, new editions, microfilms, CD-ROMs and Internet transcriptions. None of this applies to researching from books alone: there are excellent bibliographies, to take one example, of British gardens (Desmond 1984), and researching the meaning of artefacts in the landscape will always demand seeking works of reference in library catalogues.

Two useful general publications are Adkins (1988), which is a guide to all the libraries in the United Kingdom, and Downs (1981), a bibliographic guide to resources such as library catalogues. Beyond this, records from central government departments are kept in the Public Record Office at Ruskin Avenue, Kew, Richmond, Surrey, TW9 4DU, and a *Guide* to its contents is available. Also important is the British Library, which holds the most extensive collection of printed books, manuscripts and maps in the UK. The other copyright libraries – at Edinburgh, Aberystwyth, Oxford and Cambridge – are also sources of vital records, and all have useful computerized catalogues available on the Internet. Finally, every county has its own record office specializing in local material such as parish registers, es-

tate records, county court papers and the like. Many counties have published bibliographic guides to available material (see also Richardson 1986). For historians of geographical thought, an especially useful bibliographic starting point is Sitwell (1993), whilst more inclusive ones are provided by Taylor (1930, 1934). For historical geographers, the Historical Geography Research Group has published some useful bibliographic material (Finlay 1981; Whyte and Whyte 1981; Shaw 1982; Mills and Pearce 1989), which can be obtained from its publication editor, Dr Philip Howell, Department of Geography, Downing Place, Cambridge, CB2 3EN. Broader searches can be conducted using the three *Short-title Catalogues* which list all works in the English language from the birth of printing to 1800 (Pollard and Redgrave 1976–91; Wing 1982–98; British Library 1990).

Analysing Historical Sources

Given that you have found data relevant to answering your historical-geographical question, how do you go about analysing this material? The answer to this question will vary to some extent according to the type of data you have collected and the type of question you are asking. If you have large quantities of simple data (say, parish records of births, marriages and deaths), you can use quantitative statistical techniques to discover family size, life expectancy and the like for a past society. Perhaps more frequently, your material will be both more limited and less quantitative, at which point close and careful reading, be it of a book or a landscape, is the first necessity. In fact, this is needed just as assuredly for quantitative material: historians have to be sensitive to the patchy survival of evidence from the past, and even where we are confident that a dataset is reasonably complete, and therefore amenable to statistical analysis, its *meaning* in terms of the light it sheds on a past society and its geography cannot be disclosed merely by the generation of a body of statistical information. Primary data are not in and of themselves history; understanding a society demands that insight be extracted from information.

Historically sensitive reading of primary data seeks to understand the information more fully as a product of its times, and thereby to reveal what it tells us of a past society and its geography. To do this historical geographers place their primary sources – their *texts* – in a whole series of *contexts*, thereby gaining a richer picture of their meaning (Skinner 1969). For example, biographical context is important: who wrote a book or designed a garden or painted a landscape is a key to unravelling their interpretation of a now past society and its geography. To understand *The Name of the Rose*, for example, it is highly relevant to know that its author, Umberto Eco, is a distinguished scholar of medieval aesthetics, and as such his tale is based on extensive knowledge of the age in which it is set. Similarly, the nature of a past society is key background information for the study of any historical material. As an example, historical demography uses English parish records to reconstruct past populations, but these records only cover members of the Church of England with any accuracy, such that other faiths go unrecorded. From the historical geographer's viewpoint, this social context – that religious affiliation drove data collection – is of particular importance since there were regions where non-Anglicans were particularly numerous, as was the case in Lancashire with its strong Roman Catholic community. The social context behind the data collection, then, demands that geographical sensitivity is used in assessing the reliability of historical statements based on parish records. As this example reminds us, even the most comprehensive historical records were normally collected for some pragmatic purpose, not to give future generations an impartial view of the age in which they were collected. As such, caution is a key quality in the historian as he or she as-

sesses the worth of his or her data sources. In a similar vein, historians trying to read their evidence, be it in a book or a landscape, need to beware, because the meanings of key terms in language and of physical objects in the landscape change over time. Reading data historically is never a straightforward activity. Furthermore, historically valid questions can sometimes be in principle unanswerable, since there are no data sources available. Here, however, historical imagination can come into play in how extant sources are used to answer questions tangentially. For example, whilst direct data about the geography of agriculture in early modern England are extremely patchy, arable areas had seasonal marriage regimes (with a rush of marriages after the harvest season), where pastoral areas did not. As such, historical geographers can reconstruct agrarian geography from the aforementioned parish registers, which tell us (with limitations) when marriages occurred (Kussmaul 1990). Finally, one needs to look at comparative material; in other words, at primary sources of a similar variety to that in which you are interested. For example, to understand the meaning of an eighteenth-century geography book, you need to see it in the light of other geography books of the time. This is an often sobering experience, as geography books were frequently just cribbed from earlier ones, such that what seems to be an author's viewpoint about a place is nothing of the sort, but instead a mere plagiarism. It is chasing up texts and their innumerable contexts which leaves the historical geographer trekking across landscapes and scurrying around libraries, coming upon forgotten features and books banished by time.

The Relevance of Researching Historical Geography

Finally, what is the outcome of research in historical geography and its wider relevance to geography? The result of any piece of historical research can never be the truth but only what the evidence obliges us to believe (Oakeshott 1933). For historians and historical geographers are engaged essentially in a process of inference, moving from limited data to a picture of a society. Fresh evidence, and extant evidence interpreted in the light of new questions or different contexts, can and will lead to differing and equally valid conclusions in historical geography. But your piece of research, if it tracks down appropriate data sources for the question you have in mind, reads those sources with historical sensitivity and places them carefully in the contexts from which they emerged, can result in a greater understanding of a past society, of how it operated over space, of its interactions with its environment or of what it understood geography as a subject to be about. It should be emphasized that the aim of historical ge-ography is *understanding* rather than *explanation* (Langton 1988). Where scientific explanation develops hypotheses to generate laws of general applicability, the aim of historical enquiry is to move from a question about a past society to a better understanding of that society which is non-transferable to other societies and claims no universal validity.

Beyond this greater understanding of past geographies for their own sake, such an enquiry also has relevance to geography in the present day in two ways. First, historical-geographical enquiry can lead us to rethink key concepts in geography such as the region or landscape and what we mean by those taken-for-granted terms. Second, and more importantly, historical geography can give us a comparative perspective, questioning what we take geography to be and how our society organizes itself geographically from the perspective of another and different society, one from the past. This can create a certain healthy scepticism about ourselves and our world: across time, space and space-time, conceptions of geography and modes of organization in space have varied; our conceptions and modes are not better, only subsequent. If we can learn this,

the upshot of researching historical geography might just be to produce human geographers who are humane citizens.

References

Adkins, R.T. (ed.) 1988: *Guide to Government Departments and Other Libraries*. London: British Library Science Reference and Information Services.

British Library 1990: *The Eighteenth Century Short-title Catalogue*. London: British Library [a CD-ROM version was also produced in 1991].

Cannadine, D. 1989: *The Pleasures of the Past*. London: Collins.

Clarke, K. 1999: *Between Geography and History: Hellenistic Constructions of the Roman World*. Oxford: Clarendon.

Desmond, R. 1984: *A Bibliography of British Gardens*. Winchester: St Paul's Bibliographies.

Downs, R.B. 1981: *British and Irish Library Sources: A Bibliographic Guide*. London: Mansell.

Eco, U. 1983: *The Name of the Rose*. London: Picador.

Finlay, R. 1981: *Parish Registers: An Introduction*. Norwich: GeoAbstracts.

Kussmaul, A. 1990: *A General View of the Rural Economy of England, 1538–1840*. Cambridge: Cambridge University Press.

Langton, J. 1988: The two traditions of geography, historical geography and the study of landscapes, *Geografiska Annaler*, 70B, 17–25.

Mills, D. and Pearce, C. 1989: *People and Place in the Victorian Census: A Review and Bibliography of Publications based substantially on the Manuscript Census Enumerators' Books, 1841–1911*. Bristol: Historical Geography Research Group.

Oakeshott, M. 1933: *Experience and its Modes*. Cambridge: Cambridge University Press.

Pollard, A.W. and Redgrave, G.R. 1976–91: *A Short-title Catalogue of Books Printed in England, Scotland and Ireland, and of English Books Printed Abroad: 1475–1641*, 2nd edition. London: Bibliographical Society.

Richardson, J. 1986: *The Local Historian's Encyclopaedia*. New Barnet: Historical Publications.

Shaw, G. 1982: *British Directories as Sources in Historical Geography*. Norwich: GeoAbstracts.

Sitwell, O.F.G. 1993: *Four Centuries of Special Geography: An Annotated Guide to Books that Purport to Describe all the Countries in the World*

Published in English before 1888, with a Critical Introduction. Vancouver: University of British Columbia Press.

Skinner, Q. 1969: Meaning and understanding in the history of ideas, *History and Theory*, 8, 3–53.

Taylor, E.G.R. 1930: *Tudor Geography, 1485–1583*. London: Methuen.

Taylor, E.G.R. 1934: *Late Tudor and Early Stuart Geography, 1583–1650*. London: Methuen.

Whyte, I.D. and Whyte, K.A. 1981: *Sources for Scottish Historical Geography: An Introduction*. Norwich: GeoAbstracts.

Wing, D.G. 1982–98: *A Short-title Catalogue of Books Printed in England, Scotland, Ireland, Wales and British America, and of English Books Printed in Other Countries, 1641–1700*, 2nd edition. New York: Modern Language Association of America.

Further Reading

Two collections of essays by historical geographers give a good insight into methodological approaches to research in history as they have been taken up by historical geographers. These are A.R.H. Baker and M. Billinge (eds), *Period and Place: Research Methods in Historical Geography* (Cambridge: Cambridge University Press, 1982) and A.R.H. Baker, J.D. Hamshere and J. Langton (eds), *Geographical Interpretations of Historical Sources: Readings in Historical Geography* (Newton Abbot: David and Charles, 1970). Additional material is to be found in D. Brooks Green (ed.), *Historical Geography: A Methodological Portrayal* (Savage, MD: Rowman and Littlefield, 1991); M. Morgan, *Historical Sources in Geography* (London: Butterworth, 1979); W. Norton, *Historical Analysis in Geography* (London: Longman, 1984); and M. Pacione (ed.), *Historical Geography: Progress and Prospect* (London: Croom Helm, 1987).

For direct access to debates amongst historians concerning how to practise the subject, traditional views might profitably be investigated in G. Elton, *The Practice of History* (Glasgow: Fontana, 1967); R. Shafer, *A Guide to Historical Method* (Homewood, IL: Dorsey Press, 1969); and L. Gottschalk, *Understanding History: A Primer of Historical Method* (New York: Knopf, 1969); whilst more recent approaches are well treated in L. Jordanova, *History in Practice* (London: Arnold, 2000). A neat summary overview is J. Tosh, *The Pursuit of History: Aims, Methods, and New Directions in the Study of Modern History*, 3rd edition (Harlow: Longman, 2000).

45

Geographical Ethics: Reflections on the Moral and Ethical Issues Involved in Debate and Enquiry

Tim Unwin

Imagine that you are undertaking research for your dissertation. You are interviewing young people about their experiences of place, and in the course of these interviews someone tells you that they are being abused by one of their relatives. In accordance with what you have read about research methods, you had previously assured everyone that what they told you would be in confidence, and that they would not be identifiable in anything you wrote or said. What would you do?

How would you go about resolving this question? Try discussing it with your friends, and examine how you all seek to disentangle the issues involved (for ways in which geographers have recently engaged with such issues, see Valentine 1999 and the collection of papers on the ethical dimensions of working with young people in *Ethics, Place and Environment*, 4(2), especially Aitken 2001). Such reflection on what is right and wrong, and the actions that we take in response to how we answer such questions, are the concern of ethics. While the example above may be unusual, we are all involved in making

ethical decisions every day of our lives. Likewise, we all have values, we judge some things and actions as better than others; we consider some places ugly, and others beautiful. These decisions reflect our engagement with the norms and expectations of the societies in which we live. Crucially, humans have the capacity to say what *should* be, and to try to shape the world to reflect these claims. Indeed, as David M. Smith (2000: 1) has emphasized, '[i]t is this capacity to reason about the normative aspects of life, beyond the pursuit of mere physical survival, that most clearly differentiates humankind from other creatures'.

It is useful to distinguish between three commonly confused, and much debated, terms: values, morals and ethics (for a more detailed discussion in a geographical setting, see Smith 2000). None is easy to define. In general usage, *values* are seen as the aspects of life that individuals consider desirable or worthy, and that guide their actions. Frequently, values are considered as being opposed to facts (Proctor 1999). In such a context, values are the domain of the *normative*, or what should be, while facts are the realm of the *positive*, what is. This usage of the word 'values' shades into the meaning of morals. *Morals* are the rules that people follow; what they think is good and bad, right and wrong. *Ethics*, in turn, is the systematic reflection on moral questions; the philosophical discourse on morals. Hence the term 'moral philosophy' is often used to refer to ethics (Hinman 1994).

Two broad types of ethics may be identified: *professional ethics*, or reflections on geographical practice (see Hay 1998); and the philosophical or theoretical dimensions of ethics, which are often divided into *descriptive ethics* (the characterization of existing moral practices and beliefs), *normative ethics* (concerned with the solution of moral problems relating to such matters as social justice, inequality and environmental change) and *metaethics* (the examination of ethical reasoning itself) (see, for example, Proctor 1999; Smith 2000). These distinctions frame the account that follows.

The Context of Ethical Considerations in Geography

Two interrelated series of influences have led to the rising prominence of ethical considerations in geography during the 1990s and early twenty-first century. First, there has been considerable growth in concern within European and North American society about moral and ethical matters, typified by the increase in debate over environmental and human rights agendas. This has not only been reflected in the activities of radical or fringe groups, but is now increasingly being incorporated into some central economic and political institutions. In July 2001, for example, the *Financial Times* launched its ethical index FTSE4Good (http://www.ftse4good.com), reflecting the increased importance that financial institutions now attribute to ethical investments. In part, such concerns reflect a growing awareness of the failure of the model of science that had come to domi-

nate 'western' society during the second half of the twentieth century. While this model is reasonably powerful at describing what exists in the world, its failure to consider its own ethical foundations means that it cannot provide society with satisfactory solutions to the much more difficult questions about how to act on such information. Increasingly it has been recognized that the claim of science to be value-free is nothing but an illusion. Science, like any other form of discourse, is imbued with the values and prejudices of the societies that produce it. As Habermas (1978: 67) has so forcibly commented, 'by making a dogma of the sciences' belief in themselves, positivism assumes the prohibitive function of protecting scientific enquiry from epistemological self-reflection. Positivism is philosophical only insofar as is necessary for the immunization of the sciences against philosophy'. As he goes on to maintain, '[t]he positivistic attitude conceals the problems of world constitution. *The meaning of knowledge itself becomes irrational* – in the name of rigorous knowledge' (Habermas 1978: 68–9).

While geographers have in part responded to these wider social trends, a second set of influences can be seen as growing out of the intellectual traditions of the discipline itself. Moral questions concerning the place of humans in the world were of some considerable significance for scholars in antiquity, as well as in the eighteenth and nineteenth centuries (Unwin 1992). However, the positive stance adopted by many geographers in the 1950s and 1960s, with its emphasis on quantification and spatial science, relegated normative concerns very much to a secondary position. Moreover, the subsequent radical opposition, drawing heavily on Marxism, tended to focus much more on critique than it did on the advocacy of alternative normative models of society. Surprisingly few geographers, for example, were prepared to advocate what they believed the world *should* be like. Some did voice opposition on moral grounds to questions of social inequality (Harvey 1973), but little attention was explicitly paid to the ethical foundations of the discipline. Notable exceptions were Yi-Fu Tuan's (1986, 1989) examination of the ways in which morality is imagined and experienced in different places and times, and David M. Smith's (1977, 1979) advocacy of a welfare approach to geography.

During the 1990s, an increasing number of geographers (see, for example, Driver 1988; Sack 1997; Corbridge 1998; Proctor 1998) began to address moral geographies more directly. Sack (1997: 24), for example, stressed that geography is fundamental for an understanding of ethical issues. As he argues, '[t]hinking geographically heightens our moral concerns; it makes clear that moral goals must be set and justified by us in places and as inhabitants of the world'. Moreover, as Smith (2000: 14) has emphasized, there is 'one metaethical issue of such obvious geographical interest that it cannot be bypassed: that of relativism'. At one level, this is fundamentally concerned with the character of truth itself: whether it is a relative or universal concept. How we answer this question plays a very significant part in determining the way in which we embark on geographical enquiry. If we adopt the former stance, our task is to come ever closer to an understanding of what that truth is. For example, many physical geographers are concerned to identify *the* universal or absolute laws that will explain processes in the physical world. In contrast, most human geographers tend to adopt a relativist stance, noting that what is accepted as truth in any given society closely reflects the social and cultural contexts in which that truth is produced (for an attempt to resolve these different positions, see Walzer 1994).

A focus on difference has also brought geographers into a much more overt consideration of the ethical implications of other aspects of their discipline, notably on the moral significance of distance. In his examination of development ethics, Corbridge (1998: 35) thus emphasizes that 'citizens and states in the advanced industrial world have a responsibility to attend to the claims of distant strangers'. He goes on to assert that development ethics 'is not just about questions of transnational justice and positionality; it is also about the construction of plausible alternative worlds and practical development policies' (Corbridge 1998: 35). Likewise, Silk (1998) has drawn attention to important issues surrounding the significance of distance for care, highlighting the complexity of assumptions all too often ignored when we use phrases such as 'nearest and dearest'. A further area of moral enquiry with which geographers have recently begun to engage more directly has been concerned with the difference between the human and non-human worlds (see, for example, Welch and Emel 1998; Low and Gleeson 1999), and particularly on the moral values associated with different biocentric, ecocentric or geocentric views of the world (Lynn 1998).

Professional Ethics

Turning to matters of professional ethics, there is some considerable debate about the creation and use of ethical codes or guidelines. While guidelines are generally designed to encourage good practice, formal codes can become constraining (Hay 1998). Nevertheless, along with many other professional bodies (see, for example, American Sociological Association 1989) geographers have in recent years sought to establish ethical statements that can provide guidance on what is deemed to be acceptable behaviour by members of the profession. Among the most comprehensive of such statements for geographers was that endorsed by the Council of the Association of American Geographers in 1998 (see Association of American Geographers 1999). In its preamble, this stresses that the diversity of research undertaken by geographers makes it impossible to generate a comprehensive ethical statement that will be appropriate for all geographers, and it emphasizes that the principles it sets forth should be seen primarily as starting points for ethical consideration. The statement focuses on six main areas of ethical engagement:

1 Professional relations with one another (including avoiding discrimination and harassment, sustaining community and promoting fairness in hiring).
2 Relations with the larger scholarly community (concentrating on attributing scholarship, evaluating scholarship and self-plagiarism).
3 Relations with students (addressing instructional content, pedagogical competence, training students with funded research and confidentiality).
4 Relations with people, places and things (covering project design, ethical behaviour during field research, and reporting and distributing results).
5 Relations with institutions and foundations that support research (focusing on funding research, and the use of results from funded research).
6 Relations with governments (addressing government research support and employment).

While not all of the statement is of immediate relevance to your learning and research, the guidelines are well worth reading for the scope of the practical ethical issues that they address. They highlight four principles that are of particular relevance to your debate and enquiry:

• The importance of honesty: results should not be fabricated, and geographers should not plagiarize the work of others.
• The impact of field research: geographers should consider the effects of their research on people, places, flora, fauna and environments prior to embarking on the research and should seek to minimize any potential damage arising from the research.
• The golden rule: people, places and things should be treated in the same way

that researchers would like their own selves, places and possessions to be treated.
• Returning results: research results should be given in an accessible form to those with whom the research was undertaken.

It is essential to note from these that ethical considerations do not just apply in the context of human geography, but are equally applicable to research on and in the physical environment. Even taking an ice core on a glacier or digging a soil section have significant ethical implications that need to be considered before the research is undertaken.

As well as such guidelines, there is a growing body of literature written by geographers that addresses specific practical ethical dimensions of geographical enquiry. This can beneficially be examined at the early stages of planning a research project. Hay (1998), for example, proposes a helpful set of prompts for moral contemplation and action, grouped under the five headings of free and informed consent, confidentiality, minimizing harm, cultural sensitivity and feedback to participants. These closely mirror the concerns reflected in the Association of American Geographers' (1999) statement, and provide a useful checklist for those embarking on geographical research. Hay goes on to stress the importance of encouraging undergraduate and postgraduate students to acquire a grasp of the complexity of moral enquiry through engaging with a series of case studies that highlight particular ethical conundrums. Given the diversity of geographical enquiry, and the difficulty of ever reaching agreement on a definitive code of ethical practice for geographers, he stresses the importance of encouraging geographers to learn to think ethically.

Geographical Ethics

This brief overview has stressed that ethical considerations enter geographical enquiry at both a practical and theoretical level. As with any academic discipline, there are important ethical considerations that need to be borne in mind when undertaking empirical research; indeed, geographers in their immediate interactions with the physical and socio-cultural world around them need to be more aware than most of the ethical implications of their research practices. However, geographers' preoccupations with fundamental moral questions concerning social justice, the identity and meaning of place, our interactions with the non-human physical world, and the significance of distance, all mean that there is much to be gained from a deeper exploration of the interface between geography and ethics.

References

Aitken, S.C. 2001: Fielding diversity and moral integrity, *Ethics, Place and Environment*, 4, 125–9.

American Sociological Association 1989: *Code of Ethics*. Washington, DC: American Sociological Association.

Association of American Geographers 1999: Association of American Geographers' statement on professional ethics, *Ethics, Place and Environment*, 2, 258–66.

Corbridge, S. 1998: Development ethics: distance, difference, plausibility, *Ethics, Place and Environment*, 1, 35–54.

Driver, F. 1988: Moral geographies: social science and the urban environment in mid-nineteenth century England, *Transactions of the Institute of British Geographers*, 13, 275–87.

Habermas, J. 1978: *Knowledge and Human Interests*. London: Heinemann.

Harvey, D. 1973: *Social Justice and the City*. London: Arnold.

Hay, I. 1998: Making moral imaginations: research ethics, pedagogy and professional human geography, *Ethics, Place and Environment*, 1, 55–75.

Hinman, L.M. 1994: *Ethics: A Pluralistic Approach to Moral Theory*. Fort Worth, TX: Harcourt Brace.

Low, N. and Gleeson, B. 1999: *Justice, Society and Nature: An Exploration of Political Ecology*. London: Routledge.

Lynn, W.S. 1998: Contested moralities: animals and moral values in the Dear/Symanski debate, *Ethics, Place and Environment*, 1, 223–42.

Proctor, J.D. 1998: The social construction of nature: relativist accusations, pragmatist and critical realist responses, *Annals of the Association of American Geographers*, 88, 352–76.

Proctor, J.D. 1999: Introduction: overlapping terrains. In J.D. Proctor and D.M. Smith (eds), *Geography and Ethics: Journeys in a Moral Terrain*, London and New York: Routledge, 1–16.

Sack, R.D. 1997: *Homo Geographicus: A Framework for Action, Awareness and Moral Concern*. Baltimore: Johns Hopkins University Press.

Silk, J. 1998: Caring at a distance, *Ethics, Place and Environment*, 1, 165–82.

Smith, D.M. 1977: *Human Geography: A Welfare Approach*. London: Arnold.

Smith, D.M. 1979: *When the Grass is Greener: Living in an Unequal World*. Harmondsworth: Penguin.

Smith, D.M. 2000: *Moral Geographies: Ethics in a World of Difference*. Edinburgh: Edinburgh University Press.

Tuan, Yi-Fu 1986: *The Good Life*. Madison, WI: University of Wisconsin Press.

Tuan, Yi-Fu 1989: *Morality and Imagination: Paradoxes of Progress*. Madison, WI: University of Wisconsin Press.

Unwin, T. 1992: *The Place of Geography*. Harlow: Longman.

Valentine, G. 1999: Being seen and heard? The ethical complexities of working with children and young people at home and at school, *Ethics, Place and Environment*, 2, 141–55.

Walzer, M. 1994: *Thick and Thin: Moral Argument at Home and Abroad*. Notre Dame, IN: Notre Dame University Press.

Welch, J. and Emel, J. (eds) 1998: *Animal Geographies: Place, Politics and Identity in the Nature–Culture Borderlands*. London: Verso.

Further Reading

An excellent introduction to research ethics in geography, incorporating useful practical examples and guidance, is Hay (1998), while a thought-provoking collection of essays by some of the leading geographers concerned with ethics is found in J.D. Proctor and D.M. Smith (eds) *Geography and Ethics: Journeys in a Moral Terrain* (London and New York: Routledge, 1999). The most accessible text on the interface between geography and ethics is David M. Smith's *Moral Geographies* (2000). The journal *Ethics, Place and Environment* includes numerous examples of the ways in which geographers have recently sought to engage with ethics and moral questions con-cerning place and the environment. *Environmental Ethics* is a scholarly interdisciplinary journal dedicated to the philosophical aspects of environmental problems.

Part IV
Geography in Context

Geography and geographers do not exist in isolation. The history of the subject shows how there have been numerous exchanges with natural sciences, social sciences and the humanities. Over the decades geography has borrowed ideas and theories, as well as passed them on to other researchers. In many instances, a distinct and separate geography is hard to find. Some people may find this disturbing, but for the majority what matters is that geography provides a base from which to ask searching and difficult questions about the physical and social worlds, as well as the relations between them.

In this section we place geography in its wider intellectual context. David Livingstone's short but informative history of the subject reveals continuities and changes from the Greek and Roman worlds. He argues that the 'geographical tradition', as he calls it, has evolved and adapted to different social and intellectual environ-ments. Rather than define the *nature* of ge-ography, he shows how it has meant differ-ent things to different people at different times and places, though all the while main-taining a finite number of themes and ques-tions. Two further chapters discuss how geography shares some of the major issues in natural and social sciences. In relation to the social sciences, Gary Bridge and Ali Rogers discuss issues including whether peo-ple are predictable, the impact of technol-ogy on society and the role of physical environment in world economic inequality. In relation to the natural and physical sci-ences, Heather Viles considers themes in-cluding the philosophy and methodology of science, chaos theory and Gaia. Finally, we have compiled a timeline from the mid-nineteenth century to the end of the twentieth century in order to help you place developments in geography in their social and historical contexts.

46

A Brief History of Geography

David N. Livingstone

Geography has meant, and still means, different things to different people. For some it conjures up images of faraway places and intrepid explorers going where none has gone before. For others, the geographer is regarded as the person with an encyclopaedic knowledge of the longest rivers, the highest mountains, the largest cities and so on – a sort of talking atlas invaluable for television quiz shows but useful for little else. For still others geography is the subject that deals in charts and globes; history is about chaps, it is said, geography about maps. In all likelihood today's professional geographers would reject all these commonplace notions as definitions of their discipline and provide their own explanation of just what geography is all about.

I do not intend to adjudicate these disparate claims. All of them – and doubtless many others – are valid interpretations of geography to one degree or another. Instead my intention is to look at what people *have taken geography to be* over the years and to trace the evolution of what I like to call 'the geographical tradition'. Accordingly I have no desire to defend any particular definition of geography, as many historians of the subject have done; rather, I just want to consider some of the different ways people have thought about it down through the ages.

In order to confront this task I propose to identify some ten different discourses – conversations, if you will – in which geography has been engaged. Certainly my list is not exhaustive. Nor does it have to be. For the heart of my argument is simply that geography changes as society changes, and that the best way to understand the tradition to which geographers belong is to get a handle on the different social and intellectual environments within which geography has been practised. Some of the topics that we shall explore will undoubtedly seem bizarre, or exotic, or quaint to modern eyes; but if we are to take history seriously we will have to learn to understand past geographies in *their own contexts* without subjecting them to twenty-first-century judgements.

To the Ends of the Earth

The story of geography, like the history of many sciences, has frequently been traced back to the Greek and Roman worlds, to figures like Thales, Anaximander, Herodotus, Strabo, Ptolemy and a dozen or more others. Their contributions – frequently of a mathematical character – did much to advance geographical theory. But it was through the explorations of Muslim scholar-travellers like ibn-Batuta and ibn-Khaldun, and the voyages of the Scandinavians, the Chinese and medieval Christian adventurers that first-hand knowledge of the world began to contribute to geographical lore. Eventually the European explorers of the fifteenth and sixteenth centuries helped to transform these earlier fragmentary gleanings into a more or less coherent body of knowledge about the terrestrial globe.

Indeed, it could be argued that the voyages of discovery, so called, made a vital contribution to the development of science in the West. Many of these seafarers, for example, saw themselves as involved in world-scale experiments to test the accuracy of Renaissance concepts inherited from the ancient classical world. This is not to say, of course, that they all thought of themselves as proto-scientists; many were just lustful for adventure on the high seas or greedy for the untold riches of exotic kingdoms. But the information they gathered helped challenge the scholarly authorities of the day by demonstrating that people *did* inhabit the southern hemisphere or that there were varieties of plant and animal that just did not fit into Aristotle's taxonomy. Besides all this, the whole business of navigation required sophisticated technological and scientific skills to determine a ship's position at sea and, more important, to chart the way back to safe havens. So it is not surprising that the navigational institute that Prince Henry the Navigator established at Sagres in the early fifteenth century – and which drew together experts in cartography, astronomy and nautical instrumentation – has been seen as a crucial early move in the development of western science. The names of Diego Cão, Bartholomew Dias, Vasco da Gama, the Cabots, Christopher Columbus, Francis Drake and Ferdinand Magellan – to name but a few – thus all occupy as important a niche in the early annals of modern geography as the republication of Ptolemy's *Geography* in 1410.

Of course geography's engagement with exploration did not come to an end in the fifteenth century. Voyages of reconnaissance continued to expand geographical knowledge of the globe throughout later centuries, and special mention might be made of the eighteenth-century journeys of James Cooke and Joseph Banks into the South Pacific and the nineteenth-century circum-

navigations of such naturalists as Charles Darwin and Thomas Henry Huxley. At the same time, the significance of scientific travel was being championed by men like Alexander von Humboldt, Henry Walter Bates and Alfred Russel Wallace through their own explorations of the Far East and South America. Indeed, the Royal Geographical Society, which did so much to advance overseas exploration in the Victorian era, continues to sponsor expeditions of this sort right up to the present day. Moreover, geographers have continued to speak of expeditions in other contexts: expeditions into the urban jungle, ethnic ghettos and other such 'threatening' environments. The vocabulary of exploration thus continues to capture the spirit of certain aspects of the geographical tradition. My argument here is simple: geography has always been closely associated with the exploring instinct.

Geography is Magic!

Even while new geographical knowledge was challenging accepted scholarly traditions there were ways in which geographical lore continued to confirm long-held beliefs. Thus, just as other nascent sciences were deeply implicated in various magical practices, so too was geography. This is plain, for example, in the early development of modern astronomy. Much interest in the stars was stimulated by astrological concerns and among the earliest Copernicans there is evidence of a continuing interest in that enterprise. Kepler, for example, cast his own horoscope every day, and in so doing he was far from unique. Aside from this the belief that various plants possessed hidden occult powers that could be harnessed for medicinal powers led to important pharmacological and chemical findings. Moreover, the writings of such giants of the scientific revolution as Bacon and Newton reveal a substantial interest in such seemingly arcane practices.

Geography, of course, was no less identified with astrology and natural magic than these other fields of discourse. Numerous early writers on geography, like William Cunningham, Thomas Blundeville, John Dee, and Thomas and Leonard Digges, involved themselves in various aspects of magic. For some, like Dee, the key lay in the mystical significance of number – the celestial and terrestrial worlds were held together in certain mathematical relationships in such a way that changes in one directly influenced the other. For others, like the Digges, astrology was of first importance and their early meteorological efforts were all of a piece with astrological knowledge; to them weather forecasting required acquaintance with the significance of celestial changes in the moon, the stars and the planets. For still others, notably Jean Bodin and Cunningham, the diversity of the world's peoples and cultures was closely bound up with which sign of the zodiac governed the particular region they inhabited.

No doubt this chapter in the history of geography will seem utterly bizarre to modern eyes. But it would be mistaken to ignore it, or suppress it, as historians of geography have all too frequently done, because it demonstrates the role of apparently non-rational discourse in the evolution of the discipline. Moreover, this geographical interest in the mystical has continued to manifest itself right up to the early twenty-first century. Recent work has shown various mystical elements in the history of the modern conservation movement – in late nineteenth-century and early twentieth-century figures like Francis Younghusband and Vaughan Cornish, for example – and that strain of thought which spiritualizes, even divinizes, nature continues to be with us to the present day.

A Paper World

The knowledge explosion occasioned by the European voyages of exploration soon brought new cartographic challenges and accomplishments. To be sure, the science of cartography was not born in the sixteenth century. Portolano sea charts had been circulating for long enough around the Mediterranean, and of course there already existed numerous symbolic depictions of the world in the form of various Mappaemundi. But now whole new worlds had to be reduced to paper and that brought new challenges. Gerard Mercator solved some of the mathematical problems associated with transferring a sphere to a flat surface with his famous map projection. Soon a series of Dutch and Belgian cartographers such as De Jode, Jodocus Hondius and Petrus Plancius splendidly mapped the progress of overseas discovery. Closely associated with these accomplishments was the development of surveying skills and instruments, and so instrument making was frequently one of the craft competences of the early cartographer.

Map making, of course, was as artistic a practice as it was scientific. Frequently maps were elaborately decorated and skilfully executed, so much so that they frequently became *objets d'art* in their own right. Besides, the whole cartographic impulse in painting is nowhere more clearly manifest than in the Dutch art of the seventeenth century. And this serves to remind us of the early associations between geography and humanistic endeavours.

In the following centuries geography's links with cartography have continued to be maintained. The progress of the Ordnance Survey's work in nineteenth-century Britain was regularly reported at the Royal Geographical Society; geographers frequently involved themselves in the thematic mapping of drift geology, soils, disease, populations and so on; now in our own day geographers maintain this tradition when they turn to remote sensing and computer mapping. The mapping drive has thus always been strong in geography; so much so that Carl Sauer believed that, if a geographer was not fascinated by maps to the extent of always needing to be surrounded by them, then that was a clue that he or she had chosen the wrong profession.

A Clockwork Universe

In the wake of the mechanical philosophy that came to dominate science in the seventeenth century, there were numerous efforts to retain the integrity of religious discourse in the face of the apparently naturalistic implications of a mechanistic world picture. One of the most common strategies, defended by men like Newton and Boyle, was to argue that the world was essentially like a grand clock, comparable with that at Strasbourg, and that by investigating the world machine scientists were interrogating the very mind of the Great Designer. This logistic move was to play a key role in the evolution of the geographical tradition. Numerous writers during the period of the Enlightenment developed a style of natural history called physico-theology. Regarding the world as teleologically designed and providentially controlled, they interpreted the world environment as a functioning revelation of divine purpose. In the writings of Thomas Burnet, John Ray, John Woodward, William Derham, as later in the works of William Paley, the world's geography – its physical and organic forms – was seen as pointing beyond itself to nature's God.

Of course these practitioners of natural theology differed frequently among themselves on both detail and strategy; but between them they delivered to history a vision of nature as a holistic system, a sort of ecological picture, that emphasized the interrelationships and interdependencies among organisms and environment. Here the image of warfare between science and religion turns out to be something of a historical fiction. Indeed, there were geographers like Bartholomaus Keckermann in Germany (author of *Systema Geographicum*) and Nathanael Carpenter in England (author of *Geography Delineated Forth*) whose commitment to the theology of the Reformation encouraged them to *reject* ecclesiastical authority in matters of science and to argue for the liberation of science from scholastic censure.

This particular intellectual trajectory continued to inform geographical thought over the next centuries. In the nineteenth century Karl Ritter exemplified the same stance, and the Ritterian vision was propagated in the United States by his disciple-devotee Arnold Guyot. Besides these there is much evidence of teleological thinking in the works of Mary Somerville and David Thomas Ansted in England, and Matthew Fontaine Maury and Daniel Coit Gilman in the United States. Indeed, H.R. Mill, writing in 1901, was entirely correct when he noted that teleological modes of reasoning were 'tacitly accepted or explicitly avowed by almost every writer on the theory of geography'. Even more recently, the self-same teleological vision comes through in the writings of the Dutch geographer De Jong. Here geography continues to operate as the handmaiden to theology.

On Active Service

If, as we have just seen, geography could subserve theological ends, its services to external interests did not stop there. Throughout the nineteenth century it was frequently cast as the *aide de camp* to militarism, imperialism, racism and doubtless a host of other 'isms'. Maps, it was long known, were as vital implements of warmongering as gunnery, and it is no surprise that institutional geography first flourished in military schools. Indeed, the prehistory of the Ordnance Survey can be traced back to military needs during the Jacobean era, while in the twentieth century geographers like Isaiah Bowman played their parts in America's involvement with post-war European reconstruction.

By the same token British expansion overseas aroused a renewed interest in geography for its functional purposes. At the inaugural meeting of the Royal Geographical Society of London in the early 1830s the need for such a society was defended on the grounds that geography was vital to the imperial success of Britain as a maritime nation. Accordingly there was, and continued to be, considerable debate in British – not to mention German and American – geography on the subject of acclimatization because the question of white adaptation to the tropical and subtropical worlds was of pressing international significance. Here geographers worked closely with medical experts to delineate the significance of climatic factors. Indeed, in so doing they kept alive an ancient tradition, rejuvenated by Montesquieu, that explained the cultural in terms of the natural.

Besides this there were certain aspects of geographical theory ripe for manipulation. Environmental determinism – a doctrine emphasizing the moulding power of physical conditions – could be used for a range of purposes. Some found in it justification for a racial ideology; indeed, racial questions were commonplace in geography texts around the turn of the twentieth century and in some cases long after that. Others saw in it a doctrine with strategic potential. Halford Mackinder, for example, outlined a theory of world political power that crucially depended on the control of a particular piece of territorial space in the Old World. Friedrich Ratzel in Germany erected an organic theory of the state on his notion of *Lebensraum*, urging that the character and destiny of a *Volk* were umbilically tied to a definite area or *Raum*. In the United States the Ratzelian viewpoint was propagated by Ellen Semple, who used it to chart the necessitarian course of American history, while Ellsworth Huntington turned to climate as the great mainspring of civilization. In all of these, as in the stop-and-go determinism of Griffith Taylor, the constitutive links between geographical theory and social outlook are clearly displayed. This is not to say, of course, that geographical determinism as a precept was *just* social ideology writ large. But it *is* to recognize that there is a *social* history of geographical ideas as well as a purely *cognitive* one.

The Regionalizing Ritual

Even while environmental determinism in one form or another was spreading like wildfire among professionalizing geographers, there were those who insisted on the capacity of human culture to transform its natural milieu rather than remaining in nature's deterministic grip. In Britain H.J. Fleure emphasized the importance of human agency in modifying environment and thus turned away from the conventional concentration on natural regions towards the significance of transitional zones of culture contact down through history. Moreover, even those like A.J. Herbertson, in whose geography the concept of natural region occupied a strategic place, nevertheless recognized the subtle interplay of environment, heredity and consciousness in producing the geographical patterns of human diversity across the face of the globe. For both the idealist strain in Lamarckian evolution – an evolutionary model stressing the significance of life-force and will – was of crucial importance. Another strain of environmentalist critique was forthcoming from a different, though related, conceptual source around the turn of the twentieth century, namely the vibrant tradition of French cultural geography associated with Vidal de la Blache. For Vidal and the Vidalians, environment was to be seen not as a determinative force but rather as a limiting factor setting limits on cultural possibilities. Possibilism, as this doctrine was styled, also emphasized the science of human regions because it was in specific physical milieux that distinctive *genres de vie* – modes of life – found expression.

A third strand of determinist criticism emanated from Carl Sauer and the Berkeley school of cultural geography in the United States. Here inspiration was derived less from evolutionary biology than from cultural anthropology and can be traced back to the seminal influence of the anthropologist Franz Boas. Boas had begun his academic career as a physical geographer but turned to anthropology when his work among the

Inuit led him to question environmental determinism. The mild cultural relativism that he came to espouse was mediated to Sauer through anthropological colleagues at Berkeley, and Sauer built on these foundations as he emphasized the importance of residual material culture as historical artefacts of cultural diversity.

Whatever the differences in approach, all these geographers shared a conception of geography as a study of regions. And this brand of geography received its benediction in Richard Hartshorne's influential monograph *The Nature of Geography*, in which he argued his apologetic case from a partisan review of historical – and in particular German – sources. Thus the notion of geography as the 'regionalizing ritual' provided a paradigm that still governs much geographical work, whether in the qualitative contributions of writers of regional personality or in the more quantitative emphasis of the practitioners of regional science.

The Go-between

Alongside these efforts to delineate for geography a piece of cognitive territory – a sector of conceptual space in the academic scheme of things – there were those who were rather more inclined to stress its functional role. Frequently the case was made that geography was the integrating discipline *par excellence* that kept the study of nature *and* culture under one disciplinary umbrella. W.M. Davis, for example, otherwise remembered for his elucidation of the cycle of erosion, nevertheless felt that physical geography was incomplete without ontography, its human counterpart. This go-between function was valuable in a number of contexts. For one thing it was appealed to to justify geography as a coherent and independent academic discipline both in Britain and the United States. Indeed, Halford Mackinder in Britain found this to be the only foundation on which geography as a causal science could be built. In the

United States Isaiah Bowman championed the same view.

Besides this, geography's bridging role between nature and humanity frequently took the form of a strenuous engagement with questions of resources. In America the roots of this geographical tradition go back to such figures as Nathaniel Southgate Shaler and George Perkins Marsh, and later J. Russell Smith, whose contributions were resurrected by early twentieth-century geographers seeking the recovery of a tradition of environmental sensitivity. For some this emphasis led to a historical reassessment of 'man's role in changing the face of the Earth'; for others the needs of the future fostered an engagement with environmental systems analysis or with ecological energetics in the attempt to model the changing human–nature interface. In our own day, as the environmental crisis has bitten even more deeply, geographers like Timothy O'Riordan and Andrew Goudie have done much to keep this tradition at the forefront of geographical discourse. Moreover, so far as institutional identity is concerned, it is noteworthy that university and college geography is not infrequently housed in schools of environmental studies.

A Science of Space

If some identified geography's essence in its focus on regional integration, there were those who found the emphasis on the particularity of places lacking in methodological rigour. To them, all the talk of bridging the gulf between the sciences and the humanities seemed little more than academic-political rhetoric, and the idea of regional personality frankly unscientific. Fred Schaefer spearheaded the attack with his article on 'Exceptionalism in geography', published in the *Annals of the Association of American Geographers* in 1953. Schaefer's critique was designed to transform geography into a true science by urging that it become a law-seeking explanatory discipline concerned with universal laws, not regional specifics, or, as he put it, 'exceptions'. Schaefer's paper, it is commonly believed, heralded the introduction of logical positivism into the discipline, and its curriculum was defended in William Bunge's *Theoretical Geography* of 1962 and David Harvey's *Explanation in Geography* published at the end of the decade. And thus was born the idea of geography as a science of spatial distribution – locational analysis, as it was frequently styled – and soon various theorems seeking to explain the location of economic behaviour were introduced to geography by figures such as W.L. Garrison in America and Peter Haggett in Britain. In particular the earlier economic theorizing of Von Thünen, Alfred Weber, Walter Christaller and August Lösch soon began to receive an airing in the discipline.

Along with this definition of geography as spatial science came the paraphernalia of scientific knowhow, and thus geography received its newest initiation into scientific method and statistical technique. Not, of course, that geography had been utterly innocent of quantification hitherto. The roots of geography as a mathematical practice can be traced back at least to the period of the scientific revolution in the seventeenth century, and doubtless before that. Nor does it mean that all geography was quantified; plainly many areas of the tradition remained statistically immune. Still, positivism did make substantial inroads into geographical theory and practice from the 1950s and a variety of reasons for geography's relatively late baptism in positivist philosophy have been put forward. A Marxist-converted Harvey believes it represented, at least in America, a strategic attempt by geographers to escape the political suspicion falling on social science in the post-McCarthy era by retreating into the safety of number-crunching. At least as compelling, I think, would be an explanation that takes seriously the perceived need for geographers to accrue to themselves a set of craft competences which bolstered their professional vested interests in creating a spatial *science*.

Statistics Don't Bleed

Whatever the causes of geographical quantification may have been, recent decades have witnessed a sequence of attacks on positivism from different perspectives. From the radical side comes the complaint that the whole quantitative procedure is ideologically laden from the start. The argument here is that by keeping geography as just a sort of spatial calculus, a geometric technique for depicting distributions, fundamental questions of justice and political involvement are simply – and too comfortably – ruled out of court. Accordingly, various contemporary radical geographers see themselves in a geographical lineage stretching back to figures such as Elisée Réclus, Peter Kropotkin and Karl Wittfogel who strenuously advocated social engagement. In this scenario, and it has to be admitted that it is far from unified, there is something of an emphasis on the determinative role of economic structure. Whether investigating the significance of residential segregation, the vicissitudes of the world economic system or the historical change from feudalism to capitalism, this same *motif* regularly reasserts itself.

From another perspective, there are those humanistic geographers who insist that the quantitative tabulation of economic data and other activities has dehumanized geography by ignoring, not to say suppressing, human agency. Statistics are simply not made of flesh and blood. Whole acres of human experience – fear, imagination, emotion – are left out of the picture. And these geographers have seen it as their task to keep the geographical world open to the artistic side of its history by their interrogation of literary texts and their championing of subjectivism in the subject. Yi-Fu Tuan's meditations on 'topophilia' and 'topophobia', David Ley's excursion into the mind of the inner-city ghetto, Leonard Guelke's turning to Collingwood's idealist philosophy of history are just some of the currents to have swept through the discipline recently. Again partisans are quick to point out that this is not a wholly new departure: some claim that the earlier behavioural geography of J.K. Wright, David Lowenthal and William Kirk accorded a key role to subjective experience, while others, ignoring Vidal's natural science aspirations for *géographie humaine*, speak of the revivification of the Vidalian tradition.

cultural and epistemological pluralism now seems inevitable. Fragmentation of knowledge, social differentiation and the questioning of scientific rationality have all coalesced to reaffirm the importance of the particular, the specific, the local. And in this social and cognitive environment a geography stressing the centrality of place is seen as having great potential. Once again the constitutive nature of the relationship between geography's internal domain and external context is clearly evident.

Everything in its Place

These respective emphases on the role of social structure and human agency in accounts of geographical phenomena have most recently led some to wonder whether explanatory privilege ought to be accorded to either side of the equation. In the attempt to find a way out of the impasse, some geographers have turned to the theory of 'structuration' advanced by the Cambridge sociologist Anthony Giddens. This account of social formation and transformation highlights the interplay of both forces: human beings find themselves in structural circumstances not of their choosing, but through the exercise of their own agency can do something to bring about change. The never-ending ebb and flow of agent–structure intercourse provides the engine power of social transformation. Where geography enters the picture is in the need to 'earth' this general model of historical change. Just how the interplay of social structure and human agency falls out is evidently different from place to place and depends crucially on the particular arena of encounter. Hence geographers – arguing for the prime significance of locale – increasingly call for the geographizing of social theory.

What has given further encouragement to this renewed emphasis on the significance of place is a whole series of philosophical and social developments. The details need not concern us, save to note that the idea of

Geographical Conversations

Little needs to be said in conclusion. My argument throughout has been that the geographical tradition, like a species, has evolved as it has adapted to different social and intellectual environments. Geography, as was noted at the beginning, has meant different things to different people at different times and in different places. It has employed different vocabularies to suit different purposes – from magic and theology to science and art. Sometimes these discourses have been in conflict; at other times they have been mutually reinforcing. Sometimes the conversations have admitted a range of geographers; sometimes only a select group were allowed to take part. Either way, what is important is that in telling the story of the tradition to which geographers belong there needs to be a recognition of the integrity of each of these diverse discourses in their own terms. Otherwise the history and future of geography will be enslaved to partisan apologists who wish to monopolize – even hijack – the conversation in order to serve their own sectarian interests.

Further Reading

This chapter is derived from D.N. Livingstone, *The Geographical Tradition: Episodes in the History of a Contested Enterprise* (Oxford: Blackwell, 1992). The history of geography is further explored in

D.N. Livingstone and C.W.J. Withers (eds), *Geography and Enlightenment* (Chicago: University of Chicago Press, 1999); R. Mayhew, *Enlightenment Geography: The Political Languages of British Geography, 1650–1850* (London: Macmillan, 2000); and C.W.J. Withers, *Geography, Science and National Identity: Scotland Since 1520* (Cambridge: Cambridge University Press, 2001).

Two books on geography and its relationships with exploration and imperialism are A. Godlewska and N. Smith (eds), *Geography and Empire* (Oxford: Blackwell, 1994); and F. Driver, *Geography Militant: Cultures of Exploration and Empire* (Oxford: Blackwell, 2001).

On physical geography in particular, see K.J. Gregory, *The Changing Nature of Physical Geography* (London: Arnold, 2000); and K.J. Tinkler, *A Short History of Geomorphology* (London: Croom Helm, 1985).

On geography and its relations with the earth sciences in general, including evolutionary biology, plate tectonics and ecology, see P.J. Bowler, *Fontana History of the Environmental Sciences* (London: Fontana, 1992).

47

Geography and the Natural and Physical Sciences

Heather A. Viles

Science, man's greatest intellectual adventure, has rocked his faith and engendered dreams of a material Utopia. At its most abstract, science shades into philosophy; at its most practical, it cures disease. It has eased our lives and threatened our existence. It aspires, but in some very basic ways fails, to understand the ant and the Creation, the infinitesimal atom and the mind-bludgeoning immensity of the cosmos. It has laid its hand on the shoulders of poets and politicians, philosophers and charlatans. Its beauty is often apparent only to the initiated, its perils are generally misunderstood, its importance has been both over- and underestimated, and its fallibility, and that of those who create it, is often glossed over or malevolently exaggerated.

(Silver 1998: xi)

As Brian Silver eloquently and engagingly describes in his book *The Ascent of Science*, over recent decades the natural and physical sciences have both produced some remarkable achievements in fields such as cosmology and biology (such as the Human Genome Project) and also come under great attack for being expensive, arrogant, fallible and biased. Science (for short) has also increasingly been asked to provide answers to major world challenges such as how to increase world food production, and how to predict climatic change and variability over the twenty-first century. As in previous centuries, scientists have developed a vast range of new ideas and theories, most of which have been the subject of debate and discussion both within and outside the wider scientific community. Furthermore, scientists across many diverse disciplines have also continued to debate and develop issues of philosophy and methodology. Geographers (both physical and human) have contributed

to many of these issues and debates as scientists and also as geographers; the very subject matter of geography is at the heart of many of the debates. In this chapter I focus on five areas of debate involving science and scientists which have affected, and in turn been affected by, geography and geographers.

Science Wars: Attacking the Privileged Position of Science

The rise of science has been almost inexorable. From the seventeenth century to the 1960s modern science developed from a hobby of gentlemen (and less commonly recorded, women) into a powerful force for change, dominated by high-budget, big projects producing major technological advances. Along the way, scientists have answered some of the great questions and produced some fundamental insights into the behaviour of life, the universe and (almost) everything. Confidence in science probably peaked sometime between the late nineteenth and mid-twentieth centuries, when the positive advances spurred on by scientific progress seemed invincible. An extreme view on the primacy of science which should strike fear into any geographer today, dating from 1954, was given by the socialist scientist J.D. Bernal, who wrote:

> The transformation of nature, along the lines indicated by the biological sciences, will be undertaken with the use of heavy machinery, including possibly atomic energy. All the river basins of the world can be brought under control, providing ample power, abolishing floods, droughts, and destructive soil erosion, and widely extending the areas of cultivation and stock raising ... Beyond this lie possibilities of further extending the productive zone of the world to cover present deserts and mountain wastes and making full use of the resources of the seas. (Bernal 1954, quoted in Gillott and Kumar 1995)

Dissenting voices, however, have suggested that science is fallible and can be wrong and dangerous, and also that scientific knowledge should not be given any higher profile or credence than other belief systems (Gillott and Kumar 1995; Sardar 2000). Early attacks on science focused on its failure to solve problems, and its linked ability to create problems. Rachel Carson's classic book *The Silent Spring*, originally published in 1962, provides a good example of the decreasing confidence in science during the latter half of the twentieth century. Carson, a marine biologist, wrote of the harmful effects that DDT and other synthetic substances created by scientists were having on the environment. In the case of DDT, as well as in worries about many other applications of scientific discoveries, it is not perhaps science itself which is being criticized, but rather the application of science as technology through the agencies of government and industry.

More recently, attacks on science have expanded to provide a critique of its privileged status, from sociologists, feminists and post-colonial theorists. Thus, David Bloor and Barry Barnes published *Scientific Knowledge* in 1996, which tackles the sociology of scientific knowledge and shows how 'theory-laden' the reporting of many scientific results is. Other sociologists take a more extreme and counter-intuitive view, such as the constructionists Bruno Latour and Steve Woolgar, whose 1979 book *Laboratory Life: Social Construction of Scientific Facts* looked at the history of a particular scientific fact and illustrated the role of funding, personalities and technology in the identification of TRF(H), or thyroptropin releasing factor (hormone). When such authors describe observations as theory-laden or socially constructed, they mean that the available evidence 'underdetermines' scientific truth, i.e., it is not sufficient to explain why a certain theory or hypothesis is accepted. In part, as these authors claim, truth is a geographical problem, relating to how a proposition or observation literally travels from the laboratory to the outside world. Feminists have also attacked science for its masculinist viewpoint. Sandra Harding (1986) questions whether science can be reformed so as to include feminist concerns, whether a distinctively feminist standpoint on science is inherently preferable, or whether science should be transcended in some postmodern fashion. Post-colonial critics have also attempted to combat the westernized view of science and its history, for example by uncovering the contribution of Indian and Islamic civilizations to mathematics and examining the presence of racism in supposedly neutral and objective scientific practices.

Scientists have been quick, not surprisingly, to defend their endeavour. Several scientists have produced books outlining the key attributes and approaches of science and have revisited the norms of scientific behaviour proposed by Robert K. Merton in 1942 (i.e., communalism, universalism, disinterestedness and scepticism) (Ziman 1984; Lee

2000). Others have gone on the attack, showing the absurdity (as they view it) of other approaches to knowledge rather than defending science. In 1996 the physicist Alan Sokal had a paper published in *Social Text*, a cultural studies journal, under the title 'Transgressing the boundaries: towards a transformative hermeneutics of quantum gravity'. In it he presents the (patently absurd) argument that unifying quantum mechanics and general relativity produces a postmodern science. The episode, known as the 'Sokal hoax', made scientists laugh at the fallibility of social scientists, but perhaps did nothing to redress the criticisms of scientists as arrogant and of scientific knowledge as not deserving any privileged status.

Geographers occupy a special position within this multi-faceted debate about the role of science. Many geographers would describe themselves as scientists, and many geographers of various persuasions are involved in the dissemination and application of scientific ideas in a range of ways. For example, geographers are involved in studying the causes of desertification (e.g. Fullen and Mitchell 1994) and are also heavily involved in trying to develop schemes which manage the problem more effectively (e.g. Goudie 1990; Agnew and Warren 1996). Desertification, as a problem, has both social and environmental causes. Managing desertification requires both scientific and socio-cultural understanding, and thus the primacy of scientific knowledge cannot be taken for granted as a way of providing effective solutions. Furthermore, our scientific knowledge is not always adequate, that is to say, we do not always fully understand the nature and causes of desertification. As Thomas and Middleton (1994) put it, there is a 'myth' of desertification. The myth of desertification may be just as important as the science of desertification in trying to manage the resulting problems. Desertification, and the wider problem of land degradation, provide an ideal opportunity to study the role of science in causing and solving environmental problems.

Geography departments also offer a microcosm of the academic world, bringing together as they do a wide range of scientists and social scientists in (usually) one building. Departmental seminars provide a useful insight (which all geography students should make use of as part of their course) into the current relationships between science and other forms of knowledge. A cultural geographer talking to an audience half-full of pollen analysts, climatologists and fluvial hydrologists will get used to the instinctive criticisms of small sample sizes, qualitative methodologies and lack of utility of the findings. On the other hand, a fluvial hydrologist presenting a paper on the latest model of nitrate pollution down a small waterway will face a barrage of questions about how this helps to solve the problem.

Chaos and Complexity

In 1963 Edward Lorenz built a simple model of atmospheric behaviour using three independent first-order differential equations. This simple model produced unexpectedly complex results. In 1982 Benoit Mandelbrot's book *The Fractal Geometry of Nature* came out. These two events, as well as a whole host of other developments taking place in many different fields of science around the world, contributed to the rise of non-linear dynamical thinking as a powerful force within science. As Brian Silver puts it, 'One of the most delightful meetings in science has been between chaos and fractals' (Silver 1998: 233). More recently, chaos and fractals have been joined by complexity (Nicolis and Prigogine 1989; Lewin 1993). For many writers, chaos theory can be seen as a critique and replacement of conventional modes of scientific thinking and analysis. Instead of equilibrium, steady state and predictable behaviour, the world can be seen as functioning in unpredictable ways. Simple deterministic systems can give rise to highly complex non-linear behaviour (such as the simple climate model Lorenz built).

Conversely, complex systems can show ordered behaviour (such as the regular patterns formed by many turbulent flows in sedimentary environments). Chaos and complexity ideas have brought with them exotic terminology such as strange attractors and self-organized criticality. They have been proposed as a way of getting rid of reductionist tendencies in science, as they encourage links between different scales and types of science and discourage narrow focus on cause–effect links. Chaos and complexity also have wide applications within various fields of science and social science. On the other hand, there have been many criticisms of chaos and complexity as being interesting but not leading to any major or useful insights.

Geographers have been entwined with the emergence of chaos in science in many ways. On the one hand, the subject matter of geography (such as predator–prey relations, distribution of earthquakes and big cities, coastlines and climatic systems) has provided one of the most fertile testing grounds of the development of non-linear dynamical thinking. One of Mandelbrot's classic explanations of fractals relates to the measurement of the length of a coastline (Mandelbrot 1967). Pattern formation in nature, as well as highly complex behaviour, is clearly apparent in many geographical phenomena, from sand dunes through to urban morphology (Batty and Longley 1994). On the other hand, geographers have been enthusiastic in their application of non-linear dynamical concepts to such problems as soil formation, cave passage networks and the location of polar bears (respectively in Phillips 1993; Laverty 1987; Ferguson et al. 1998). Chaotic behaviour and the unpredictability of chaotic systems is of key relevance to those geographers trying to untangle the natural and human and environmental consequences of climatic change and variability. Indeed, notions of change and complexity have become firmly embedded in many attempts to explain geographical behaviour (in contrast to equilibrium and

linear ideas in the past). However, it remains to be seen whether the adoption of non-linear dynamical systems thinking to human systems is anything more than a superficial analogy. Alan Sokal's hoax ridiculed the use of mathematics and quantum theory by social theorists.

Philosophy and Methodology of Science

Allied to concerns over the status of scientific knowledge, and also linked to the rapid development of science within recent years, has been a re-evaluation of the scientific method and of the underpinning philosophies. Scientific methods have been developed in order to provide some sort of standardized frameworks for the creation of scientific knowledge. The conventional viewpoint was that science is predicated on positivist principles, and deals with the formulation, testing and refinement of hypotheses with a view to developing theories, laws and predictions. Positivism arose from the ideas of sociologist Auguste Comte (1789–1857), and flowered most notably in the early twentieth century as logical positivism expounded by a group of philosophers (with a strong scientific bent) led by Moritz Schlick. Logical positivism gave great emphasis to observation and measurement, and had a basic underlying principle that propositions were only true if they could be verified either logically or by observation. Verification has proved to be a thorny issue, and Karl Popper proposed that propositions (or hypotheses or theories) could only be falsified (i.e., disproved) rather than verified (Popper 1959). His position, referred to as critical rationalism, was also widely welcomed among philosophers of science, as well as some physical geographers (Haynes-Young and Petch 1986).

Though positivism and critical rationalism provide the bedrock of conventional scientific method, there are many alternative positions. The debates among them are often quite complicated, with several writers defining and using terms in different ways. Often the debates are rather removed from the practice of science, as scientists tend not to indulge in philosophical speculation, but they are an important element of understanding how science relates to other forms of knowledge. One alternative to positivism is 'instrumentalism', which takes the view that a proposition is true if the results that follow from it are useful (see Silver 1998: 503). Scientists often utilize such a pragmatic approach. Another alternative is realism, although there are several different versions of it. Jeff Lee, who is a geographer, has written a useful book on the scientific endeavour in which he provides a simple threefold division of post-positivist approaches to science based on the work of Rhoads and Thorn (1994), i.e., post-positivist empiricism, relativism and realism. Post-positivist empiricism acknowledges that some parts of theories cannot be verified directly by experimentation but are nevertheless cornerstones of scientific progress. Thus, according to this way of thinking truth is at least partly theory-laden. In contrast, relativism, like constructionism, emphasizes the role of social factors in the development of scientific ideas. Thomas Kuhn's famous book *The Structure of Scientific Revolutions*, published in 1962, presents a version of this approach, illustrating how science works within 'paradigms' or shared visions of how the world works up until the existing paradigm becomes unable to cope with a new development, and then a revolution occurs and a new paradigm replaces it. A more extreme position is taken by anarchist philosopher Paul Feyerabend in his book *Against Method*, published in 1975. He claims that there are no grounds for believing that any method, including science and witchcraft, is preferable to any other. There are no rules the breaking of which cannot be shown to have produced sound results. Finally, realism acknowledges that even though science may not ever reach absolute truth, it can get close, as there is a real world out there separate to our perceptions of it. Theories or laws are therefore always adequate approximations.

Many scientists view such philosophical and methodological concerns as being irrelevant to their work. Different areas of science also experience different difficulties with method. In much of modern physics, for example, theory has outpaced experimentation and areas such as superstring theory, which combines general relativity with quantum mechanics, have been claimed to be untestable (Gillott and Kumar 1995). In geography, positivism and its near neighbours have been largely (although by no means totally) rejected by human geographers over the past couple of decades. Physical geographers, however, still profess to follow 'a or the scientific method', even though many do not appear to know exactly what that is. Something of a debate has emerged in physical geography, especially amongst geomorphologists, about the role of realist and other approaches to the development of theory within the subject. In 1990 Keith Richards wrote an editorial for the journal *Earth Surface Processes and Landforms* which put the case for a realist approach to geomorphology rather than the critical rationalist or Popperian tradition. Some debate has ensued over the interpretations of both realism and critical rationalism proposed by Richards and of the role of realism in guiding geomorphological enquiry (e.g. Bassett 1994; Rhoads 1994; Harrison and Dunham 1998).

The Nature of Evolution

Changing ideas over the nature of evolution have, even since before the publication of Charles Darwin's *The Origin of Species* in 1859, exerted a major influence on the development of science (and indeed other spheres of endeavour such as social sciences). Ideas on evolution, and more broadly on the changing nature of communities and environments over long timespans, have continued to develop over recent decades and to have widespread influence. Niles Eldredge, a palaeontologist, has expressed the nature of evolutionary theory as being a continuing dialogue and debate between what he calls ultra-Darwinians (usually geneticists) and naturalists with contrasting perspectives. As he puts it:

> Ultra-Darwinians emphasise continuity through natural selection and the primacy of active competition for reproductive success as the prime mover underlying absolutely all evolutionary phenomena. Naturalists, in contrast, see the complex biotic world as composed of discrete entities. Discontinuity is as important as continuity in depicting the real, natural world. (Eldredge 1995: 7)

Ultra-Darwinian viewpoints are represented by the ideas of Richard Dawkins amongst others, whilst naturalists have a champion in Stephen Jay Gould. Both authors have been prolific in writing accessible books which present evolutionary ideas to a wide audience. Stephen Jay Gould, along with Niles Eldredge, wrote a paper in 1972 which introduced the 'punctuated equilibrium' view of evolution that has been much debated ever since. The basic idea at the heart of punctuated equilibria is that evolutionary change seems to concentrate in particular episodes (linked with the origin of new species) rather than accumulating gradually and steadily over time.

Similar ideas have emerged in other areas of the earth and environmental sciences, and geography has been quick to absorb evolutionary ideas. In geomorphology, for example, Denys Brunsden sees geomorphic change as acting as waves of activity interspersed with periods of nothing much happening (Brunsden 2001). Indeed, Ian Douglas and co-workers go as far as to use the term 'punctuated equilibria' to describe the alternating periods of rapid change and stasis within forested geo-ecosystems in the Danum Valley, Sabah (Douglas et al. 1999). There are clear convergences between the concepts of punctuated equilibria and those of self-organized criticality and the shift from chaotic to ordered behaviour. The tendency of geographers to borrow concepts and ideas from other scientific disciplines and to read widely among other scientific literatures has led to a number of different applications of evolutionary ideas and metaphors.

Gaia

Discussion of the division between living and non-living things, and the links between them, has lasted for centuries. During the second half of the twentieth century, stemming largely from pioneering work carried out by the great Russian scientist Vernadsky, a new, highly controversial view has been developed by James Lovelock and co-workers. In a series of books, Lovelock has outlined the view that life regulates the environment on Earth and that, following this argument to its logical conclusion, the Earth can be regarded as an organism (e.g. Lovelock 1979, 2000). This concept of a self-regulating life-force has been given the title Gaia and has been the subject of much debate amongst scientists and those interested in the environment. A whole bandwagon of environmental mumbo jumbo has developed alongside the Gaia hypothesis, which, as some critics have argued, might not have appeared if 'instead of being named after a Greek goddess, the theory had been called coordinated interactive non-linear dynamics in the terrestrial bio- and geospheres' (Silver 1998: 447). Gaia also purports to be a scientifically valuable idea with practical applications, and James Lovelock has more recently coined the term geophysiology to cover attempts to understand and 'mend' the functioning of the Gaian system.

Geographers are concerned with much that is at the heart of Gaia, and indeed many physical geographers, ecologists and geologists find much of interest at the interface between living and non-living systems (Myers 1990). There have been complaints that physical geography has neglected the sense of the physical world as enchanted, or possessing spiritual values. The discipline more or less missed out on the wave of environmental concern in the 1970s, perhaps in part because of the preoccupation with being a hard, numerical science (Simmons 1990). Few geographers, however, are happy to jump completely onto the Gaian bandwagon, especially as many of the ideas are seemingly untestable.

Conclusions: The Relations Between Geography and Science

It is clear that many geographers view themselves as scientists, while a large percentage do not. It is also clear that developments in scientific knowledge and its results continue to have a major impact on both the subject matter of geography and the ways in which we study it. There has always been an interplay of concepts and ideas between the 'hard' and social sciences, and the start of the twenty-first century is no different in this respect. Geography is poised at the meeting point between natural and physical sciences and the social sciences, and thus is in a key position to both assist and benefit from this flow of ideas. Human–environment relations, one of the core areas of modern geography, provide a fertile area for the development of scientific theories and products.

References

Agnew, J. and Warren, A. 1996: A framework for tackling drought and land degradation, *Journal of Arid Environments*, 33, 309–20.

Bassett, K. 1994: Comments on Richards: the problems of 'real' geomorphology, *Earth Surface Processes and Landforms*, 19, 273–6.

Batty, M. and Longley, P. 1994: *Fractal Cities*. London: Academic Press.

Brunsden, D. 2001: A critical assessment of the sensitivity concept in geomorphology, *Catena*, 42, 99–123.

Carson, R. 1965: *The Silent Spring*. Harmondsworth: Penguin.

Douglas, I., Bidin, K., Balamurugan, G., Chappell, N.A., Walsh, R.P.D., Greer, T. and Sinun, W. 1999: The role of extreme events in the impacts of selective tropical forestry on erosion during harvesting and recovery phases at Danum Valley, Sabah, *Philosophical Transactions, Royal Society of London B*, 354, 1749–61.

Eldredge, N. 1995: *Reinventing Darwin: The Great Evolutionary Debate*. London: Weidenfeld and Nicolson.

Eldredge, N. and Gould, S.J. 1972: Punctuated equilibria: an alternative to phyletic gradualism. In T.J.M. Schopf (ed.), *Models in Palaeobiology*, San Francisco: Freeman and Cooper, 82–115.

Ferguson, S.H. et al. 1998: Fractals, sea ice landscape and spatial patterns of polar bears, *Journal of Biogeography*, 25, 1081–92.

Fullen, M.A. and Mitchell, D.J. 1994: Desertification and reclamation in North Central China, *Ambio*, 23, 131–5.

Gillott, J. and Kumar, M. 1995: *Science and the Retreat from Reason*. London: Merlin Press.

Goudie, A.S. (ed.) 1990: *Techniques for Desert Reclamation*. New York: Wiley.

Harding, S. 1986: *The Science Question in Feminism*. Milton Keynes: Open University Press.

Harrison, S. and Dunham, P. 1998: Decoherence, quantum theory and their implications for the philosophy of geomorphology, *Transactions, Institute of British Geographers*, 23, 501–14.

Haynes-Young, R.H. and Petch, J. 1986: *The Nature of Physical Geography*. London: Harper and Row.

Laverty, M. 1987: Fractals in karst, *Earth Surface Processes and Landforms*, 12, 475–81.

Lee, J.A. 2000: *The Scientific Endeavor: A Primer on Scientific Principles and Practice*. San Francisco: Benjamin Cummings.

Lewin, R. 1993: *Complexity: Life at the Edge of Chaos*. London: Phoenix.

Lovelock, J. 1979: *Gaia: A New Look at Life on Earth*. Oxford: Oxford University Press.

Lovelock, J. 2000: *Gaia: The Practical Science of Planetary Medicine*. Stroud: Gaia.

Mandelbrot, B. 1967: How long is the coast of Britain? *Science*, 156, 637.

Mandelbrot, B. 1982: *The Fractal Geometry of Nature*. San Francisco: W.H. Freeman.

Myers, N. 1990: Gaia: the lady becomes ever more acceptable, *Geography Review*, 3, 3–5.

Nicolis, G. and Prigogine, I. 1989: *Exploring Complexity*. New York: W.H. Freeman.

Phillips, J.D. 1993: Chaotic evolution of some coastal plain soils, *Physical Geography*, 14, 566–80.

Popper, K. 1959: *The Logic of Scientific Discovery*. London: Hutchinson.

Rhoads, B.L. 1994: On being a 'real' geomorphologist, *Earth Surface Processes and Landforms*, 19, 269–72.

Rhoads, B.L. and Thorn, C.E. 1994: Contemporary philosophical perspectives on physical geography with emphasis on geomorphology, *Geographical Review*, 84, 90–100.

Richards, K. 1990: 'Real' geomorphology, *Earth Surface Processes and Landforms*, 15, 195–7.

Sardar, Z. 2000: *Thomas Kuhn and the Science Wars*. Cambridge: Icon Books.

Silver, B.L. 1998: *The Ascent of Science*. New York: Oxford University Press.

Simmons, I. 1990: No rush to grow green, *Area*, 22, 384–7.

Thomas, D.S.G. and Middleton, N.J. 1994: *Desertification: Exploding the Myth*. London: Arnold.

Ziman, J. 1984: *An Introduction to Science Studies: The Philosophical and Social Aspects of Science and Technology*. Cambridge: Cambridge University Press.

Further Reading

Bak, P. 1997: *How Nature Works*. Oxford: Oxford University Press. A readable introduction to self-organized criticality from an expert.

Bauer, B.O. 1999: On methodology in physical geography: current status, implications and future prospects, *Annals, American Association of Geographers*, 89, 677–778. This paper introduces a collection of papers on methodological issues in physical geography which discusses the utility of chaos theory to geography, as well as a host of other issues.

Phillips, J.D. 1999: *Earth Surface Systems: Complexity, Order and Scale*. Oxford: Blackwell. A useful review of non-linear dynamical systems ideas and their application to geomorphology.

Volk, T. 1998: *Gaia's Body: Towards a Physiology of Earth*. A readable explanation of Gaian ideas from an enthusiast.

Internet Resources

- After The Sokal Affair and *Impostures Intellectuelles* is a website devoted to the Sokal hoax and science wars: http://www.math.tohoku.ac.jp/~kuroki/Sokal/index.html

48

Geography and the Social Sciences

Gary Bridge and Alisdair Rogers

The relationship between geography and the social sciences has been both productive and selective, with strong connections forged around common themes existing alongside weak links, misunderstanding and indifference. Taking the long view of the discipline, as David Livingstone does in chapter 46, we can see that geography has often exchanged ideas and concepts with the full range of social sciences. In the late nineteenth century this included anthropology and sociology, and into the twentieth century economics figured prominently. In recent decades political economy and cultural studies can be added. A full account of these exchanges is beyond our scope. Here we focus on just five key debates or problems that have continued through social science in the past century. In each instance we describe the issue and then critically examine the responses and contributions made by geography. In many, but not all, cases, we show how geographers have not simply reacted to developments outside the discipline, but have also extended these debates through original insights.

Big Structures and Small People

> Men make their own history, but
> they do not make it as they please;
> they do not make it under self-
> selected circumstances, but under
> circumstances existing already,
> given and transmitted from the past.
> (Karl Marx, *The Eighteenth
> Brumaire of Louis Bonaparte*, 1852)

This often-cited observation by Karl Marx expresses one of the most enduring problems across the social sciences. While, on the one hand, as individuals we believe that our actions stem from our personal intentions and motivations, on the other, we also suspect that things go on 'behind our back', or that larger forces are at work in shaping our lives (Giddens 1984).

This issue preoccupied the major social scientists of the nineteenth century, notably Karl Marx, Max Weber and Emile Durkheim. They wrote from different perspectives – Marx the philosopher/economist and radical intellectual, Weber concerned with historical sociology and Durkheim an analytical sociologist and a conservative. They differed over the degree to which what happens in society can be explained by the actions of individuals who are said to compose that society, or whether other, larger, forces are at work. This is the distinction between micro and macro explanations. It affects what we think 'the social' and 'society' might be. It also relates to just what kind of science of society you can have. And above all, it relates to the possibilities of our individual and collective action changing that society for the better.

Despite their different political and intellectual ideas, both Marx and Durkheim emphasized the macro forces that provide a context for individual action. For Marx this was economic forces 'in the last instance' (in Engels's famous phrase). Marx argued that it was possible to identify certain 'laws of motion' of capitalism (such as capitalist competition and accumulation for accumulation's sake) that drove the economy and shaped society. Individual actions often counted for very little against these macro technical/economic imperatives that not only surrounded individual action, but also marked different socio-historical periods.

Durkheim also stressed the importance of macro factors but in his case they were explicitly sociological. He thought that certain properties were emergent from society as a whole: that any society was more than the sum of its parts. These emergent properties were 'social facts' that were in some senses invariant and stood apart from individuals. In his most famous example he asked himself the question, 'Is suicide a social phenomenon?' From comparative suicide rates Durkheim concluded that suicide could be explained as a social and not an individual act. The social fact that explained suicide was the degree to which individuals were integrated in their society. In nineteenth-century France this was manifest in relatively higher rates for members of the Protestant community, which was more individualistic, than for members of the Catholic community, which had closer social ties.

The work of Max Weber (1864–1920) is often portrayed in contrast to Durkheim. Weber was an advocate of methodological individualism: to understand how society works, we must first understand how individuals work. He argued that to understand any action, we must first understand the individual's intention in the action, as well as what the action means socially. No analysis of society is possible without an understanding ('*verstehen*') of individual motivations – trying to see the world from the individual's point of view (the spur of much qualitative ethnographic research). Overall, Weber's work shows careful attention to both micro and macro factors in sociological explanations without ever fully integrating them.

In many ways this lack of integration is explicable in disciplinary terms. At the end of the nineteenth century the emerging discipline of sociology sought to establish itself with its own object of knowledge and modes of analysis. These were formed in opposition to the micro foundations of the marginal revolution in economics which meant that sparing assumptions about human rationality and instrumental action could be applied to all kinds of supply and demand situations. In contrast, the pioneers of sociology (such as Weber and Durkheim) wanted to set society apart from economy by arguing that there were properties in societies that existed beyond individual decision making (such as Durkheim's social facts) or that micro analysis of individuals had to have a much richer and more nuanced view of them as sociological actors (hence Weber's insistence on understanding the meanings as well as intentions of actions).

This lack of integration persisted in social science throughout the twentieth century. Different sociological paradigms came from either micro or macro assumptions. Thus Talcott Parsons's comprehensive theory of action that dominated Anglo-American sociology mid-century gave greatest weight of explanation to certain structural (macro) features of society that ensured its reproduction (and was therefore known as structural functionalism). Symbolic interactionism started its analysis of society with (micro) gestures between individuals in order to build up an idea of social communication. Here society was built out of the sociological competences of interacting individual actors. These involved elements of self-presentation, etiquette and ploys in conversation (see Goffman 1959). This micro analysis was challenged for being naive about larger social structures that impinge on the context of interaction. These contrasts were also apparent in anthropology. On one side were the structuralist and functionalist explanations proposed by Radcliffe-Brown and Lévi-Strauss, searching for the larger cultural rules underpinning both modern and traditional societies. These contrasted with the small-scale ethnographies favoured by many anthropologists in the field.

If we believe that our actions are determined by larger forces outside our control, then we may be inclined to adopt a passive model of the actor in social sciences. But if we place too much faith in micro-level explanations, we are likely to underestimate the deep-seated conditions of social life that perpetuate injustice, inequality, racism and so forth. How to reconcile micro and macro explanations has therefore been not just an idle exercise for scholars, but also an important claim about the power of politics or collective action.

One of the more recent attempts to combine macro and micro elements in the explanation of society is structuration theory, most notably the version outlined by British sociologist Anthony Giddens (Giddens 1984). Rather than the dualism of micro/macro explanations, he argues for a duality of structure and agency. According to this theoretical synthesis, structures can enable as well as constrain action in the sense that the taken-for-granted routines that reproduce social life are also consolidated social competences for further action. Some of these routines are intensely geographical. Giddens draws heavily on abstract ideas of time-geography to understand the daily time-space paths of individuals where the rules of social reproduction meet the competences of 'coupling points' of social interaction. Structuration theory also had an impact on geography, notably in the work of Derek Gregory on regional historical geography (Gregory 1982) and Allan Pred (1986) on the modern transformation of Swedish society. At the same time, the approach has been criticized for being unable

adequately to integrate observations on individuals with larger social forces.

The links between geography and structuration theory are an exception. The micro/macro debate that has dogged social science in general and sociology in particular over the last 100 years did not find the same register in geography. To be sure, geography was swept by the quantitative revolution of the 1960s involving spatial models with implicit assumptions about rational economic decision making (based on micro foundations), but the analytical/methodological split between micro and macro explanations for social phenomena was never so hotly debated in geography as it was in social science more generally. In part this is because geographers have always had a keen sense of the relationships between the society and its physical/biological environment. Whereas sociology had to establish itself on the terrain of social science and the influential discipline of economics (formally political economy), geography traditionally looked to the physical sciences and the power of the physical and biological environment to shape human behaviour. In geography's past this has manifested itself in various forms of determinism – environmental, racial, spatial and cultural.

It is therefore often difficult to judge just what kind of actor geographers are assuming within these different paradigms from the late nineteenth century onwards. In his call for a humanistic approach to geography in the 1970s, David Ley (in Ley and Samuels 1978) sought to capture a holistic view of actors and their lifeworlds to rescue the idea from its reduced existence as an atom in the spatial models of the 1960s. The later 'cultural turn' gave another set of imperatives to the actor (see chapter 24 by Peter Jackson in this volume), and the more recent influence of post-structuralism has meant that the idea of the actor and agency becomes diffuse in networks of dispersed and weak subjectivity.

The Rational Actor

A second and related key question is whether humans are explicable and predictable in the same way as phenomena in the physical sciences. If they are, then the social sciences can use the same techniques of investigation and analysis as the physical sciences. This was certainly the aim of positivism from Auguste Comte's initial proposals in the wake of the French Revolution onwards. The most powerful idea in maintaining these assumptions in social science has been that human beings are rational actors. Economists have pared down this idea to utility maximization. This assumes that individuals have preferences for things in the world, that they are able to rank these preferences (apples above oranges, for instance) and that they choose in a way that is consistent with these preferences. These narrow utilitarian assumptions of rational choice theory have had an enormous impact, most obviously on economics but also on political science and sociology. The claim is that these simple micro foundations explain larger social and economic phenomena.

One especially influential model of the social consequences of individual rational choice is the prisoner's dilemma, which seems to demonstrate the real-world paradox that people will choose a less rewarding outcome for themselves because of temptation to get even greater gains by selfish behaviour, that they are unable to trust and cooperate with others, and that they assume that others will do the same. A variant of this type of behaviour is shown by the 'tragedy of the commons' (Hardin 1968), where the gain for an individual farmer in adding another cow to the common land exceeds the loss to that individual incurred by overgrazing and loss of quality of the pasture. If all farmers think like this, then the common land is overgrazed and in the end all suffer. These assumptions seem to capture a whole range of social situations, from attitudes to taxation, to traffic jams, urban sprawl and to responses to the threat of global warming.

The prisoner's dilemma game is in fact trivial because the way that the incentives (or payoffs) are set up means that the rational chooser need not take the rationality of the other player into account. In these parametric situations the actor is choosing in an environment that is non-intentional (deciding whether to take an umbrella is a choice against the weather) or where aggregated choices can be considered to comprise a parameter (such as consumer demand in marginalist economics or class action according to certain sociologists – for example Goldthorpe 1998). In the last case the consolidated choices that constitute the parameter can be seen as similar in their effects to a structure – hence the potential power of rational choice to provide micro foundations for macro phenomena (Schelling 1960, 1978). Even those most enthusiastic about this *parametric rationality* call for its softening in social situations. Important here are Herbert Simon's ideas that people 'satisfice' rather than optimize their utilities and that their knowledge of alternatives and therefore their rationality is bounded, rather than being based on perfect information.

Assumptions of parametric rational decision making were very much implicit in the spatial models of location theory dominant in the 1960s and 1970s (see, for example, Haggett et al. 1977). More prominent in this modelling were the formal spatial and geometric features rather than their behavioural assumptions. This also applied to behavioural geography where optimizing decisions by individuals were translated into distance minimization in spatial behaviour. Simon's ideas of satisficing and bounded rationality were important, for example in Wolpert's (1964) study of spatial decision making among Swedish farmers. However, behavioural geography got caught in the pincers of the critique of its overly narrow view of human consciousness from humanistic geography and its positivist foundations by the burgeoning Marxist geography. Rational choice geography was therefore killed off in its parametric form before certain advances in game theory began to impact on the rest of social science.

The advances in game theory in the 1970s and 1980s were to do with how situations might be analysed where the decision making is interdependent and the rationality of the other player is a crucial component of choice (strategic situations). Getting what you want now in part depends on how others decide and your own anticipation of those decisions depending on how you think they will act. Settled ways of behaving might be found if an actor is content not to change a decision given what s/he thinks others will choose, and so on for all the actors in the interaction (the so-called Nash equilibrium; see Binmore 1992). Adaptations of *strategic rationality* start to shade into questions of consciousness and interdependence. Some rational choice theorists have argued that such settled behaviour could be the basis of social norms. Robert Sugden (1989), for example, gives the case of the creation of an elaborate social code for the collecting of driftwood from a beach. Others have criticized the idea that strategic interdependence can form the 'cement of society' (Elster 1989). Nevertheless, the idea that social norms can arise spontaneously as a rational response to a novel situation poses a deep threat to more traditional sociological approaches that assume norms to be inherited and taken-for-granted. Strategic assumptions have also become common in political science over such issues as voting behaviour and the positioning of political parties to capture the median voter, collective bargaining and other dilemmas of collective action. It has even proved challenging for political theory, from the rationalist assumptions of Rawls's theory of justice (Rawls 1971) to the example of strategic reasoning over 'morals by agreement'.

As we suggested earlier, rational choice geography was eclipsed before the insights of strategic rationality had gained their immense influence on social science. Rational choice theory (RCT) is now a major ap-proach in psychology, sociology and politi-cal science to the extent that there are strong objections to it as a form of economic im-perialism. Its absence in geography since the 1960s means that some of the questions that RCT poses remain unanswered – such as whether we can afford to dispense with the assumptions that humans are predictable. To be sure, governments setting rates of taxation, or trying to reduce CO_2 emissions by using tariffs and other incentives, can-not!

Towards a Network View of Society

The foundation of the modern social sci-ences was laid in the late nineteenth cen-tury in Europe and America. They each took as a fundamental unit of analysis the 'na-tion-society-state', wrapping the three terms into a single entity as if it were the natural basis for understanding human life. Thus, 'realist' political science studied states, leav-ing international relations to explain the re-lations between these states. Sociology understood society to mean something con-tained by a nation/state. Macro economics considered national economies, later con-tributing to the belief that they could be regulated by governments, while trade theory covered exchanges between national economies. Anthropologists' 'tribes' were, in some ways, smaller versions of the same thing, inspected in isolation from the mod-ern world. But like nation-state-society, they were instances of an organic conception of society as bounded, functioning coherently, and explicable in terms of what could be observed within that society.

In the past two decades an alternative approach has emerged around the idea of networks. To speak of a single network ap-proach would be misleading, however. In-stead, across the social science disciplines from sociology to business management, migration studies and the history of science, a number of distinct modes of thinking with networks have developed. In their contexts, network is variously treated as a metaphor, method, theory and paradigm. But all the approaches share two broad characteristics. First, networks are regarded as alternatives to 'regions', or bounded notions of social and economic life such as nations or firms. Networks focus on the capacity of individuals or firms to organize relations and transactions across boundaries. They are also contrasted with hierarchies and markets, suggesting a looser, more flexible and adaptable form of social organization for companies, ethnic groups or terrorists. Second, networks are often regarded as an intermediate concept between macro and micro approaches (Powell and Smith-Doerr 1994). They enable analysts to focus on the form and content of social relationships. Social network analysis seeks to reveal the form and nature of social interaction in order to establish the structural principles on which that interaction occurs (see Mitchell 1969). One particularly influential piece of research that captures the flavour of this work was Mark Granovetter's (1973) counter-intuitive argument about the 'strength of weak ties', in which he argued that dense networks (where everybody knows each other) were less useful in finding a job than loose networks, with lots of acquaintances that acted as links to other networks – and sources of information about jobs. In most of this work the spatial aspects were at best implicit and the connections between social network analysis and geography were weak.

Examples of network thinking abound. Michael Mann's magisterial two- (soon to be three-) volume historical sociology from ancient times to the present, *The Sources of Social Power* (Mann 1986, 1993), replaces the 'unitary, closed system' of society with 'multiple overlapping and intersecting sociospatial networks of power' (1986: 1), suggesting that societies are much more messy and interconnected with one another than is generally depicted in social science. Identities are more fluid and comprised of networks of power that vary across space and time – with simultaneous tendencies to stretch out and fold in. Concentrating on the late twentieth century alone, sociologist Manuel Castells has produced a three-volume argument on the information age that centres around the idea that we are living in a network society (Castells 1996). Networking logic, partly entwined with information technology, permeates governments, firms, protest groups, criminal organizations and other forms of association. Within geography, Doreen Massey (1993) has argued that the idea of 'power networks' is a more useful way of understanding people's experience of agency in space and time rather than some common experience of time-space compression. Networks with a range of global and local connections overlap in the city to help create a progressive sense of place – which she describes in terms of the transnational atmosphere of her own neighbourhood of Kilburn in north London.

Alongside these grand theories there is a growing field of empirical research lying on the boundaries between economics, sociology, management studies, economic geography and ethnic and migration studies. The role of networks in economic activity was given a much more general application with the rise of socio-economics (see especially Smelser and Swedberg 1994). Drawing on an older tradition of institutional and historical analysis of economics, this approach suggested the importance of sociological and cultural influences 'embedded' – to use Granovetter's (1985) phrase – in economic activity, rather than relying on the abstract assumptions of microeconomics. One important strand of this work concerns how a good deal of contemporary economic activity is networked rather than hierarchical. Flexible specialization and just-in-time production rely on strong reliable networks of supply from subcontractors to the core firm. Geographers, such as Allen J. Scott (1988), have explored the spatial implication of these network relations and how they produce new industrial districts of clustered activities in more dispersed metropolitan settings. There are also networks of trust between business people that are global in scope. Yeung and Olds (1999) apply these ideas to Chinese business networks, critically assessing the claims that the economic success of overseas Chinese capital is related to the qualities of so-called *guanxi* networks. At the other end of the spatial scale, trust exists in networks of reciprocity between people that share common activities and place-based community. These ties can give regional differences in economic organization and performance, most famously analysed in a comparison of northern and southern Italy by Robert Putnam. Place-based loyalties act as forms of 'social capital' which, according to Putnam (1993), has major consequences for democracy and civic life. He gives the example of one man who donated a kidney to another based on the fact that they went bowling together. Putnam argues, however, that American citizens are now more often 'bowling alone', indicative of a decline in associational life (Putnam 2000).

In all these discussions of networks the unitary identity of the individual in the network has been assumed, even if identity in some senses emerges from the overlapping networks of which the individual is a part. A more radical reading of network effects dispenses with the idea of a unitary identity (either of individuals or social wholes). Allied to the post-structuralist critique of the idea of the self-conscious thinking subject, actor network theory sees human subjects not as nodes in the network but rather that human subjectivity is dispersed through the network, alongside the actions of non-human actors (such as animals or bacteria), machines (computers for instance) and texts. Human subjectivity is part of the overall network effect. These network connections are long or short, involving action-at-a-distance. Geographers such as Nigel Thrift and Sarah Whatmore have employed such theories to enquire into financial capital, food chains and the intertwining of social, natural and technological hybrid objects (see Murdoch 1997).

Geographers have been increasingly involved in the debate on networks – as empirical/analytical tool, metaphor and theoretical approach. So far discussions have ranged freely from the micro to the macro, from local to global, and network effects that explain everything – time, space, power, identity. A good deal of work remains to be done, for instance, in bringing together the burgeoning debate on scale in geography (see Delaney and Leitner 1997) with an understanding of how different forms of power are realized in networks.

Representation, Image and Reality

Where social science and the humanities meet there has arisen a set of loosely connected, but also profound, issues related to representation, image and reality. On the one hand, the line between image and reality is increasingly regarded as arbitrary, elusive or even non-existent. On the other hand, representations of all kinds – images and texts – have become central to an understanding of society, giving rise to new methodologies and techniques (see chapter 43 by John Morgan in this volume). Just as the significance of representation is recognized, so its relation to the world of things becomes harder to pin down. New scholarly disciplines such as cultural studies and media studies are responses to these developments, and their impact has also been felt throughout human geography, for example in the 'cultural turn' (see chapter 24).

The distinction between image and reality is but one of several related dichotomies, including the distinctions mental/physical and metaphorical/material. In each case, the former term is generally regarded as less substantial, less scientific or less trustworthy than the latter. Good social scientific knowledge is often assumed to proceed from a direct relation to material or concrete reality. The rest is fiction, entertainment and distraction, or worse still, deception. As

Nigel Thrift (2000: 371) states, 'perhaps the most important recurring motif in work on images has been suspicion'.

Since at least the late 1960s developments in society and technology, coupled with shifts in intellectual fashion, have brought questions of representation to the fore. It is not difficult to see why. In the West at least, we live in a world saturated with media images and signals of all kinds, from television to computer screens, advertisers to spin doctors, and lifestyle magazines to 24-hour news stations. Image is everything, it seems sometimes. Distinguishing image from reality feels harder.

A number of key theorists have tried to make sense of this world, including two thinkers inspired by the civil unrest in France during the late 1960s, during which there was an explosion of new political and theoretical ideas. For Guy Debord (1932–94), a central figure in a radical group of thinkers known as the Situationist International, capitalism had reached a new level. The production and consumption of things had been replaced by the production and consumption of images, forming a 'society of the spectacle'. The only way to resist capitalism was through creating 'situations', moments of disorder or transgression that combined politics, art and play. Jean Baudrillard (b. 1929) also began by exploring the enlarged role of the symbolic value of commodities, the 'sign'. He also argued that capitalist society was now organized around symbolism and not production, but then took this idea further to claim that we are living in a new era of simulation. Signs increasingly take on a life of their own. Copies become more real than reality. As a result, 'the boundary between image or simulation and reality implodes, and with it the very experience and ground of "the real" disappears' (Best and Kellner 1991: 119). The distinctions between news, information, politics and entertainment also dissolve, rendering conventional politics and understanding irrelevant. In later writings, Baudrillard proposes that images, signs, models and simulations have so escaped our control that they dominate us and leave us with no alternative but to give up the illusion of being subjects altogether. Here he shades into the realm of science fiction and few critics are willing to follow him. None the less, Baudrillard offers a disturbing diagnosis of our joys, fears and uncertainties at being unable comfortably to differentiate image and reality.

What Baudrillard and others have identified is one aspect of postmodernism, understood as both a condition and a philosophy that recognizes deep changes in our world and the way we can make sense of it. Allied with various post-structuralist philosophies, postmodernism is sceptical about, and critical of, the power of reason to understand and change the world (see Peet 1998: ch. 6). But, as Denis Cosgrove (1990) has suggested, metaphor and image were also considered central to human understanding in the *pre*-modern world of the European Renaissance. Then as now, 'images no longer illustrate, reflect or disguise a reality existing below themselves, rather they present themselves as simulacra, constitutive of their own reality' (1990: 353).

This theme of images and representations, including texts, less as mirrors of reality and more as participants in its construction now runs throughout geographical research. It may imply that reality should be approached in terms of the relations between and among texts and images, a condition of 'intertextuality'. Meanings are not securely anchored in hard facts, but also formed 'horizontally', as it were, in the realm of signs. It may also imply what anthropologists have described as a 'crisis of representation', a marked unsettling of the assumption that writing is an innocent or transparent bearer of meaning. Instead, it is now commonplace to acknowledge how much power in various guises – e.g. colonialism and masculinism – is implicated in the production and circulation of accounts.

Because geography has always emphasized imagery of all kinds – for example maps, photographs, topographic sketches and now geographical information systems – the subject has taken on board many of these ideas and generated distinct areas of inquiry. The field of landscape studies, long a traditional concern of geography, has become expanded and transformed by the infusion of new ideas from cultural studies, art history and literary theory. James Duncan (1995) identifies two main strands. One treats landscape in a picturesque or painterly fashion. It places landscape painting and landscape gardening within the social and political context, moving seamlessly between aesthetics and economy, class and taste, all the while suspending the distinctions between material and metaphorical aspects of landscape. This strand also introduces metaphors from theatre and drama, and considers landscapes as texts to be read in relation to other cultural texts. A second strand focuses more on the ways that cultural and political values are embedded in and communicated through landscape. In both approaches, the concept of landscape has become a way of holding image and reality in a combined and productive tension.

Several other areas of geographical inquiry also contribute to the rethinking of representation. As conceived by Gearóid Ó Tuathail (1996), critical geopolitics updates traditional concerns for statecraft and geography for the televisual age by subjecting various geopolitical 'scripts' to critical interpretation. It addresses how foreign policy intellectuals, through their publications, speeches and maps, strive to shape how the world's political geography is represented. These scripts are not mirrors of some 'real' politics; they are interventions in how the world of states and nations, good and evil, is imagined. This was no more apparent than in the Gulf War. Closely related to geopolitics is the critical interpretation of cartography, including both maps and mapping, pioneered by the work of J. Brian Harley (1989). By introducing post-structuralist themes, Harley is not simply concerned with how maps 'lie', but also how cartographic

practices and conventions construct the effect of truth and, in so doing, render certain accounts of the world apparently normal or natural. Other geographers, for example Bartram and Shobrook (2000), have drawn upon Baudrillard's concepts of simulation and hyperreality to understand the social and symbolic construction of nature – in their case, the Eden Project (see chapter 14 by Noel Castree in this volume).

The Eden Project is the perfect example of the ways in which the comforting distinctions we depend upon to make our way in the world, between original and copy, simulation and reality, authentic and contrived, nature and artefact, may become confused. Such confusion is the source of both creative stimulation and corrosive anxiety. Through research in landscape, critical geopolitics, cartography, nature and other fields, geography is making an original and challenging contribution to a key theme in the social sciences and humanities.

Space, Technology and Environment

Since the last decade or so of the twentieth century there has been a confluence of economic, social and technological changes that has posed new challenges for understanding society. Different but related aspects of these changes are captured by such terms as 'globalization', 'the new economy', 'the knowledge economy', 'cyberspace' and 'digital living'. An unprecedented period of economic growth in the United States persuaded some economists that the constraints of the old industrial economy had been swept away by the forces of the information revolution. They predicted an era of growth without the past problems of inflation and unemployment. Certain economists and political scientists also debated the end of the nation-state in the wake of globalizing flows of capital, trade and information (see chapter 18 by Henry Wai-chung Yeung in this volume). Futurologists and gurus of the computer age foresaw science fiction-like transformations in the relations between bodies, minds and technologies that might point towards a post-human world of artificial and disembodied intelligence. To some, these new technologies will usher in a nightmare world of increased electronic surveillance, loss of personal privacy and autonomy, and huge disparities of wealth and freedom. Others foresee a utopia of ever-widening communication undermining state control and establishing new virtual communities founded on the full diversity of human interests and passions.

The social scientific response to these transformations can be classified under at least two broad headings. The first, termed 'substitution and transcendence' by Stephen Graham (1998), points towards a radically reduced significance of space, place and location in human activity. In contrast, 'co-evolution' approaches attempt to analyse the new relationships between technology and geography in less reductionist ways.

The initiative in making sense of these changes has been seized by a new breed of intellectuals, writers and commentators operating on the borders of social science, management studies and the media. Indeed, they are part of the phenomenon itself, apostles and interpreters of the information age. Their books often imply the disappearance of geography as a factor in human affairs. Titles such as *The Borderless World, Living on Thin Air, The Death of Distance, The Weightless World* and *The End of Geography* summon up an ethereal world in which land, location and place no longer seem to matter. Either electronic flows and spaces will directly substitute for material ones, or they will enable individuals and firms to transcend the material world. To take but one example, Frances Cairncross's *The Death of Distance* (1997) looks towards a near future when the costs of telecommunications will fall to zero, thereby freeing economic activity from its industrial-age spatial constraints. If this is true, the information age may well spell the end of geography as a discipline!

The alternative, 'co-evolution' approach is characterized by a scepticism towards technological determinism and a preference for analysing technological change in its social, historical and geographical contexts. This position does not deny that there are significant developments in work, life and entertainment. The vital importance of new information and communication technologies to world development has been acknowledged by the UNDP's *Human Development Report* (1999), and the World Bank's *World Development Report 1999* focuses on 'knowledge for development'. Major research initiatives, such as the Economic and Social Research Council's Virtual Society? programme, have begun to analyse the social science of electronic technologies with a view to separating out the myth and hype from the reality (see http://www. virtualsociety.sbs.ox.ac.uk). Such approaches are, however, more open-minded about the multiple causes and possible futures of economic and technological change.

Ongoing work in geography has made a substantial and significant contribution to this kind of understanding of the complex and varying relationships between space, knowledge, economy and technology (see Graham 1998 for a review). The diffusion of technological innovations and their impacts on social and spatial inequalities is a well-established field of geographical research, beginning with the work of Swedish geographer Torsten Hägerstrand in the 1950s. The so-called digital divide, or the uneven access to the Internet, mobile telephony and digital technologies in general, has express geographical dimensions at global, national and urban scales. There has been exciting research in the field of virtual reality, for example (see Crang et al. 1999), as well as innovatory attempts to map cyberspace (see Dodge and Kitchen 2001).

But even in geographical studies there are significant differences and controversies over the role of space and the physical environment. While some research implies too weak a concept of geography, other arguments appear to propose too much significance for geography.

The first position can be detected in the so-called 'new economic geography' (NEG). Certain economists, notably Harvard's Paul Krugman and Michael Porter, have revived some of the classic location theories of Von Thünen, Lösch and Weber, along with ideas from international trade theory, and developed them to address core issues such as the spatial agglomeration of industry, regional specialization, the emergence of industrial districts and clusters of firms. These issues have direct interest to governments trying to understand how they can regenerate ailing local economies and/or capture a share of mobile capital investment (Krugman 2000). The new economic geography is beginning to be taken very seriously by planners and policy makers. But it faces stiff criticism from geographers themselves. As Ron Martin (1999) has argued, NEG relies too much on stylized and abstract models, in which real places only enter the reckoning to confirm mathematical equations. He points out that their models might be quite good at establishing why firms cluster, but not why they do so in one place rather than another. He calls for greater dialogue between geographers, whose research is more institutional, qualitative, discursive and contextual – i.e., they talk to people – and the economists, whose research is often more rigorous and exact. It is true to say that economic geography has developed as one of the leading fields of the discipline.

By contrast, some economists and historians have suggested too strong a role for geography. Two bestselling books published in the 1990s revived the debate on environmental determinism which geographers had assumed was settled in the 1930s (see chapter 14 in this volume). Jared Diamond, a professor of physiology, wrote *Guns, Germs and Steel* (1997) to explain how differences in the physical geographical environment, and not race or culture, could explain current global inequalities in wealth. A key part of the argument is that the Old World had

more wild grasses suitable for cultivation and animals suitable for domestication. People in Eurasia evolved in close proximity to their animals, developing immunity to the diseases that so devastated the New World when Europeans arrived. The east–west orientation of the major physical features of Eurasia allowed agricultural innovations to diffuse more easily than in the Americas, where features are more aligned north–south against the grain of climatic regions. A Harvard professor of history and economics, David Landes, similarly argued in *The Wealth and Poverty of Nations* (1998) that the distribution of climate, soils and disease can partly account for the rise of Europe to global economic and political power. A culture of innovation and enterprise also helped, argues Landes.

Geographers have always been suspicious of claiming a major role for climate, soils and physical factors for differences in development because such ideas were often used to justify imperialism and racism in the decades before the First World War. Further, accounts that explain differences in gross national product (GNP) by using GIS analyses of climate and access to sea-based trade – for example Mellinger et al. (2000) – run the risk of downplaying colonialism and the subsequent global political inequalities that have affected terms of trade and development. As Richard Peet (1999) writes in his review, Landes has ignored virtually all geographical research over the past 30 or more years that has argued against single-factor explanations of human geographical differences.

The interesting thing about these debates is that disciplines outside geography are taking geographical ideas very seriously in the important questions of world and regional development. But at the same time, the work of actual geographers has often been overlooked, perhaps because it does not fit simple historical accounts and economic models. A balance must be struck which recognizes the significance of space, place, location and the physical environment alongside social and cultural processes, while rejecting simple reductionist or deterministic explanations which imply that geography is destiny.

References

Bartram, R. and Shobrook, S. 2000: Endless/endless nature: environmental futures at the fin de millennium, *Annals of the Association of American Geographers*, 90, 370–80.

Best, S. and Kellner, D. 1991: *Postmodern Theory: Critical Interrogations*. Basingstoke: Macmillan.

Binmore, K. 1992: *Fun and Games*. Lexington, MA: D.C. Heath.

Cairncross, F. 1997: *The Death of Distance*. Cambridge, MA: Harvard Business School Press.

Castells, M. 1996: *The Rise of the Network Society*. Oxford: Blackwell.

Cosgrove, D. 1990: Environmental thought and action: pre-modern and post-modern, *Transactions of the Institute of British Geographers*, N.S. 15, 344–58.

Crang, M., Crang, P. and May, J. 1999: *Virtual Geographies: Bodies, Space and Relations*. London: Routledge.

Delaney, D. and Leitner, H. 1997: Special issue on the political geography of scale, *Political Geography*, 16(2), 93–185.

Diamond, J. 1997: *Guns, Germs and Steel: A Short History of Everybody for the Last 13,000 Years*. London: Jonathan Cape.

Dodge, M. and Kitchen, R. 2001: *Atlas of Cyberspace*. Boston: Addison Wesley.

Duncan, J. 1995: Landscape geography, 1993–94, *Progress in Human Geography*, 19, 414–22.

Elster, J. 1989: *The Cement of Society*. Cambridge: Cambridge University Press.

Giddens, A. 1984: *The Constitution of Society: Outline of a Theory of Structuration*. Cambridge: Polity.

Goffman, E. 1959: *The Presentation of Self in Everyday Life*. Garden City, NY: Doubleday.

Goldthorpe, J. 1998: Rational action theory for sociology, *British Journal of Sociology*, 49, 167–92.

Graham, S. 1998: The end of geography or the explosion of place? Conceptualizing space, place and information technology, *Progress in Human Geography*, 22, 165–85.

Granovetter, M. 1973: The strength of weak ties, *American Journal of Sociology*, 78, 1360–80.

Granovetter, M. 1985: Economic action and economic structure: the problem of embeddedness, *American Journal of Sociology*, 91, 481–510.

Gregory, D. 1982: *Regional Transformation and Industrial Revolution: A Geography of the Yorkshire Woollen Industry*. London: Macmillan.

Haggett, P., Cliff, A.D. and Frey, A. 1977: *Locational Analysis in Human Geography*, 2 vols. London: Arnold.

Hardin, G. 1968: The tragedy of the commons, *Science*, 162, 1243–8.

Harley, J.B. 1989: Deconstructing the map, *Cartographica*, 26, 1–20.

Krugman, P. 2000: Where in the world is the 'new economic geography'? In G.L. Clark, M.P. Feldmann and M.S. Gertler (eds), *The Oxford Handbook of Economic Geography*, Oxford: Oxford University Press, 49–60.

Landes, D. 1998: *The Wealth and Poverty of Nations*. New York: W.W. Norton.

Ley, D. and Samuels, M. (eds) 1978: *Humanistic Geography: Problems and Prospects*. London: Croom Helm.

Mann, M. 1986, 1993: *The Sources of Social Power*, 2 vols. Cambridge: Cambridge University Press.

Martin, R. 1999: The 'new economic geography': challenge or irrelevance? *Transactions of the Institute of British Geographers*, N.S. 24, 387–91.

Massey, D. 1993: Power geometry and a progressive sense of place. In J. Bird, B. Curtis, T. Putnam, G. Robertson and L. Tickner (eds), *Mapping the Futures*, London: Routledge, 87–132.

Mellinger, A.D., Sachs, J.D. and Gallup, J.L. 2000: Climate, coastal proximity, and development. In G.L. Clark, M.P. Feldmann and M.S. Gertler (eds), *The Oxford Handbook of Economic Geography*, Oxford: Oxford University Press, 169–94.

Mitchell, J.C. 1969: The concept and use of social networks. In J.C. Mitchell (ed.), *Social Networks in Urban Situations*, Manchester: Manchester University Press, 1–50.

Murdoch, J. 1997: Towards a geography of heterogeneous associations, *Progress in Human Geography*, 21, 321–37.

Ó Tuathail, G. 1996: *Critical Geopolitics*. London: Routledge.

Peet, R. 1998: *Modern Geographical Thought*. Oxford: Blackwell.

Peet, R. 1999: Review of *The Wealth and Poverty of Nations* by David Landes, *Annals of the Association of American Geographers*, 89, 558–60.

304 Gary Bridge and Alisdair Rogers

Pred, A. 1986: *Place, Practice and Structure: Social and Spatial Transformations in Southern Sweden, 1750–1850.* Cambridge: Polity.

Putnam, R., with Leonardi, R. and Nanetti, R. 1993: *Making Democracy Work: Civic Traditions in Modern Italy.* Princeton, NJ: Princeton University Press.

Putnam, R. 2000: *Bowling Alone.* London: Simon and Schuster.

Rawls, J. 1971: *A Theory of Justice.* Cambridge, MA: Belknap Press.

Schelling, T. 1960: *The Strategy of Conflict.* Cambridge, MA: Harvard University Press.

Schelling, T. 1978: *Micromotives and Macrobehaviour.* New York: W.W. Norton.

Scott, A.J. 1988: *Metropolis: From the Division of Labour to Urban Form.* Berkeley: University of California Press.

Smelser, N. and Swedberg, R. (eds) 1994: *The Handbook of Economic Sociology.* Princeton, NJ: Princeton University Press.

Sugden, R. 1989: Spontaneous order, *Journal of Economic Perspectives*, 3, 85–97.

Thrift, N. 2000: Image. In R.J. Johnston, D. Gregory, G. Pratt and M. Watts (eds), *The Dictionary of Human Geography*, Oxford: Blackwell, 371–2.

Wolpert, J. 1964: The decision process in spatial context, *Annals of the Association of American Geographers*, 54, 537–58.

Yeung, H.W.-C. and Olds, K. (eds) 1999: *The Globalisation of Chinese Business Firms.* London: Macmillan.

Further Reading

A good account of the potential and pitfalls of borrowing ideas and concepts from the social sciences is given by J. Agnew and J.S. Duncan, 'The transfer of ideas into Anglo-American human geography', *Progress in Human Geography*, 5 (1981), 42–57.

A very accessible introduction to game theory is provided by A. Dixit and S. Skeath, *Games of Strategy* (London: W.W. Norton, 1999).

For an application of rational choice to sociology and social theory, see J. Coleman, *Foundations of Social Theory* (Cambridge, MA: Belknap Press, 1990).

For a review of network approaches, see G. Bridge, 'Mapping the terrain of time-space compression: power networks in everyday life', *Environment and Planning D: Society and Space*, 15 (1997), 611–26; P. Dicken, P. Kelly, K. Olds and H.W.-C. Yeung, 'Chains, networks, territories and scales: towards a relational framework for analysing the global economy', *Global Networks*, 1 (2001), 89–112; and W.W. Powell and L. Smith-Doerr, 'Networks and economic life', in N.J. Smelser and R. Swedberg (eds), *The Handbook of Economic Sociology* (Princeton, NJ: Princeton University Press, 1994), 368–402.

For the central, but accessible, statement of actor network theory, see B. Latour, *Science in Action* (Milton Keynes: Open University Press, 1987).

A good introduction to the themes of image and reality is provided by M. Crang, 'Image–reality', in P. Cloke, P. Crang and M. Goodwin (eds), *Introducing Human Geographies* (London: Arnold, 1999), 54–61.

On landscape see D. Cosgrove and S. Daniels (eds), *The Iconography of Landscape* (Cambridge: Cambridge University Press, 1988); and T. Barnes and J.S. Duncan (eds), *Writing Worlds* (London: Routledge, 1992).

On economic geography and its relations with economics, including the new economic geography, see T. Barnes and E. Sheppard (eds), *A Companion to Economic Geography* (Oxford: Blackwell, 2000); and G.L. Clark, M.P. Feldmann and M.S. Gertler (eds), *The Oxford Handbook of Economic Geography* (Oxford: Oxford University Press, 2000).

49

(Some) Spaces of Critical Geography

Lawrence D. Berg

The term 'critical geography' is a relatively new one in the lexicon of human geographers. One cannot find critical geography, for example, in the latest online version of the *Oxford English Dictionary*. Similarly, the term is absent from all but the most recent (fourth) edition of the highly influential *Dictionary of Human Geography* (Johnston et al. 2000). Notwithstanding the recent arrival of the term in Anglo-American geography, we should not be fooled into thinking that critical geography is a new approach. Critical approaches have a long history in the disciplinary practices of human geographers across a range of locales.

In the United States, for example, critical geographers can be seen as part of a long tradition of 'radical geographers', whose scholarship developed as part of the new social movements (e.g. civil rights, women's rights, gay rights) of the 1960s and 1970s. In the UK, critical geography has links to the development of Marxist geography and socialist-feminist geographic thought among the New Left in the 1960s and 1970s. The term critical geography gained wide usage in the UK only in the mid-1990s, as a result of left-oriented geographers' varied responses to the Shell Oil corporation's sponsorship of the Royal Geographical Society (with the Institute of British Geographers). Shell Oil has been linked with the repressive military regime in Nigeria that was re-sponsible for the 'judicial murder' of Ogoni activist Ken Saro-Wiwa, and one response to this set of events was the formation of the Critical Geography Forum (Gilbert 1999). In contrast, the term critical human geography (*kritisk samfundsgeografi*) has been used for at least 20 years by a significant proportion of human geographers in the Nordic countries (Denmark, Finland, Norway and Sweden). Nordic human geographers have been meeting at an annual conference for critical geography (Nordiske Symposium for kritisk samfundsgeografi) since the very early 1980s. Likewise, geographers working in Aotearoa/New Zealand have a long history of critical work (Berg and Kearns 1997), especially as it relates to attempts to decolonize geographic knowledge in the contradictory spaces of a colonial settler society whose dominant inhabitants – white settlers from the UK – can be seen as both colonizer and colonized (Berg and Kearns 1998). Similar arguments can be made for geographic work done in places like Australia (see Morris 1991; Anderson and Jacobs 1997) and South Africa (see Crush 1994).

As these brief comments suggest, there is a multitude of approaches that academics can choose in order to be critical geographers today. Critical human geography thus involves a wide-ranging and rapidly changing set of ideas and practices within human geography. Moreover, critical geography overlaps with and incorporates older traditions of 'radical geography' (Painter 2000) and other forms of social and spatial critique practised across a range of locations. It is appropriate, therefore, to think of critical *geographies* rather than a single critical geography. Drawing on earlier work with a colleague (Morin and Berg 1999), I want to suggest that in coming to understand the varied practices of critical geographers, one should focus on the emplaced *geographies* of specific disciplinary histories. In that way, one can – in part at least – avoid the construction of implicit stories of linear progress in critical geographic thinking. In order to counter notions of progress and improvement – whereby the latest critical theories are seen as 'cutting edge' whilst earlier theories are past their use-by date – I want to introduce readers to the idea of writing a *geography* of critical geographies. As part of this brief introduction to critical geographies, I suggest that certain theoretical approaches in critical geography have specific emplaced histories. In this way, we might avoid, or at least reduce the tendency to construct, a monolithic critical geography that implicitly suggests that certain critical approaches are less sophisticated than others. Instead, I suggest that different critical theories can be read as strategic reactions to, and as constituted within, specific sets of geographically contingent social relations (Morin and Berg 1999).

The 'Critical' in Critical Geography?

As Caroline Desbiens and Neil Smith (1999) have argued, critical geographers have until now resisted trying to outline what a 'critical' geography might stand for. Such resistance to defining critical geography stems primarily from concerns that such definitions may be exclusionary. Critical geographers have tended to be interested in forging coalitions across a range of critical approaches and spaces rather than defining who can or

cannot participate as a critical geographer (although this is not always the case). Notwithstanding their concern for a politics of inclusion, critical geographers draw upon notions of being both 'radical' and 'critical' in their understanding of what it means to do critical geography. With this in mind, it is useful to understand what these terms mean before we can know how they have been incorporated in geographic thinking:

> **Critical:** [...] 2. Involving or exercising careful judgement or observation; nice, exact, accurate, precise, punctual. Now Obs. (or merged in other senses); 3. a. Occupied with or skilful in criticism. b. Belonging or relating to criticism. c. critical theory [...] a dialectical critique of society (esp. of the theoretical bases of its organization) associated with the leaders of the Institute for Social Research at Frankfurt (the Frankfurt School). (*Oxford English Dictionary*)
> **Radical:** [...] 1. b. Of qualities: Inherent in the nature or essence of a thing or person; fundamental. 2. a. Forming the root, basis, or foundation; original, primary. [...] 3. a. Going to the root or origin; touching or acting upon what is essential and fundamental; thorough; esp. radical change, cure. [...] 3. e. Characterized by independence of, or departure from, what is usual or traditional; progressive, unorthodox, or revolutionary (in outlook, conception, design, etc.). (*Oxford English Dictionary*)

It is not implausible to suggest that most, if not all, the above characteristics would apply, to some degree or another, to most critical geographers. Critical geographers are supposed to exercise careful judgement, they should be skilful in criticism, and they draw upon critical theory (broadly conceived) as a basis of such criticism. It is perhaps this latter point that is most important here, for the term critical geography draws on the idea of utilizing some form of critical theory, that is, theory that focuses on a critique of the foundational bases of the organizational structures of society. In this sense, critical geographers are interested in focusing their critiques on the *origins* of problems, and they do so in a manner that reflects a commitment to progressive thought and action. To be a critical geographer means, in part at least, to draw on a range of theories and approaches that rub against the grain of taken-for-granted or 'common-sense' understandings of human social and spatial relationships.

Critical geographers have also been able to agree, for the most part at least, on a range of critical theories and practices that contribute to critical geography. We can gain some clues as to what critical geographers consider to be 'critical' through examining some of the editorial statements of journals that are committed to publishing critical and radical geographic scholarship. The first English-language journal of radical geography, *Antipode*, which began publication in 1969, 'publishes articles that offer a radical (Marxist/socialist/anarchist/antiracist/feminist) analysis of geographical issues and whose intent is to contribute to the *praxis* of developing a new and better society'. These are some of the original approaches that underpinned radical geography as it developed in the 1960s and 1970s. More recently, there have been some additions to the 'list' of what might be considered appropriately 'critical' theories in geography. These are reflected in the newest addition to the line-up of critical geography journals, an online critical geography journal called *ACME*, which supports anarchist, anti-racist, environmentalist, feminist, Marxist, post-colonial, post-structuralist, queer, situationist and socialist perspectives. According to the editors of *ACME*:

analyses that are critical and radical are understood to be part of the praxis of social and political change aimed at challenging, dismantling, and transforming prevalent relations, systems, and structures of capitalist exploitation, oppression, imperialism, neo-liberalism, national aggression, and environmental destruction. (http://www.acme-journal.org)

The journal *Environment and Planning D* has a similar set of objectives, although they tend to be defined in a broader sense. It is committed to publishing 'interpretations [that] move across theoretical spectrums, from psychoanalysis to political economy. The journal editors are equally committed to the nitty gritty of practical politics and the abstractions of social theory'.

While these editorial statements illustrate some of the common themes that resonate with critical geographers – their commitment to social justice, for example, and their interest in acting in the world in ways that serve to contest unequal social relations – they do not illustrate the contested character of critical geography today. As Joe Painter (2000: 127–8) suggests, there are at least four key themes that animate current debates in critical geography: the relationship between theory and practice; politics inside and outside the academy; questions of positionality and reflexivity; and attempts to internationalize critical geography. Critical geographers are thus interested in a broad range of issues, as evidenced by the wide-ranging substantive geographies discussed by critical geographers who have contributed to this volume (for example, see chapters 14, 17, 22 and 23 in this volume by Castree, Page, Lees and Dwyer).

Given the brevity of this chapter, I cannot discuss all of the themes that interest critical geographers. Rather, I will focus on the question of internationalizing critical geography, and some of the debates that arise in relation to the dominance of Anglo-American geographic knowledge in the development of an international critical geography. The discussion will, of course, spill over to issues of positionality, reflexivity, theory production, activism and politics. My hope is that by focusing on the internationalization of critical geography, my discussion will elucidate some of the specific power-laden geographical contexts that critical geographers work within and thereby lead us to think about the *geographies* of critical geography.

Thinking Critically *and* Geographically

Critical geographers' commitments to social justice and transformative politics are informed by and in turn inform their use and development of critical social theory (see Painter 2000). Critical geographers are thus 'critical' because they understand that the foundations of current taken-for-granted geographies are in need of significant change. One key component of critical geography is a commitment to *praxis*, the development and use of *theory* in order to *act* in ways that contribute to the transformation of unequal and oppressive social and spatial relations.

Notwithstanding such commitments to social change, there would appear to be little agreement among critical geographers on how to effect such change. A recent epistolary exchange between a number of 'differently situated social actors', geographers from Canada, Denmark, Mexico, the UK and the United States (Katz et al. 1998) – all of whom seemed to have different theoretical understandings and empirical concerns – illustrates some of the fraught relationships between various branches of critical geography. Geographers from Canada, Denmark and the United States drew upon critical cultural studies (part of what has been called the 'cultural turn' in geography) for analyses of issues relating to place and the politics of identity in America and Europe (see letters from Berg, Morin and Simonsen in Katz et al. 1998). In contrast, a colleague from Mexico was quite opposed to such 'postmodern' approaches. Instead, she advocated the necessity of adopting a Marxian political economy to understand the geographies of economic deprivation and neo-liberal hegemony in Latin America (see letters from Uribe-Ortega in Katz et al. 1998). These kinds of disagreements are not uncommon, and they stem in part from the diversity of geographically contingent social relations within which different critical geographers work.

(Hierarchical) Spaces of Radical and Critical Geographies

At the same time, however, we must recognize that critical geographers are not free from the social and spatial relations that they wish to analyse (Harvey 1984; Berg 2001). Rather, geographers are as much a product of specific social, cultural and geographical constructions as the taken-for-granted geographies they wish to contest and transform, and this affects their analyses (and their disagreements with each other). Such geographies are always constructed within relations of power in which some spaces (and ways of understanding them) are effaced by dominant and hegemonic geographies. With this in mind, we need to recognize that, notwithstanding a sincere desire to be inclusive, critical human geography remains, for the most part, dominated by Anglo-American approaches (Slater 1992; Berg and Kearns 1998; Katz et al. 1998; Minca 2000; Painter 2000; Gregson et al. 2001).

Indeed, as a colleague and I have argued elsewhere (Berg and Kearns 1998), Anglo-America is the constitutive referent for philosophical and theoretical reflection. By this we mean that many geographers take for granted that geographic knowledge is produced *by* Anglo-Americans *for* Anglo-Americans. With few exceptions, the key authorities of what are constituted as the important debates and central positions in 'Geography' are Anglo-American. While geographers from the peripheries are allowed to participate in such debates, they are rarely able to set the agenda or frame the boundaries of what can and cannot be known (see also Slater 1992). This is not to suggest that the issues that concern critical geographers in America or Britain are unimportant, for this is clearly untrue. Instead, what I want to suggest is that to understand critical geography – and, I would argue, to BE good critical geographers – we all need to be more aware of the wider international context within which critical geographies have been developing for some time now.

Discussion: Critical Geographies in Geographical Context

Let me illustrate some of my concerns with a brief discussion from my own familiar context, Aotearoa/New Zealand, a place where geographers have been working for some time to 'decolonize' geographic thinking. Such a process of decolonization, ironically, operates within a paradoxical relationship to post-colonial studies. I say this relationship is paradoxical because, on the one hand, the goal of post-colonial studies is decolonization (of knowledge and space), while on the other hand, post-colonial theory has tended to operate in ways that recolonize the space of academic knowledge production (see Spivak 1988).

Post-colonial theory has been taken up with keen interest by critical geographers across a wide range of sub-disciplines, but it has been particularly influential in historical geography. This was very evident at the 1992 International Conference of Historical Geographers, from which many of my colleagues from the southern hemisphere had returned feeling marginalized because they were, apparently, not up to speed with the latest in metropolitan post-colonial theory. In describing their responses to that conference, a number of colleagues had suggested that they had the feeling that as inhabitants of the (theoretical) neo-colonies, they were not appropriately 'post-colonial' subjects. Given that for many years a number of these people have been involved in the fight to decolonize both the epistemological and the 'real' material spaces of colonial societies like Australia and Aotearoa/New Zealand, the ironies of this marginalizing post-colonialism were all too clear.

In the case of Aotearoa/New Zealand, for example, I was fortunate to work with a doctoral supervisor, Professor (now Dame) Evelyn Stokes, who was not particularly interested in post-colonial or any other 'critical' theory. Yet her intellectual endeavours as a key critic of white settler land policies

and as a member of the Waitangi Tribunal (a permanent Royal Commission of Inquiry) – have resulted in the return of thousands of hectares of land confiscated from Maori people and reparations payments of millions of dollars to Maori *iwi* (tribes). Her work must be seen as pivotal in the decolonization of Aotearoa/New Zealand. Certainly, it has led to radical transformations of the relationship between Maori and *Pakeha* (white New Zealander) peoples in Aotearoa/New Zealand.

Although I am fairly certain that Professor Stokes would see herself as anti-colonial, I doubt she would ever call herself a post-colonial geographer. None the less, I know of few critical geographers who, given the full story of her work, would deny her the moniker 'critical'. Interestingly, taken from its context and put in another place (the Vancouver Historical Geography conference), her work was not seen as critical enough (not 'post-colonial enough'). But my primary argument throughout this chapter is that critical geography cannot be divorced from its geographical context. Accordingly, one of the defining issues for critical geography in future will be how well we begin to accept different emplaced understandings and perspectives on what it means to be 'critical geographers', and how well we begin to understand and draw upon the varied geographies of critical geography.

References

Anderson, K.J. and Jacobs, J.M. 1997: From urban Aborigines to Aboriginality and the city: one path through the history of Australian cultural geography, *Australian Geographical Studies*, 35, 12–22.

Berg, L.D. 2001: Masculinism, emplacement and positionality in peer review, *Professional Geographer*, 53, 511–21.

Berg, L.D. and Kearns, R.A. 1997: Constructing cultural geographies of Aotearoa, *New Zealand Geographer*, 53, 1–2.

Berg, L.D. and Kearns, R.A. 1998: America Unlimited, *Environment and Planning D: Society and Space*, 16, 128–32.

Crush, J. 1994: Post-colonialism, de-colonization, and geography. In A. Godlewska and N. Smith (eds), *Geography and Empire*, Oxford: Blackwell, 333–50.

Desbiens, C. and Smith, N. 1999: The International Critical Geography Group: forbidden optimism? *Environment and Planning D: Society and Space*, 18, 379–82. Available online at: http://www.envplan.com/html/d1704fst.html

Gilbert, D. 1999: Sponsorship, academic independence and critical engagement: a forum on Shell, the Ogoni dispute and the Royal Geographical Society (with the Institute of British Geographers), *Ethics, Place and Environment*, 2, 219–28.

Gregson, N., Simonsen, K. and Vaiou, D. 2001: On writing (across) Europe: writing spaces, writing practices and representations of Europe. Unpublished MS available from authors.

Harvey, D. 1984: On the history and present condition of geography: an historical materialist manifesto, *Professional Geographer*, 36, 1–11.

Johnston, R.J., Gregory, D., Pratt, G. and Watts, M. (eds) 2000: *The Dictionary of Human Geography*, 4th edition. Oxford: Blackwell.

Katz, C., Bakker, K., Berg, L.D., Morin, K., Page, B., Pratt, G., Simonsen, K., Swyngedouw, E. and Uribe, G. 1998: Lost and found in the posts: addressing critical human geography, *Environment and Planning D: Society and Space*, 16, 257–78. Available online at: http://www.envplan.com/html/d1603ed.html

Minca, C. 2000: Venetian geographical praxis, *Environment and Planning D: Society and Space*, 18, 285–9.

Morin, K.M. and Berg, L.D. 1999: Emplacing current trends in feminist historical geography, *Gender, Place and Culture*, 6, 311–30.

Morris, M. 1991: Afterthoughts on Australianism, *Cultural Studies*, 6, 468–75.

Painter, J. 2000: Critical human geography. In R.J. Johnston, D. Gregory, G. Pratt and M. Watts (eds), *The Dictionary of Human Geography*, 4th edition, Oxford: Blackwell, 126–8.

Slater, D. 1992: On the borders of social theory: learning from other regions, *Environment and Planning D: Society and Space*, 10, 307–27.

Spivak, G.C. 1988: Can the subaltern speak? In C. Nelson and L. Grossberg (eds), *Marxism and the Interpretation of Culture*, Urbana: University of Illinois Press, 271–313.

Further Reading

Desbiens and Smith (1999), Gilbert (1999), Katz et al. (1998), Minca (2000) and Painter (2000) together provide an excellent introduction to the debates around critical geography.

Internet Resources

Websites

- Cincinnati Mini-conference on Critical Geography: http://geog-www.sbs.ohio-state.edu/cinciconf/
- Critical Geography Forum Online: http://www.mailbase.ac.uk/lists/crit-geog-forum
- East Asian Regional Conference in Alternative Geography: http://econgeog.misc.hit-u.ac.jp/earcag/index.html
- Geo-Critica: Online Sources for Critical Geography: http://www.ub.es/geocrit/menuuk.htm
- Geographic Perspectives on Women Specialty Group (of the AAG): http://www.online.masu.nodak.edu/divisions/hssdiv/meartz/gpow/gpow.htm
- International Critical Geography Group http://econgeog.misc.hit-u.ac.jp/icgg/index.html-ssi
- Nordic Critical Geography Conference http://www.geo.ruc.dk/Nordkrit/Home.htm
- People's Geography Project: http://www.peoplesgeography.org
- Political Ecology Society: http://www.library.arizona.edu/ej/jpe/eco~1.htm
- Sexuality and Space Specialty Group (of the AAG): http://www.frc.csm.cc.md.us/soc richardr/SaSSG.htm
- Socialist Geography Specialty Group (of the AAG): http://marcod.tripod.com/geography sgsg.html

Listservs

- Critical Geography Forum: CRIT-GEOG FORUM@JISCMAIL.AC.UK
- Disabilities and Geography List: GEOGABLE@LSV.UKY.EDU
- Geography and Feminism List: GEOGFEM@LSV.UKY.EDU
- Sexuality and Space List: SXSGEOG@LSV.UKY.EDU
- Socialist Geography List: LEFTGEOG@LISTSERV.ARIZONA.EDU

Online critical geography journals

- *ACME: An Online Journal for Critical Geographies*: http://www.acme-journal.org
- *Journal of Political Ecology*: http://www.library.arizona.edu/ej/jpe/jpeweb.html
- *Journal of Psychogeography and Urban Research*: http://www.psychogeography.co.uk

Other critical geography journals

- *Antipode: A Radical Journal of Geography*: http://www.blackwellpublishers.co.uk/journals/ANTI/descript.htm

- *Environment and Planning D: Society and Space*: http://www.envplan.com/epd/epd_current.html
- *Gender, Place and Culture: A Journal of Feminist Geography*: http://www.tandf.co.uk/journals/carfax/0966369X.html
- *Social and Cultural Geography*: http://www.tandf.co.uk/journals/routledge/14649365.html

50

A Chronology of Geography, 1859–1999

Alisdair Rogers and Heather A. Viles

Historical Events	Events and Publications in Geography

1860s

1859 Charles Darwin, *Origin of Species*	**1859** d. Carl Ritter and Alexander von
1861–5 US Civil War	Humboldt
1866 E. Haeckel coins term 'Oecologie'	**1862** John Wesley Powell loses right arm at
1867 Karl Marx, *Capital vol. 1*	Shiloh
Meiji restoration, Japan	**1864–5** Piotr Kropotkin's Siberian expeditions
1869 Suez Canal opens	**1864** George Perkins Marsh, *Man and Nature*
	1869 Powell's first Grand Canyon expedition

1870s

1870–1 Franco-Prussian War, Paris Commune	**1871** Elisée Réclus captured fighting for Paris
1871 Stanley finds Livingstone at Ujiji	Commune and exiled to Switzerland
1872 Yellowstone National Park designated	**1874** Prussia decrees geography a university
1872–6 HMS *Challenger*'s oceanographic cruise	discipline
1873 International Meteorological Organization	**1874–5** Grove Karl Gilbert's study of Henry
(IMO) founded	Mountains
1874 First Impressionist exhibition, Paris	**1875** Francis Galton's first weather map
1875 Bell patents telephone	published in a newspaper
1876 Battle of Little Big Horn	James Croll, *Climate and Time*
1878 First telephone exchange	**1876** E. Ravenstein, 'Laws of migration'
	Kropotkin flees Russia
	Alfred Russel Wallace, *The Geographical*
	Distribution of Animals
	1877 Thomas H. Huxley, *Physiography*
	1878 William M. Davis to Harvard

1880s

1880 Development of seismograph	**1880** A.R. Wallace, *Island Life*
1882 Britain occupies Egypt	**1881** F. Ratzel, *Anthropogéographie vol. 1*
d. Charles Darwin	**1882** Archibald Geikie, *Textbook of Geology*
1883 Krakatau volcano erupts	**1883** Franz Boas's expedition to Baffin Island
1884 Greenwich Mean Time established as	Edouard Suess, *Das Antlitz der Erde (The*
world standard	*Face of the Earth), 3 vols*
1884–5 Conference of Berlin carves up Africa	**1880s** Term 'geomorphology' coined by US
among European states	Geological Survey

Historical Events	*Events and Publications in Geography*
1887 Queen Victoria's golden jubilee	**1885** *c.*94 geographical societies around the
Herz identifies radio waves	world
Trafalgar Sq. riots, London	P. Kropotkin, 'What geography ought to
1888 William II, Emperor of Germany	be'
1889 Paris Exhibition	**1886** Ratzel to chair at Leipzig
	1887 Halford J. Mackinder to readership at
	Oxford
	1888 Geography at Cambridge
	National Geographical Society, USA
	1889 b. Carl O. Sauer

1890s

1893 World's Columbian Exposition, Chicago	**1890** G.K. Gilbert, *Lake Bonneville*
1895 Roentgen discovers X-rays	*Annales de Géographie* founded, Paris
Lumière brothers develop cinema	Ellen C. Semple studies under Ratzel in
1896 Becquerel discovers radioactivity	Leipzig
1898 Start of German naval build-up	**1892** John Muir helps found Sierra Club
Spanish–American War	**1893** *The Geographical Journal* founded
Term 'tropical rainforest' (*tropische*	Frederick J. Turner, *Frontier in American*
Regenwald) coined by A.W.F. Schimper	*History*
1899 Second Boer War	**1894** E. Réclus, *Universal Geography*
	Albrecht Penck, *Morphology of the Earth's*
	Surface
	1896 F. Ratzel, 'The territorial growth of states'
	1898 Paul Vidal de la Blache to chair at
	Sorbonne
	1899 Oxford, first UK geography department
	W.M. Davis, *The Geographical Cycle*

1900s

1900 Sigmund Freud, *The Interpretation of*	**1903** Chicago, first US geography graduate
Dreams	programme
First pan-African conference	P. Vidal de la Blache, *Tableau de la*
Planck's quantum theory	*géographie de la France*
1901 T. Roosevelt, US President	A.W.F. Schimper, *Plant Geography on a*
Marconi's first radio signals across Atlantic	*Physiological Basis*
1903 Wright brothers' first flight	**1904** d. F. Ratzel
1905 Albert Einstein, 'Special theory of relativity'	H.J. Mackinder, 'The geographical pivot
1906 San Francisco earthquake	of history'
Japanese defeat Russian army and navy	Association of American Geographers
1909 Model T Ford	(AAG) founded in Philadelphia
Oil drilling in Iran by British	Francis Younghusband expedition to
Andrija Mohorovicic discovers Moho	Tibet
discontinuity in Earth's crust	**1907** Griffith Taylor lectures in geography in
	Sydney
	Ellsworth Huntington, *The Pulse of Asia*
	1909 Alfred Weber's industrial location theory
	Walery von Lozinsky coins term
	'periglacial'
	E. de Martonne, *Traité de géographie*
	physique

| *Historical Events* | *Events and Publications in Geography* |

1910s and First World War

1911 Atomic structure explained by Rutherford
Amundsen reaches South Pole
Post-impressionist exhibition, London
1912 Sinking of the *Titanic*
1912–13 Balkan wars
1914 Panama canal opens
A.E. Douglass publishes early
dendrochronological work
1914–18 First World War
Einstein's general theory of relativity
1916 Pollen analysis technique first used by
Lennart von Post
1917 Russian Revolution
1919 Alcock and Brown fly Atlantic
Versailles Peace settlement

1910–22 Mackinder elected to Parliament
1911 *Annals of the Association of American
Geographers* founded
Ellen Semple, *Influences of Geographic
Environment*
Many geographers in military intelligence
1914 P. Kropotkin, *Mutual Aid*
1915 Patrick Geddes, *Cities in Evolution*
First Canadian geography course,
University of British Columbia
Alfred Wegener, *The Origin of Continents
and Oceans*
1916 F.E. Clements, *Plant Succession*
1918 d. P. Vidal de la Blache
Isaiah Bowman and Semple contribute to
US preparation for Paris peace talks

1920s

1922 James Joyce, *Ulysses*
T.S. Eliot, *The Waste Land*
Max Weber, *Economy and Society*
1923 France occupies Ruhr
1925 First Surrealist exhibition, Paris
1926 General Strike, UK
1927 Stalin comes to power, USSR
Transatlantic telephone links
1929–33 Great Depression

1921 Semple first woman president of AAG
I. Bowman joins Council on Foreign
Relations
First full geography course in Canada,
University of Montreal
1922 d. P. Kropotkin
1923 Sauer to Berkeley
J.H. Bretz, 'The channelled scablands of
the Columbia plateau'
1924 Harold Jeffreys, *The Earth: Its Origin,
History and Physical Constitution*
1925 C.O. Sauer, 'The morphology of
landscape'
1928 Griffith Taylor leaves Australia because of
his views on environmental determinism
1929 V.I. Vernadsky, *La Biosphère*

1930s

1930 Pluto discovered
Turbo jet engine patented by Frank
Whittle
1930s Purges in Germany and USSR
Dust Bowl, USA
1931 K. Popper, *The Logic of Scientific Discovery*
Japan occupies Manchuria
1932 Splitting of atom, Cockcroft and Walton
Southern Oscillation first documented by
Sir Gilbert Walker
1933 Hitler named German chancellor
1933–45 F.D. Roosevelt, US President
1933–41 New Deal in USA

1931–5 British Land Utilization Survey
under L. Dudley Stamp
1931 C. Warren Thornthwaite, 'The
climates of North America
according to a new classification'
1933 Institute of British Geographers
(IBG) founded
A. Weber attacked by Hitler Youth,
Alfred Hettner prevented from
publishing by Nazis
Bowman chair of National Research
Council, USA
Walter Christaller, *Central Places in
Southern Germany*

Historical Events	*Events and Publications in Geography*

1934 C. Raunkiaer classification of life forms
1935 Nylon invented
Ecosystem concept developed by A.G. Tansley
1936 J.M. Keynes, *The General Theory of Employment, Interest and Money*
Television broadcasts, UK
1936–9 Spanish Civil War
1937 Japan invades China

1934 d. W.M. Davis (having published 615 papers, including 33 in his 80s)
I. Bowman, *Geography in Relation to the Social Sciences*
C. Daryll Forde, *Habitat, Economy and Society*
1935 Fllip Hjulstrom develops empirical curve relating stream flow velocity to sediment movement
1936 H. Clifford Darby, *Historical Geography of England before AD 1800*
Marion Newbigin, *Plant and Animal Geography*
1937 Alex du Toit, *Our Wandering Continents*
1938–9 Richard Hartshorne in Germany and Austria
L. Dudley Stamp and Stanley Beaver, *The British Isles*
1939 Richard Hartshorne, *The Nature of Geography*

1940s and Second World War

1939–45 Second World War
1940s Development of radiocarbon dating
1941 Milutin Milankovitch's ideas on climatic cycles first published in Serbia
1944 Bretton Woods Conference
Term 'palynology' first used by Hyde and Williams
1945 United Nations charter
International Monetary Fund and World Bank
1947 Indian independence
Marshall Aid plan
1948 Apartheid in South Africa
1949 Division of Germany
People's Republic of China formed
UK Government establishes the Nature Conservancy Council
Aubreville coins the term 'desertification'

1939–45 Geographers recruited to war effort for intelligence, training, air photography, meteorology etc.
Christaller works on planning for occupied Poland
1941 R.A. Bagnold, *The Physics of Blown Sand and Desert Dunes*
H. Jenny, *Factors of Soil Formation*
1942 Gilbert F. White, *Human Adjustment to Floods*
1944 Richards, *The Tropical Rain Forest*
1945 William Kirk at fall of Mandalay, Burma campaign
New Zealand Geographical Society founded
R. Horton, 'Erosional development of streams and their drainage basins'
1946 Arthur Holmes, *Principles of Physical Geology*
1947 d. H.J. Mackinder
1948 Harvard geography closes
AAG and American Society for Professional Geographers merge
d. I. Bowman
1949 A. Leopold, *A Sand County Almanac*
G.K. Zipf, 'Principle of least effort'

1950s

1950 World Meteorological Organization (WMO) formed

1951 Canadian Association of Geographers founded

Historical Events	Events and Publications in Geography
1950–3 Korean War	**1951** L.D. Stamp appointed director of
1953 Identification of DNA double helix by	World Land Use Survey
Crick and Watson	Stanley Wooldridge and Gordon
Mt Everest summit reached	East, *The Spirit and Purpose of*
McCarthy investigations, USA	*Geography*
E. and H. Odum, *Fundamentals of Ecology*	**1952** A. Strahler, 'The dynamic basis of
North Sea storm surge affects England and	geomorphology'
the Netherlands	**1953** English translation of Walther
1954 Start of Vietnam War	Penck's *Die Morphologische Analyse*
1954–62 Algerian War of Independence	F.K. Schaefer, 'Exceptionalism in
1954 FORTRAN devised	geography'
1956 Hungarian uprising put down by Soviet	Torsten Hägerstrand, 'Innovation
troops	diffusion as a spatial process'
Stalin denounced in USSR	**1954** A. Lösch, *The Economics of Location*,
Elvis Presley	English translation
1957 Treaty of Rome leading to the European	**1955** Start of seminars in mathematical
Economic Community (EEC)	statistics under E. Ullman and W.L.
Sputnik in orbit	Garrison at University of
1957–9 International Geophysical Year	Washington, Seattle
1958 Fifth Republic, France	**1956** W.L. Thomas, *Man's Role in*
Great Leap Forward, China	*Changing the Face of the Earth*
Silicon chip invented	**1957** H.T. Odum, 'Trophic structure
1959 North Sea gas discovered	and productivity of Silver Springs,
	Florida'
	1958 Richard Chorley to Cambridge
	Institute of Australian Geographers
	founded
	C.S. Elton, *The Ecology of Invasions*
	by Animals and Plants
	L. Croizat, *Panbiogeography*
	1959 Torsten Hägerstrand visits
	University of Washington

1960s

1960 W. Rostow, *The Stages of Economic Growth*	**1960s** Early developments of GIS
Oral contraception commercially available	**1960** Anuchin's *Theoretical Problems in*
First specialized meteorological satellite	*Geography* creates fierce debate in Soviet
launched	geography
1961 Vostok 1, first manned space flight	M.G. Wolman and J.P. Miller,
Berlin Wall built	'Magnitude and frequency of forces in
1962 Cuban Missile Crisis	geomorphic processes'
The Beatles	**1960** Walter Isard, *Methods of Regional Analysis*
R. Carson, *Silent Spring*	**1961** IBG creates study groups
First geodetic satellite	Jean Gottmann, *Megalopolis*
1963 President J.F. Kennedy assassinated	**1962** William Bunge, *Theoretical Geography*
1964 Civil Rights Act and Great Society welfare	**1963** William Kirk, 'Problems of geography'
programmes, USA	**1964** Julian Wolpert, 'The decision process in
First weather satellite	spatial context'
Alaskan earthquake	**1965** L. Dudley Stamp knighted
1964–74 International Biological Programme	Peter Haggett, *Locational Analysis in*
(IBP)	*Human Geography*
1965 Watts riot, Los Angeles	S.A. Schumm and R. Lichty, 'Time, space
First mini-computer	and causality in geomorphology'

Historical Events	*Events and Publications in Geography*
1965 Peak in global emissions of sulphur Basic idea behind plate tectonics published in *Nature* by J. Tuzo Wilson	**1965** Donald Meinig, 'The Mormon culture region'
1966 Cultural Revolution, China Global Atmospheric Research Project (GARP) began	**1966** W. Christaller, 'Central places in southern Germany', English translation
1967 Six Day War, Middle East	**1967** AAG project in remote sensing *Regional Studies* founded
1967–70 Biafran War, Nigeria	Clarence J. Glacken, *Traces on the Rhodian* *Shore*
1968 Civil unrest in Paris, Chicago, Baltimore, Mexico City and elsewhere Prague Spring uprising in Czechoslovakia Tet Offensive, Vietnam Start of Deep Sea Drilling Project	Allan Pred, *Behaviour and Location* R. Chorley and P. Haggett, *Models in* *Geography* Robert H. MacArthur and Edward O.Wilson, *Theory of Island Biogeography*
1969 Apollo 11, first humans on moon Northern Ireland, start of civil strife De Gaulle resigns as French President R. Nixon, US President Woodstock Festival	**1968** AAG annual conference moved from Chicago to Ann Arbor in protest over Chicago police tactics against protesters Barry and Chorley, *Atmosphere, Weather* *and Climate*
	1969 *Geographical Analysis, Antipode* and *Environment and Planning A* founded d. W. Christaller David Harvey to Baltimore Detroit Geographical Expedition organized by W. Bunge D. Harvey, *Explanation in Geography*

1970s

1970 Earth Day, 22 April Bangladesh floods and famine Vostok ice-core drilling begun, Antarctica	**1971** The Socially and Ecologically Responsible Geographer founded by Wilbur Zelinsky and others
1971 Indo-Pakistan war leading to secession of Bangladesh (1972) Aswan High Dam opened British currency is decimalized	AAG committee on status of women geographers R. Abler, J. Adams and P. Gould, *Spatial* *Organization*
1972 UN Conference on the Environment, Stockholm Club of Rome, *Limits to Growth* Landsat 1 satellite launched UNESCO accepted World Heritage Convention	R. Chorley and B. Kennedy, *Physical* *Geography: A Systems Approach* **1972** P. Haggett, *Geography: A Modern Synthesis* H.H. Lamb, *Climate: Past, Present and* *Future vol. 1*
1973–4 Oil crisis	R.H. MacArthur, *Geographical Ecology:* *Patterns in the Distribution of Species*
1973 UK and Ireland join EEC Yom-Kippur War, Middle East F. Schumacher, *Small is Beautiful*	**1973** D. Harvey, *Social Justice and the City* **1974** Relevance and public policy theme of IBG annual conference, Norwich
1974 Nixon resigns as US President Labour unrest and 3-day week, UK CFCs linked to destruction of ozone layer	Yi-Fu Tuan, 'Space and place in humanistic geography' Alan Wilson, *Urban and Regional*
1975 Vietnam War ends New York City bankrupt Apollo and Soyuz dock in space Khmer Rouge seize power in Cambodia	*Modelling in Geography and Planning* R.U. Cooke and J.C. Doornkamp, *Geomorphology in Environmental* *Management*
1976 US bicentennial	**1975** d. C.O. Sauer

Historical Events	*Events and Publications in Geography*
1976 UK financial crisis and drought The Sex Pistols Tanghsan earthquake, China **1977** New York City blackout Queen Elizabeth II's silver jubilee Terrorist attacks in Italy and Germany **1978** Camp David peace accord Election of Pope John Paul II Proposition 13 cuts property taxes, California Jonestown, Guyana, mass suicide of religious cult members **1979** M. Thatcher elected as British Prime Minister Shah of Iran deposed Soviet invasion of Afghanistan James Lovelock, *Gaia: A New Look at* *Life on Earth*	**1975** *Journal of Historical Geography* founded AAG membership peaks at 7,000 **1976** David Lowenthal and Martyn Bowden, *Geographies of the Mind* Edward Relph, *Place and Placelessness* **1977** *Progress in Geography* splits to become *Progress in Physical Geography* and *Progress* *in Human Geography* David M. Smith, *Human Geography: A* *Welfare Approach* Manuel Castells, *The Urban Question*, English translation S.A. Schumm, *The Fluvial System* John Thornes and Denys Brunsden, *Geomorphology and Time* **1978** David Ley and Marwyn Samuels, *Humanistic Geography* I. Burton, R. Kates and G. White, *The* *Environment as Hazard* Derek Gregory, *Ideology, Science and* *Human Geography* **1979** AAG 75th anniversary Doreen Massey, 'In what sense a regional problem?' D. Brunsden and J.B. Thornes, 'Landscape sensitivity and change'

1980s

1980 R. Reagan, US President Solidarity formed, Poland Iran–Iraq war starts Mount St Helens erupts, Washington, USA World Conservation Strategy established **1981** *Columbia* space shuttle launched Riots in UK inner cities IBM-PC desktop introduced **1982** British–Argentina war over Falkland Islands El Chichon volcano erupts, Mexico **1982–3** Severe ENSO event **1983** HIV identified US invasion of Grenada **1984** Ethiopian famine Assassination of Indira Gandhi Bhopal gas disaster, India UK coal miners' strike Thames Barrier opened, London **1985** Gorbachev general secretary, Soviet Communist party	**1980** *Urban Geography* founded Robert Sack, *Conceptions of Space in Social* *Thought* H.J.B. Birks and H.H. Birks, *Quaternary* *Palaeoecology* **1981** University of Michigan closes geography department Women and Geography Study Group of IBG founded R. Johnston et al., *Dictionary of Human* *Geography* **1982** *Political Geography Quarterly* founded T. Hägerstrand, 'Diorama, path and project' Benoit Mandelbrot, *The Fractal Geometry* *of Nature* **1983** *Society and Space* founded M. Castells, *The City and the Grassroots* **1984** 25th conference of the International Geographical Union, Paris 50th anniversary of IBG Doreen Massey, *Spatial Division of Labour*

Historical Events	*Events and Publications in Geography*
1985 Mexico City earthquake Live Aid concert Ocean Drilling Project starts Ozone hole over Antarctica recorded **1986** Chernobyl nuclear emergency SPOT 1 satellite launched **1987** Stock market crash Palestinian intifada Montreal protocol on ozone J. Lovelock, *Gaia* **1988** Australia's bicentenary Intergovernmental Panel on Climate Change (IPCC) established Armenian earthquake Severe flooding in Bangladesh **1989** Berlin Wall comes down Downfall of Ceausescu regime, Romania Tiananmen Square uprising, China Exxon Valdez oil disaster, Alaska Hurricane Hugo affects USA and Caribbean	**1984** Women and Geography Study Group, IBG, *Geography and Gender* **1985** Ken Gregory, *The Nature of Physical Geography* **1986** BBC Domesday project GIS Localities research programme, UK, includes Changing Urban and Regional Systems initiative Chicago department downgraded to committee Andrew Goudie, *The Human Impact on the Environment* R.H. Haines-Young and J. Petch, *Physical Geography: Its Nature and Methods* **1987** National Center for Geographic Information and Analysis, USA The Earth as transformed by human action, symposium Clark University National Geography Awareness Week, USA J.B. Harley and D. Woodward, *The History of Cartography, vol. 1* **1988** NSA founds National Center for Geographic Information Science H.C. Darby knighted Social and cultural geography study group, IBG Denis Cosgrove and Stephen Daniels, *Iconography of Landscape* Jean Grove, *The Little Ice Age* Michael Dear, 'The postmodern challenge to human geography' **1989** Edward Soja, *Postmodern Geographies*

1990s

1990 German reunification Iraq invades Kuwait Nelson Mandela released, South Africa IPCC's first report published Drilling started on GRIP ice core, Greenland **1991** Gulf War Yeltsin president of Russia End of USSR War in Yugoslavia Maastricht Treaty Mount Pinatubo volcano erupts ERS 1 satellite launched Global Ocean Observing System (GOOS) established	**1990** M.P. Conzen, *The Making of the American Landscape* **1991** Geography part of UK national curriculum New words, new worlds conference, Edinburgh Henri Lefebvre, *The Production of Space*, English translation Stanley Schumm, *To Interpret the Earth* **1991/2** Over 10,000 geography students in UK higher education **1992** John Patten, MP, former geographer, Third UK Research Assessment Exercise George Perkins Marsh Institute, Clark University d. R. Hartshorne

Historical Events	Events and Publications in Geography
1992 UN Conference on Environment and Development, Rio North American Free Trade Agreement Los Angeles riots and earthquake Discovery of 'cosmic ripples' Hurricane Andrew affects USA **1993** Moscow rebellion Mississippi and Missouri floods Collapse of CD-PSI coalition, Italy Initial Operational Capacity for GPS navigation system reached **1994** South African elections Chiapas rebellion Israel–PLO agreement IRA ceasefire *Braveheart* **1995** Oklahoma bombing Collapse of Barings bank Kobe earthquake, Japan O.J. Simpson trial Brent Spar protest Rwanda massacres **1996** BSE crisis in UK Dolly the sheep cloned Taliban seize Kabul Spice Girls **1997–8** Severe ENSO event **1997** Hong Kong returned to China Tony Blair elected Princess Diana dies *Titanic, The Full Monty* Kyoto Accord on climate change *OK Computer*, Radiohead **1998** Monica Lewinsky affair General Pinochet arrested Water found on moon Severe floods in Bangladesh **1999** War in Kosovo East Timor revolt Turkish earthquake 150 million Internet users Chi Chi earthquake, Taiwan *The Slim Shady LP*, Eminem Seattle anti-WTO protest	**1992** Trevor Barnes and James Duncan, *Writing Worlds* Karl Butzer et al., 'The Americas before and after 1492' UNEP *World Atlas of Desertification* **1993–5** First Teaching Quality Assessment in Great Britain **1993** Geographical Association centenary David Pepper, *Eco-socialism* Gillian Rose, *Geography and Feminism* *Ecumene* and *Gender, Place and Culture* founded **1994** D. Gregory, *Geographical Imaginations* Anne Godlewska and Neil Smith, *Geography and Empire* Doreen Massey, *Space, Place and Gender* Royal Geographical Society and Institute of British Geographers merger **1995** David Bell and Gill Valentine, *Mapping Desires* Susan Hanson and Gerry Pratt, *Gender, Work and Space* **1996–7** 103 institutions of UK higher education offer geography courses **1996** Critical Geography Forum list founded Bruce Rhoads and Colin Thorn (eds), *The Scientific Nature of Geomorphology* RGS-IBG Special General Meeting on Shell's sponsorship of the society Jane Jacobs, *Edge of Empire* Richard Peet and Michael Watts, *Liberation Ecologies* Nigel Thrift, *Spatial Formations* Gearóid Ó Tuathail, *Critical Geopolitics* Richard H. Grove, *Green Imperialism* **1997** National Geographic channel launched globally First International Conference of Critical Geography, Vancouver David Goodman and Michael Watts, *Globalising Food* Roger Lee and Jane Wills, *Geographies of Economies* **1998** D. Meinig, *The Shaping of America, vol. 3* Michael Curry, *Digital Places* Tracy Skelton and Gill Valentine, *Cool Places* Heidi Nast and Steve Pile, *Places through the Body* Bruce Braun and Noel Castree, *Remaking Reality*

Part V
A Geographical Directory

As you may already have found from looking at earlier chapters in this book, there is a huge wealth of information available for geographers in print, and especially online. This section does not aim to be comprehensive in its coverage, but is intended rather to introduce you to the types of information available on the Internet which might be of interest to geographers. Many of the individual chapters give details of online resources relevant to their specific topics.

The directory includes sections on:

- gateway sites, which will lead you into a whole range of useful sites on specific geographical topics;
- basic geographical information, including maps and remotely sensed imagery;
- data sources, international and national, such as census data and topographical datasets;
- specific information, factsheets and reports on particular topics;
- e-publishing and discussion group opportunities for geographers;
- contacts for the major geography organizations and university departments;
- interesting miscellaneous sites.

Some sites are better than others and you may find some difficult to navigate around. When using the resources on the Internet you should always retain your critical awareness and ask yourself questions such as: How reliable is the information provided at this site? How recently has it been updated?

Despite presenting this directory as a list, we do not want you to regard this as an authoritative guide. The key virtue of the Internet for academics is that it encourages the art of browsing on an international scale, and allows us to wander virtually (often, it seems, aimlessly) around a huge range of resources. As such, it builds upon and extends the age-old art of browsing the shelves of libraries.

51

A Geographical Directory

Heather A. Viles and Alisdair Rogers

Geographical Gateway Sites

- **SOSIG (Social Science Information Gateway)**
This is one of the main and most useful gateways into geographical information on the Internet, and is part of the UK Resource Directory Network. The SOSIG site includes catalogues of Internet resources by subject, as well as a search engine for social science resources, and it offers the opportunity to create a personal account. Material for geographers is found under Geography – broken down into Economic Geography, Social Geography and GIS and Cartography – and Environmental Science and Issues. The Geography catalogue includes reports, databases, journal abstracts and contents, reference materials and almost anything you can think of.
http://www.sosig.ac.uk
http://www.sosig.ac.uk/roads/subject-listing/World-cat/geog.html
http://www.sosig.ac.uk/roads/subject-listing/World-cat/envsci.html

- **Resource Discovery Network (RDN) Internet Geographer**
This site hosts a tutorial on how geographers can make best use of the Internet. It includes key information skills and a guide on how to improve your Internet searching. The site was set up by Pete Maggs of the University of Newcastle and is one of a set of tutorials found at the RDN Virtual Training Site (http://www.vts.rdn.ac.uk).
http://www.sosig.ac.uk/vts/geographer/index.htm

- **Geo-information Gateway**
An index of online geo-information resources for geography, geology, transport, environment and related subject areas. It is hosted by the University of Leicester and provides a good jumping-off point to explore geography on the Web.
http://www.geog.le.ac.uk/cti/info.html

- **Geosource**
This site is hosted at the University of Utrecht, the Netherlands, and includes over 4,000 links to sites around the world, but particularly across Europe. It covers physical and human geography, as well as planning, geoscience and environmental science. Many of the links are aimed at school geography, and they include geographical games and quizzes.
http://www.library.uu.nl/geosource

- **The Virtual Geography Department**
The goal of the Virtual Geography Department Project is to offer high-quality curriculum materials and classroom and laboratory modules that can be used across the Internet by geography students and faculty at any university in the world, and to promote collaborative research. It includes *Resources for Geographers*, a

good place to begin exploring geography sites on the Web. It is hosted by the University of Boulder, Colorado, and mainly focuses on North America.
http://www.colorado.edu/geography/virtdept/contents.html

- **Blackwell Geography Resources**
  Many publishers now operate resource websites for subjects. Blackwell's site is comprehensive and regularly updated. It contains a list of listservs and newsgroups, as well as addresses for different national geographical associations. From here you can access a range of international sites, although there is a heavy representation of North American government organizations and related datasets. A useful set of links is provided by sub-discipline, taking you to various course outlines in different departments of geography.
  http://www.blackwellpublishers.co.uk/GEOG/default.asp

Basic Geographical Information

Geographical dictionaries, glossaries and encyclopaedias

- **AGI GIS Dictionary: Online Dictionary of GIS**
  The Association for Geographical Information (AGI) and the Department of Geography at the University of Edinburgh provide an online dictionary of GIS terms.
  http://www.geo.ed.ac.uk/aidict/welcome.html

- **EPA: Terms of Environment**
  The United States Environmental Protection Agency (EPA) provides a dictionary of thousands of terms used in its publications on the environment. Last revised in 1998.
  http://www.epa.gov/OCEPAterms/

- **Natureserve: Online Encyclopaedia of US/Canadian Wildlife**
  A directory of encyclopaedic information on the plants, animals and ecological communities of North America, including information on taxonomy, conservation status and distribution.
  http://www.natureserve.org

Maps and atlases

- **Internet Resources in Maps and Cartography**
  A whole host of links from the University of California at Berkeley Earth Science and Map Library.
  http://www.lib.Berkeley.edu/EART/MapCollections.html#internet

- **EDINA Digimap**
  This site delivers Ordnance Survey map data to the UK higher education community. It allows users to view and print maps of any location in Great Britain at a series of predefined scales.
  http://edina.ac.uk/digimap

- **National Atlas of the US**
  A map browser which allows you to draw maps with a range of layers, such as geology, population, environmental threats and agriculture.
  http://nationalatlas.gov/hatlas/natlasstart.asp

- **Digital Atlas of New York**
  Managed by William A. Bowen (California State University, Northridge), this site includes high-quality colour maps of New York City based on census data on race and ethnicity, income, education and so on.
  http://130.166.124.2/NYpage1.html

- **The Getty Thesaurus of Geographic Names**
  A searchable gazetteer of about 1 million geographical names from around the

world, including brief historical information on each place. There are 104 'Oxfords' in the world!
http://www.getty.edu/research/tools/vocabulary/tgn/index.html

- **Atlas of Cyberspaces**
  One of the very best geographic sites on the WWW and always worth a visit. It is maintained by Martin Dodge at the Centre for Advanced Spatial Analysis, University College, London. The atlas includes everything from maps of the early Arpanet through to imaginative cartographic representations of cyberspace, film stills, maps of fibre-optic cabling, satellites and much more. You can subscribe to a regular research bulletin.
  http://www.cybergeography.org/atlas/atlas.html

- **The Atlas of Mortality in Europe**
  The atlas shows sub-national patterns, 1980/1981 and 1990/1991 for levels and causes of mortality broken down by age and sex. The site allows you to call up two maps at once to compare them visually.
  http://www.euromort.rivm.nl

- **About.Com Free Blank Outline Maps**
  Here you can print off blank outline maps of all the countries of the world.
  http://geography.about.com/library/blank/blxindex.htm?once=true&

Remotely sensed imagery

- **Earth Images: NASA (National Aeronautical and Space Agency) Earth Observing System**
  A very helpful and well-organized site which links to a whole host of NASA imagery from space, as well as visualizations and animations of the Earth. The links are organized into beginners, intermediate and advanced and allow you to access simple imagery, as

well as more complex imagery from TOMS, SeaWIFS and other missions. The site also gives access to the online version of the publication from 1986 edited by N.M. Short and R.W. Blair, entitled *Geomorphology from Space*, which contains some useful imagery of landforms with associated descriptions.
http://eospso.gsfc.nasa.gov/eos_homepage/images.html

Geographical conversion tables

- **A Dictionary of Units**
  This site, run by Frank Tapson at the University of Exeter, contains everything you want to know on the definition of units of measurements and how to convert from metric to non-metric, or vice versa. It will tell you how many acres there are in a hectare, for instance.
  http://www.ex.ac.uk/cimt/dictunit/dictunit.htm

International Datasets

- **Ready, Net, Go! Archival Internet Resources**
  A very useful site which gives an index of archives around the world and provides links to many other national and international sites. Also contains some useful advice on archival studies and how to find your way around archives.
  http://www.tulane.edu/~lmiller/ArchivesResources.html

- **InfoNation**
  InfoNation is operated by the UN for schools use. The site allows you to select up to seven countries and then produce a table comparing them across a range of measures.
  http://www.un.org/Pubs/CyberSchoolBus/infonation/e_infonation.htm

- **Intergovernmental Panel on Climate Change (IPCC) Global Climate Data Distribution Centre**
Access to the latest climate change data and scenarios for use in climate impacts assessment. Users have to register with the Data Distribution Centre, but registration is free.
http://ipcc-ddc.cru.uea.ac.uk

- **Ciesin (Center for International Earth Science Information Network)**
Run by Columbia University, this site focuses on research and the production of information at the intersection of social, natural and information science. It gives access to a huge range of data and resources, including downloadable geographical data on China, gridded world population data, Landscan 2000 – worldwide population, and environmental sustainability index information for 122 countries.
http://www.ciesin.org/index.html

- **The Portal: Data and Meta-data on the Global Environment**
Hosted by the National Geophysical Data Center in the United States, this site provides online data access to peer-reviewed datasets on many different aspects of the global environment, such as topographic and vegetation data in a range of formats. Comes with useful advice on how to download the data.
http://www.ngdc.noaa.gov/seg/tools/gis/portalhome.shtml

- **RivDis: River Discharge Data from Around the World**
Part of the Distributed Active Archive Center (DAAC), run by the Oak Ridge National Laboratory in the United States, this site contains monthly discharge measurements for 1,018 stations throughout the world. The length of records is very variable, with a mean of 21.5 years, and the dataset covers the period 1807 to 1991. Raw data, summaries and plots are all available. The DAAC site also contains useful biogeochemical and ecological datasets, such as world NPP figures.
http://daacl.esd.ornl.gov/daacpages/rivdis.html

- **Penn World Table**
This is the authoritative listing of national economic data by region and country, beginning in 1950. It includes statistics on different ways of calculating national wealth, as well as related financial data.
http://pwt/econ.upenn.edu

National Datasets

- **National Statistics UK**
National Statistics (formerly the Office of National Statistics) is the gateway to official UK government data. It includes Statbase, which holds over 3,000 tables of information on the country, everything from adult literacy to ethnic group, transport and tourism. Tables can be downloaded or viewed on site in Excel. Neighborhood Statistics includes tables of data at local authority and ward levels. Note that not all UK census information or data are freely available.
http://www.statistics.gov.uk

- **UK National Air Quality Information Archive**
An official site which holds current and older data on air quality in the UK. The UK measurement network comprises over 1,800 sites of different types which record data on sulphur dioxide, nitrogen oxides, rainfall acidity, as well as a whole host of other types of air pollutant. The site also contains surveys of major emitters of pollutants, as well as a number of highly useful reports on different aspects of air quality.
http://www.aeat.co.uk/netcen/airqual

- **US Census Bureau**
  This gateway to the US census data contains information from the 2000 census in the form of tabulated data for each state, as well as information from the five-yearly economic census and the annual American Community Survey, which provides detailed information on selected communities.
  http://www.census.gov

- **AIRdata**
  The US Environmental Protection Agency collects a large range of air pollution data which are available from this easy-to-use site. Information is available on sources of pollution, levels of pollutants at specific sites, as well as maps showing regional pollution patterns.
  http://www.epa.gov/airdata

Major Organizations with Useful Resources for Geographers

- **The Union of International Associations (UIA)**
  The UIA, based in Brussels, is a directory of international organizations listed under topical headings. To use all the information on the site you must register.
  http://www.uia.org/topics/overve.htm

- **United Nations Development Programme (UNDP)**
  The UNDP publishes the influential *Human Development Report*, which assesses trends and patterns of human development in more than just economic terms.
  http://www.undp.org

- **United Nations Economic and Social Development**
  This is the place to start exploring the considerable social and economic resources of the United Nations. It leads to the many areas covered by the UN, from population and settlement, to sustainable development, human rights and crime. From here you can access two invaluable sites for geographers, the UN Population Fund and the UN Centre for Human Settlements.
  http://www.un.org/esa/

- **United Nations Environment Programme (UNEP)**
  The UNEP site contains, amongst other things, UNEP.net, which gives a range of environmental profile information from countries around the world, from a variety of sources.
  http://www.unep.org/

- **United Nations High Commission on Refugees (UNHCR)**
  The world's main agency dealing with refugees and humanitarian assistance. The site features up-to-date news on refugee crises, as well as background reports and statistics.
  http://www.unhcr.org

- **World Bank**
  As well as information on the World Bank, its aims and meetings, this site includes good up-to-date tables of data by topic (development, health, debt etc.) and country, together with some interesting maps and short reports on development issues. It produces an annual *World Development Report*.
  http://www.worldbank.org/

- **Organization for Economic Cooperation and Development (OECD)**
  The OECD consists of 30 of the world's richest democracies. It undertakes research and produces reports on a very wide range of issues, from money laundering to biotechnology, as well as economic and labour market matters. The site contains reports and statistics, but is not easy to use.
  http://www.oecd.org.

- **Eurostat**
Eurostat is the statistical service of the European Union, available in English, French and German. Although it catalogues the full range of information about the EU and its member states, the site does not contain many free statistical tables and can be confusing to use.
http://www.europa.eu.int/comm/eurostat

- **Food and Agriculture Organization (FAO)**
The FAO of the United Nations is the principal organization concerned with food, agriculture and hunger. As well as comprehensive statistics on agricultural production and trade, there are reports on a wide range of related development issues.
http://www.fao.org

- **Intergovernmental Panel on Climate Change (IPCC)**
The authoritative site for information on climate change. It contains information on what is happening at the IPCC, as well as summaries of the major reports and meetings, including the third assessment published in 2001.
http://www.ipcc.ch/

- **International Organization for Migration (IOM)**
The IOM is the major intergovernmental body working with governments and migrants to meet the challenges of global migration. The site contains reports on their field activities in Kosovo, East Timor and elsewhere, as well as press releases.
http://www.iom.ch

- **International Telecommunications Union (ITU)**
The ITU is based in Geneva, Switzerland, and regulates international telecommunications. Their site contains some free statistics on such things as the number of cellular phones per capita in countries of the world.
http://www.itu.int/home/index.html

- **World Health Organization (WHO)**
The WHO homepage leads to reports on health and disease worldwide, including reports on the latest outbreaks of diseases.
http://www.who.int/home-page

- **The Economic and Social Research Council (ESRC)**
The ESRC is the UK government's major agency for commissioning and funding research. From here you can access the various programmes and centres that it funds. Some are of direct interest to geographers, such as the centres on social exclusion and globalization and regionalization, or the programmes on virtual and transnational communities. It gives you a good idea of what is going on in British social science.
http://www.esrc.ac.uk/

- **United States Geological Survey (USGS)**
A very wide-ranging site which provides factsheets on environmental hazards (such as floods, hurricanes and earthquakes) as well as geospatial data and a whole host of other geological and physical geographical information. The USGS has a wide network of offices and laboratories around the United States, and it can be difficult to navigate around all the different websites. Do persevere, as there is some really good material available from this site, such as video clips of landslides from California.
http://www.usgs.gov

- **National Oceanic and Atmospheric Administration (NOAA)**
Another excellent US government site, which provides much useful information on ENSO, coastal issues and other geographically important topics, as well as

access to satellite data.
http://www.noaa.gov

- **Telegeography, Inc., Washington, DC**
  Telegeography, Inc. is run by a commercial company trading telecommunications capacity. The site includes good information on telecommunications, such as global Internet statistics as well as imaginative maps.
  http://www.telegeography.com/About/about.html

- **Human Rights Watch (HRW)**
  HRW is a non-governmental organization (NGO), founded in 1978, which reports on human rights issues in all the countries of the world. It produces regular reports on current emergencies and topics such as arms control and women's rights. There is an annual report too.
  http://www.hrw.org

- **World Resources Institute**
  This thinktank, based in Washington, DC, aims to provide a comprehensive account of global environmental issues with reports, searchable tables of data and maps.
  http://www.wri.org

- **Unrepresented Nations and Peoples Organization (UNPO)**
  UNPO is an international organization founded in 1971 by nations and peoples around the world who are not represented as such in the world's principal international organizations, such as the United Nations. It has over 50 members. Based in the Hague, its reports give an alternative perspective on the world.
  http://www.unpo.org

- **Central Intelligence Agency (CIA)**
  The CIA's World Fact Book features comprehensive and easy-to-read entries for each country in the world, including information on government, economics

and demography.
http://www.cia.gov/cia/publications/factbook/index.html

E-journals

- *CyberGEO: European Journal of Geography*
  A broad geography journal, peer-reviewed, available on the Internet free of charge.
  http://www.cybergeo.presse.fr

- *Acid News*
  The newsletter of the Swedish NGO Secretariat on acid rain, available four times a year free of charge from the website, either online or on paper. Back issues also available from the website. A good source of information on acid rain research.
  http://www.acidrain.org/acidnews.htm

- *ACME: An International e-Journal for Critical Geographies*
  This journal began publishing in 2002, and is aimed at publishing critical and radical analyses of the social, the spatial and the political.
  http://www.acme-journal.org/

Geography Listservs and Discussion Groups

- **Critical Geography Forum**
  Critical Geography Forum is an international discussion group which uses electronic mail technology to allow members to share ideas, raise questions, provide answers and air opinions. There is a browsable archive of past discussions, including critical issues such as war, development and the ethics of geographic research. The discussions are lively and fairly international, although conducted entirely in English.
  http://www.jiscmail.ac.uk/lists/crit-geog-forum.html

- **GeogNet**
  GeogNet is a moderated email discussion list, primarily for the discussion of issues associated with teaching and learning in geography in UK higher education. The list moderator is Ian Livingstone (University College, Northampton). Messages can be sent to:
  GeogNet@Northampton.ac.uk

- **Geo-Network**
  A platform for discussion of the provision of Key Skills development, tailored careers guidance and general pedagogic issues within earth science undergraduate degree courses, and a medium for the dissemination of information relating to the activities of the subject centre in the area of earth sciences.
  http://www.jiscmail.ac.uk/lists/geo-network.html

- **Urban Geography**
  UrbGeog is hosted at the University of Arizona. You can join at the first address below, while the second address leads you to the archives.
  http://listserv.arizona.edu/cgi-bin/wa?SUBED1=urbgeog&A=1
  http://listserv.arizona.edu/archives/urbgeog.html

- **Landscape Research**
  A mainly UK-based listserv on interdisciplinary landscape research. From this address you can join the list.
  http://www.jiscmail.ac.uk/lists/LANDSCAPE-RESEARCH.html

- **Left Geography**
  Founded in 1995 at the University of Kentucky, it deals with socialist and radical perspectives on geography.
  http://www.qx.net/jeff/listservs/leftgeog.htm

- **GIS-UK**
  A listserv on GIS issues.

http://www.jiscmail.ac.uk/lists/gis-uk.html

Geographical Organizations

- **Royal Geographical Society (RGS)**
  The main UK association for professional geographers and merged with the Institute of British Geographers (IBG). The site contains information on the society, its events, lectures and meetings, as well as guides to funding expeditions. An important feature is the list of research groups under the RGS, each group covering a topic or field in geography, e.g. transport geography, developing areas and the British Geomorphological Research Group. Many of these groups run their own websites and publish newsletters. The site also contains a page on 'What is Geography?'
  http://www.rgs.org

- **The Geographical Association (GA)**
  The GA is the main association of UK geography teachers, and their site provides useful resources for classroom use as well as news of meetings and events.
  http://www.geography.org.uk

- **The Association of American Geographers (AAG)**
  The main association for professional geographers in the United States. The site lists geographical associations in other countries around the world, as well as information on jobs and careers in geography.
  http://www.aag.org

- **The Royal Canadian Geographic Society (RCGS)**
  The site is in French and English, and features information about the society and geography in Canada. It also leads to Jumpstation Geo, a comprehensive list of links relating to Canadian geography

and geographic resources.
http://www.rcgs.org
http://www.ccge.org/geosources/
jumpstn.htm

- **National Subject Centre for Geography, Earth and Environmental Sciences (GEES)**
This is one of 24 subject centres set up in the UK charged with disseminating good practice in learning and testing in geography. It is hosted at the University of Plymouth and features information on teaching and learning resources, as well as useful links.
http://www.gees.ac.uk/

- **The Geography Discipline Network (GDN)**
The GDN works in association with GEES (above), and provides extensive resources connected with teaching. From here you can also access the abstracts of the *Journal of Geography in Higher Education* and the *Journal of Geography*.
http://www.chelt.ac.uk/el/philg/gdn/index.htm

- **The Society of Cartographers**
The society was founded in 1964 and its membership is mainly based in the UK.
http://www.soc.org.uk

- **The British Geological Survey**
A large organization devoted to providing geological research. Its expertise includes geologists, mineralogists, engineering geologists, palaeontologists, chemists, hydrogeologists, mathematicians, biologists, computer specialists and information technologists. Useful if you want to find out more about what is going on in UK geological sciences.
http://www.bgs.ac.uk/home.html#

University Geography Departments

- **Geography Departments Worldwide**
Links to 939 geography departments in 80 countries, hosted by the Department of Geography, University of Innsbruck.
http://geowww.uibk.ac.at/geolinks

- **UK Universities and Colleges**
A complete listing of all UK universities and colleges, from which you can find individual departments of geography.
http://www.scit.wlv.ac.uk/ukinfo/alpha.html

- **UK Geography Departments**
A list of email addresses for all UK departments of geography hosted by the RGS.
http://www.rgs.org/category.php?Page=maingeography

Miscellaneous

- **The People's Geography Project**
This site defines its aims as 'to popularize and make even more relevant and useful to ordinary people the important, critical ways of understanding the complex geographies of everyday life that geographers have [developed] and continue to develop'. Managed by Don Mitchell (Syracuse University, New York), it intends to provide a radical alternative to mainstream geographical education and use geography for social justice, through such things as 'guerrilla geographies' and establishing links with grassroots organizations.
http://www.peoplesgeography.org

- **The Global Site**
This site is run by Martin Shaw at the University of Sussex, and it is an indispensable starting point for any critical understanding of globalization and related issues. It features up-to-date com-

mentary on major world events.
http://www.theglobalsite.ac.uk

- **Falling Through the Net**
  The US Commerce Department hosts a remarkable site, frequently updated, charting the so-called digital divide in the United States, i.e., the diffusion of access to information technology in social and geographical terms.
  http://www.digitaldivide.gov/

- **The Urban and Regional Regeneration Bulletin**
  Run in association with the journal *City*, this site features short articles on urban planning issues, mainly on Britain but also on other countries. There is news of forthcoming events in urban research. The site is linked with the University of Newcastle's Centre for Urban and Regional Development Studies.
  http://www.ncl.ac.uk/curds/urrb/

- **The Earth and Moon Viewer**
  From here you can observe the Earth and Moon from a specified longitude and latitude, or above a selected city, by day or night.
  http://www.Colorado.edu/geography/virtdept/contents.html

- **Volcano World**
  Track the latest volcanic eruptions round the world, view images of volcanoes and find out all you ever wanted to know about volcanoes past and present, on the Earth and other planets.
  http://volcano.und.nodak.edu/

- **International Rivers Network (IRN)**
  Established in 1985, the IRN is a non-profit-making organization which links human rights issues with environmental protection. The site contains much of interest, including online access to the IRN's publication, *World Rivers Review*.
  http://www.irn.org

- **Geo-Images**
  The University of Berkeley, California, hosts a useful website devoted to providing slides suitable for classroom and individual use on geographical topics. The images cannot be used for commercial purposes. The site is mainly about North America, with some material on Africa and Afghanistan, but it also includes pictures illustrating geomorphological features.
  http://geogweb.Berkeley.edu/GeoImages.html

- **The Cities/Buildings Database**
  The Cities/Buildings Database is a collection of over 5,000 digitized images of buildings and cities drawn from across time and throughout the world, available to students, researchers and educators on the Web. It can be searched by architect, country and city. The site is managed by Meredith L. Clausen (University of Washington), and from there you can visit many other collections of photographs.
  http://www.washington.edu/ark2

Part VI
Expanding Horizons

One of the advantages of studying geography is that it can take you out of the library, off the campus, and even out of the country. This section contains essential information on how to expand your horizons and make the best of the opportunities available. These opportunities can begin with the course itself, either through organizing your own overseas expeditions or by taking some of your course in another European university through the SOCRATES programme. There is information here to get you started on these, including useful tips on how to apply for funding from commercial donors, charities and foundations. After one course of geography you may feel that you haven't stopped learning or want to acquire more specialized and advanced skills. If so, then you may consider a Master's course. Although the UK offers a growing number of such courses in geography and related disciplines, there are also great opportunities abroad. We have asked colleagues in several mainly English-speaking countries to describe some of the opportunities for research in their country and the practical steps required. Whatever path you take after university, the chances are that you will need to produce a curriculum vitae listing your talents and achievements. The section ends with advice on how to do this.

52

Opportunities for Study Abroad: The SOCRATES-ERASMUS Programme

Fiona O'Carroll and Joe Painter

Today there are many opportunities to broaden your horizons and add an international experience to your education by undertaking part of your studies at a university in another country. For those of you rising to the challenge, this will be the experience of a lifetime.

Opportunities to study abroad for part of your degree are available within Europe and beyond. A significant percentage of placements in Europe are organized through the SOCRATES-ERASMUS programme. This programme offers many advantages, which are discussed below in more detail, and will be the focus in this chapter. However, many of the general principles (financial arrangements most notably excluded) apply to other types of study abroad. The range of placements for geography students, across all schemes, varies from one university or college to another, so you should contact your international/European office or geography department for full details. For those of you still choosing your university, open days or selection interviews provide ideal occasions to enquire about the possibilities.

The SOCRATES-ERASMUS Programme

The SOCRATES programme is funded by the European Commission's Directorate-General for Education and Culture. It began in 1995 and is now in its second phase. It covers all areas of education from school level to adult and distance learning. The higher education section of SOCRATES is called ERASMUS. The first ERASMUS programme began in the 1980s and the term 'ERASMUS student' is well recognized across Europe, though 'SOCRATES' or 'SOCRATES-ERASMUS' are sometimes used instead. ERASMUS funds activities such as student and staff exchange and curriculum development.

The programme aims to introduce a European dimension into education and thereby to improve international cooperation and the quality of education. Practical objectives include improving knowledge of European languages and promoting cooperation and mobility throughout education. Approximately 2,000 higher education institutions across Europe are involved in SOCRATES-ERASMUS, including about 200 in the UK. The programme has expanded to include a total of 31 countries (table 52.1).

Table 52.1   Countries eligible to participate in SOCRATES-ERASMUS

- Member states of the European Union: Austria, Belgium, Denmark, Finland, France, Germany, Greece, Ireland, Italy, Luxembourg, Netherlands, Portugal, Spain, Sweden, United Kingdom.
- EFTA/EEA countries: Iceland, Liechtenstein, Norway.
- EU associated states: Bulgaria, Cyprus, Czech Republic, Estonia, Hungary, Latvia, Lithuania, Malta, Poland, Romania, Slovakia, Slovenia.
- Turkey may participate from 2003.

SOCRATES-ERASMUS student exchanges are established when the collaborating universities sign an agreement. Each university will have its own 'partner' universities, many specific to particular departments or subject areas. This means that you will not have a completely free choice of university at which to study abroad. However, there are advantages to this: each department will be working with a selected number of universities and can therefore develop a good relationship with each of them. This helps to ensure a high level of personal care. Agreements are usually signed each year, so opportunities may vary from year to year. Not all universities offer SOCRATES-ERASMUS placements in geography. You should check with your international/European office or SOCRATES Departmental Coordinator for up-to-date information.

What's in it for You?

There are many advantages in participating in the SOCRATES-ERASMUS programme.

FINANCIAL

- You will not pay tuition fees at the host institution.
- If you spend a full academic year abroad you will not have to pay tuition fees to your home institution for that year. The British government will pay your fees, in recognition of the benefits of SOCRATES-ERASMUS.
- Most ERASMUS students receive a modest student mobility grant from the European Commission to help meet the additional costs of study abroad. Supplementary student loans are available through the Student Loan Company and some assistance with travel costs may be provided by Local Education Authorities (LEAs). Contact your LEA for further details.

ACADEMIC RECOGNITION

- The period abroad will be recognized as an integral part of your degree. For some students credit will be transferred to their home university through the European Credit Transfer System (ECTS). Not all universities subscribe to this scheme and some do not use it in all subject areas. However, this should not affect the recognition of the period abroad.

LANGUAGE SKILLS

- Living and working among native speakers is the most effective way to improve your language skills.

PERSONAL

- Studying and living in another country helps to develop life skills and independence, and is an opportunity to make lots of new friends.

EMPLOYABILITY

- The ERASMUS label has significant

currency amongst employers across Europe. Employers appreciate the initiative and independence demonstrated by ERASMUS students, and their excellent language skills. There are also specific benefits for geography students. If you want to use your geography in your job and are thinking of a career in planning or urban and regional development, for example, your European experience will make you more attractive to prospective employers as these professions are now heavily influenced by developments in the European Union.

GEOGRAPHICAL

- It may be possible to arrange a period of overseas fieldwork associated with your placement. Many geography students take the opportunity of an extended period of residence abroad to work on a dissertation topic related to the city or region in which they are studying.
- If you are interested in postgraduate study for a Master's degree or a Ph.D., then there may be opportunities available in your host institution which you will be well placed to take up.

Studying Geography Abroad

The *UK Guide* (UK SOCRATES-ERASMUS Council 2001) is an excellent resource for those applying to university as well as existing students. It gives details about the benefits of the programme and has entries for all participating UK institutions. It lists the subject areas and countries in which placements are available, and includes 44 universities and colleges that offer exchanges in geography and related subjects. It should be available from your local library or careers office.

One of the most interesting aspects of a year abroad is the chance to study geography as it is taught and understood in a different country. Subjects such as chemistry and engineering are broadly similar all around the world, because there is a well-integrated international scientific community in these fields that enables the rapid dissemination of new ideas and results. In philosophy and sociology, the works of major continental philosophers and sociologists are established elements of degree courses in British universities. Geography is different. In physical geography the situation is similar to that in the other natural sciences. The content of modules on geomorphology, for example, is likely to vary as much between different universities within Britain as between universities in different countries. By contrast, human geography is divided into a number of relatively distinct linguistic traditions. The human geography taught in British universities draws almost exclusively on the anglophone (English-language) tradition (for a survey, see Johnston 1997). While there is no doubting the richness and vibrancy of the anglophone tradition, it is a mistake to assume 'that human geography conducted in English somehow constitutes the "authorized" version of the discipline' (Johnston et al. 2000: viii).

The different linguistic traditions in European geography have developed in relative isolation from one another, and there is considerable variation between countries in the dominant schools of thought, the major fields of research, the principal methodological approaches, and even in what constitutes 'geography' (for a survey of the development of geography in different countries up to 1984, see Johnston and Claval 1984). These national differences are reflected in undergraduate and postgraduate courses. For example, German geography has a strong tradition of geomorphology and historical settlement studies, and an emphasis on overseas research. Although things are changing, the traditional dominance of individual professors and their specific research interests produces a degree of inertia, so German geography has not experienced the dramatic pluralization and influx of new ideas and methods evident in anglophone geography since the 1970s. By contrast, Scandinavian and Dutch geography overlaps rather more with British and American work, but is also distinctive, particularly in having stronger links with the state planning systems in those countries. As with other aspects of SOCRATES, the key here is information. Try to find out as much as possible about the kind of geography taught at the different host universities. Use the Internet and talk to others who have studied there. If you have strong preferences about the topics you wish to study, then make sure your areas of interest are available.

There are also significant national differences in teaching methods and styles. In many European countries, students go to university later and take longer to graduate than in Britain. Degree courses are often very flexible and modules can sometimes be repeated to obtain a higher mark. This means that students (and their teachers) do not always share the British obsession with passing a rigid number of modules in a year and completing a degree in a set time. Lecture classes may be very large and there may be less contact between lecturers and students than in Britain. Examining systems also vary. In some countries assessment is based on coursework, in others there may be oral exams, and so on. Again, the key to a successful year abroad is information. Try to find out in advance what to expect, preferably from someone who has been through the system.

How to Apply and Preparing to Go

ERASMUS placements generally take place in the second or third year of your degree and last from 3 to 12 months. They are a compulsory element of some degree programmes, and in such cases you will know before you arrive at university that part of your studies will take place abroad. For other courses, an ERASMUS placement is op-

tional and you can apply after you have started your degree. In this case, you should register your interest with the person responsible in your department as early as possible. Departments select students for placements according to a range of criteria. These vary from university to university but normally include academic performance, language skills and ability to cope in a different country.

Preparation is vital and you must start well in advance. It is desirable to start preparing a year or so before you depart, as this gives you plenty of time to choose your host university, get your language skills in shape, take care of the application forms, decide which courses you will take, book your flights and arrange accommodation. Your department will help you through most of this, but be prepared to do a lot of the work yourself.

There are various ways in which you can prepare linguistically, even if your degree does not include a language component. Many universities offer language modules to non-language students and language centres provide self-access materials. Many universities offer intensive language courses for their incoming ERASMUS students before the study period starts, sometimes at no cost. The European Commission sometimes provides funding for courses in the less widely spoken languages. If all else fails, certain universities (most notably in the Netherlands and Scandinavian countries) offer courses in English, so a period abroad might still be an option. However, some knowledge of the language of the host country is desirable for social reasons and to help you make the best use of library and other facilities. Enquire about the various possibilities well in advance.

We hope we have convinced you by now that a period abroad can be a wonderful life-changing experience. Throughout this chapter, we have pointed out that information is a key element of a successful period abroad. Space permits us to give you just a flavour of the opportunities available, so if you are interested do contact the people we have suggested and consult the publications and websites below. And have a good time!

References

Johnston, R.J. 1997: *Geography and Geographers: Anglo-American Geography Since 1945*. London: Arnold.

Johnston, R.J. and Claval, P. (eds) 1984: *Geography Since the Second World War: An International Survey*. London: Croom Helm.

Johnston, R.J., Gregory, D., Pratt, G. and Watts, M. (eds.) 2000: *The Dictionary of Human Geography*, 4th edition. Oxford: Blackwell.

UK SOCRATES-ERASMUS Council 2001: *The UK Guide for Students Entering Higher Education*. Camberley: ISCO Publications.

Further Reading

The UK Guide for Students Entering Higher Education. Available from ISCO Publications, 12A Princess Way, Camberley, Surrey GU15 3SP (£13.00) or in your local library or careers office.

The European Choice: A Guide to Opportunities for Higher Education in Europe and *Student Grants and Loans: A Brief Guide for Higher Education Students*. Both available free of charge from DfEE Publications Centre, PO Box 6927, London E3 3NZ.

Travelling, Studying, Working and Living within the European Union. Available free of charge from Publications Department, The European Commission, 8 Storey's Gate, London SW1T 3AT.

Higher Education in the European Community: Student Handbook. A directory of courses and institutions in 12 countries. Available from HMSO Books, PO Box 276, London SW8 5DT.

The European Union: What's it all About - Where to find out more? Available free of charge from Publications Department, Office of the European Parliament, 2 Queen Anne's Gate, London SW1T 3AT.

Education: Guide to European Organisations and Programmes. National Foundation for Educational Research, The Mere, Upton Park, Slough SL1 2DQ.

EURO Challenge: International Career Guide for Students and Graduates. Joerg E Staufenbiel, Institut für Berufs- und Ausbildungsplanung Köln, GmbH, PO Box 10 35 43, D-50475 Köln, Germany.

Internet Resources

- European Commission site – the SOCRATES programme: http://europa.eu.int/comm/education/socrates.html
- Technical Assistance Office for SOCRATES: http://www.socrates-youth.be
- UK SOCRATES-ERASMUS Council: http://www.erasmus.ac.uk/

53

How to Fund Overseas Travel and Research

David J. Nash

One of the major reasons why students choose to study geography is because they want to understand the world around them a little better. Luckier undergraduates get the opportunity to visit exotic places as part of field trips, but for some this is not enough and they want to set off on their own to explore the globe. This often involves a summer spent 'travelling', but it may also include research that forms the basis of an undergraduate project or dissertation. Doing overseas fieldwork is incredibly rewarding but it can be very expensive, and it is likely that fundraising will be one of your biggest headaches. However, don't let this put you off. There are several excellent publications offering advice on fundraising, and even more individuals and organizations who are prepared to give you money (or support in kind). In fact, you will probably be surprised at how many organizations provide funding, often asking for little in return apart from an acknowledgement and a report on what you have done with their money.

The Planning Stage

Before you even begin to think about seeking funding, you need to consider three key issues: *time*, your project *image* and your *budget*. Raising funding for any overseas travel involves a lot of thought, careful re-search, letter writing and form filling, all of which take time. Regardless of whether you plan to travel on your own or as part of a group, you will need to allow at least nine months to raise sufficient funding – maybe more if there is a large budget involved. This means that you will need to begin planning fundraising not long after you have decided that you want to travel! The reason for this is that most funding organizations have fixed application deadlines – December and January are very popular – so if, for example, you are planning to travel in July, you need to begin fundraising by October of the preceding year. To put this in perspective, if you are an undergraduate thinking of travelling overseas between your second and third year, you need to start planning your trip before you have even finished the first year of your degree.

The next stage is to consider your image. Before any organization is going to give you money, you need to convince them that what you want to do is worthwhile and that you are competent to undertake the trip. It is important that you market yourself effectively as this will help your application stand out from the others that organizations receive. You could, for example, create your own headed notepaper, with a snappy logo and project name. If you are travelling as a group you might consider including the title of your university or educational institution or the name of an appropriate patron or sponsor. Bear in mind, however, that you will need to get written permission to use these names or titles in publicity material. A professional-looking brochure or prospectus including information about your trip is almost essential. This should be informative without being excessive, but must include the following: some background to your project, a statement of your aims and objectives, a map, a brief profile of yourself and any other group members (detailing your relevant experience), a budget, details of any support and host country contacts (absolutely essential – if you are travelling as an expedition you should try to get support from your educational institution), and a contact name and address. The brochure need not be expensive – a well-designed paper pamphlet with a clear layout is more useful than something glossy but less informative. If your image is too 'slick' it may put off potential sponsors, who could think you already have too much money.

At the same time as sorting out your project image, you should also devise a realistic budget to act as a target for fundraising. This needs to include details under four subheadings:

- *Pre-trip expenses*: postage costs, phone calls, application fees for research permits, brochure production, costs of field and safety equipment, vaccinations, training courses, transport, insurance.
- *Field expenses*: subsistence, accommodation, travel, vehicle hire, fuel, freighting costs, customs charges, interpretation costs.
- *Post-trip expenses*: film processing, report preparation, writing to sponsors, postage.
- *Contingency costs*: 'emergency' costs, calculated as an additional 10 per cent of the sub-total of other expenses.

The exact budget may vary as you attract sponsorship or support in kind. You should therefore try to identify maximum and minimum costs for each item to give an idea of the lowest total income required before the trip can go ahead. An example budget for a group expedition is shown in table 53.1, which includes variations in costs as well as the factors that may influence final expenditure. In addition to the budget, you should also consider how you are going to organize your finances. You will need to keep detailed accounts – if you are travelling as a group you should nominate a treasurer to do this. It is a good idea to set up a separate bank account to keep fieldwork funds separate from your own personal funds, and try to get your bank manager on your side to provide advice on accessing money whilst overseas.

Funding Sources

Once you have an image and budget you need to consider your fundraising strategy. The range of methods for getting support is endless, but some of the more conventional approaches are highlighted here. Do not expect a single source to provide all your funding – you will need to be prepared to invest a lot of time and effort before you reach your target (table 53.2). A key to success is spending time in libraries, telephoning charities, businesses and organizations, and finding out how to apply for funds, so that you can target the most appropriate organizations. Whether you are travelling alone or in a group, the most useful initial contact point is the Expedition Advisory Centre (EAC) of the Royal Geographical Society (with the Institute of British Geographers) in London. The EAC publishes *The Expedition Planner's Handbook and Directory* and regularly updated pamphlets, entitled *Fund-raising and Budgeting for Expeditions* and *Fund-raising to Join an Expedition*, all of which are excellent guides to securing funding. There are also equivalent organizations in North America (Explorers Club and South American Explorers Club) and Australasia (Australian and New Zealand Sci-

Table 53.1 Example budget for the University of Somewhere Expedition to the Middle of Nowhere
The expedition, investigating environmental impacts of logging in the Middle of Nowhere, involved 5
students and lasted 5 weeks. It was seeking university support to pay for some administrative costs
and finance insurance.

Proposed expenditure	Maximum	Minimum	Notes
Pre-trip expenses			
Administration			
Postage/Telephone	50.00	0.00	A
Brochure/letter printing	50.00	0.00	A
Research permit application	50.00	50.00	–
Equipment			
Field	500.00	300.00	B
Film	100.00	75.00	AB
Maps	50.00	25.00	AB
Medical	100.00	75.00	B
Insurance	250.00	0.00	A
Flights (5)	3,500.00	2,500.00	C
Field expenses			
Accommodation	500.00	0.00	D
Subsistence			
(5 weeks × 5 × £10/week)	1,250.00	1,250.00	–
Transport hire/fuel	1,000.00	300.00	D
Post-trip expenses			
Film processing	200.00	150.00	AB
Report preparation	150.00	75.00	AB
Postage	50.00	0.00	A
Sub-total	7,800.00	4,800.00	–
Contingency (10% of sub-total)	780.00	480.00	–
Personal contributions	–4,000.00	–2,500.00	E
TOTAL SOUGHT	**£4,580.00**	**£2,780.00**	

A   Costs may be reduced/eliminated if expedition given university recognition.
B   Costs may be reduced if local sponsorship and/or discount is sought.
C   May be reduced if discounted flights can be obtained.
D   Costs of accommodation and local transport hire may be reduced if links with the University of the Middle of
   Nowhere are established.
E   Level of personal contributions based on approximately 50% of maximum or minimum sub-total, with each of
   the 5 group members contributing equally.

entific Exploration Society – ANZSES).

There are four main sources of funding for overseas trips. The first of these are various *grant-giving organizations and charitable trusts* who give money to support research, especially if it is of wider social or environmental significance. You will need to do some homework to identify the most appropriate sources. Many organizations have specific rules regarding who they will fund; some only support individuals or groups, some will not fund undergraduates, a few are age- or gender-specific, and some only support work in certain subject areas or applicants from

Table 53.2 Things to remember when applying for funding

- There may be a lot of money available but a lot of people are competing for it.
- Start applying early to maximize your chances, keep copies of everything, stick to application deadlines and apply to every appropriate source.
- If in doubt about your eligibility for a funding source, phone and ask.
- Persevere – it can take forever for funding bodies to process applications.
- Fundraising is time-consuming and it could distract from your studies, so spread the burden whenever possible.

(or travelling to) particular regions. Many also require applications by a specific deadline, in a prescribed format, with a supporting statement from a referee, and may invite you to an interview. Depending upon the reason you are travelling, it is worthwhile trying to obtain funding or recognition from one of the major organizations concerned with that region or subject (e.g. a national geographical, ecological or exploration society, or international organizations such as the World Wide Fund for Nature), as this may well lead to other organizations providing similar support. There are several directories that detail grant-giving organizations, including: *The International Foundation Directory* (Europa Publications), *A Guide to Grants for Individuals in Need*, *The Third World Directory* and *The Environmental Grants Guide* (all published by the Directory of Social Change), and the *Directory of Grant Making Trusts* (Charities Aid Foundation). The last-named title is especially useful as it lists organizations and charities by region and topic that they are most likely to fund.

Commercial and industrial enterprises are the second major source of assistance. Most are unlikely to give direct funding, but may give payment in kind such as discounts on equipment and services. This is a less secure approach to getting support but, with some

background work, can be successful. However, be prepared to receive sackloads of rejection letters. As with approaches to charities, you should carefully target commercial enterprises, ideally contacting named individuals within companies that are local, supply specialist equipment required for your trip, or have business links with the country you intend to visit. You can get details of appropriate companies from the *Directory of Directors* (Reed Information Services), the *Guide to Company Giving* (Directory of Social Change), international business magazines, your local chamber of commerce or the commercial attaché at the embassy or high commission of your host country. If you send off a brochure, you should include a brief letter stating who you are, the aims of your trip, who is likely to benefit, the amount you are trying to raise, how much (or what) you would like from them and, most importantly, what you can give in return. It is also worth contacting institutions such as local councils, your old school and your current educational establishment. The latter can be particularly helpful by loaning equipment or offering cheaper (or possibly free) travel insurance. Many universities have an established Exploration Society that will also be able to advise you.

Another possible source of funding is *the general public and the media*. It is possible, for example, to raise funding through the organization of events such as sponsored walks. If you have no shame, you may consider contacting local newspapers, magazines or even local radio, as a large number of individuals have received donations following this sort of public humiliation. This works especially well if you come from a small community where you can stimulate interest in your trip. If your visit is likely to attract media interest, you may be able to sell your story or photographic rights, but make sure you seek legal advice before doing so.

Finally, one of the most important sources of funding has to come from *yourself and other group members*. You must be prepared to spend as much on your trip as you would on normal living costs and, ideally, try to fix the level of personal contributions to cover up to half the total costs. This may seem a lot, but it is essential to show sponsors that you are committed to your trip. Personal financial commitments are particularly important at the early stages of planning a trip before you have received other funding. Whether you are travelling on your own or as part of a group, try to make sure that all personal contributions are paid into your project bank account as early as possible. Remember, this money may be reimbursed if you raise sufficient funding, but it may also end up being the prime source for your visit.

Now all you have to do, having identified the most appropriate funding sources, is send off large numbers of applications and wait. Hopefully, if you are sufficiently well organized and committed to your trip, you will be successful in obtaining funding. If, however, your fundraising goes less well and you have to cancel the trip, remember that you will be expected to return any funds to your donors. At the end of a trip, don't forget to say 'thank you' to everyone who has helped you as soon as you return – you never know when other undergraduates will require assistance in the future.

Useful Addresses

The majority of the publications mentioned can be obtained from the following organizations. Many are charities, so send a stamped, self-addressed envelope when requesting information.

Australian and New Zealand Scientific Exploration Society, PO Box 174, Albert Park 3206, Victoria, Australia.
http://home.vicnet.net.au/~anzses/
Charities Aid Foundation, King's Hill, West Malling, Kent ME19 4TA, UK.
Directory of Social Change, 24 Stephenson Way, London NW1 2DP, UK.
Europa Publications, 18 Bedford Square, London WC1B 3JN, UK.
Expedition Advisory Centre, RGS-IBG, 1 Kensington Gore, London SW7 2AR, UK.
http://www.rgs.org/
The Explorers Club, 46 East 70th Street, New York, NY 10021, USA.
http://www.explorers.org/
Reed Information Services, Windsor Court, East Grinstead House, East Grinstead, West Sussex RX19 1XA, UK.
South American Explorers Club, 126 Indian Creek Road, Ithaca, NY 14850, USA.
http://samexplo.org/

Further Reading

Nash, D.J. 2000a: Doing independent overseas fieldwork 1: practicalities and pitfalls, *Journal of Geography in Higher Education*, 24, 139–49.
Nash, D.J. 2000b: Doing independent overseas fieldwork 2: getting funding, *Journal of Geography in Higher Education*, 24, 437–45.

54

Applying for UK Master's Courses

John Boardman

Why Do a Master's?

There are at least four good reasons why you might wish to apply for a Master's course:

- *Vocational*: some courses lead to immediate opportunities in the job market; these usually involve a strong element of skills training such as courses focusing on geographic information systems (e.g. Edinburgh and Leicester) or are directed at a particular sector of the market (e.g. Renewable Energy and the Environment at Reading). The growth of environmental consultancies has stimulated this aspect of the educational field.
- *Reorientation*: you may feel that previous educational experience does not match your aspirations, e.g. a language degree and a desire to work in conservation. Or you may wish to study in more depth a specific field not covered in your first degree, such as environmental economics.
- *Retraining*: increasing numbers of applicants are already in a job but wish to retrain or pick up new skills.
- *Preparation for a doctorate*: many applicants see a Master's course as a stepping stone to doctoral research. Traditionally in the UK this has not been a favoured route, but university departments and Research Councils are increasingly encouraging what is known as the 1+3 route. The Master's, and the research training that it is assumed to provide, are seen as a desirable precursor to a three-year doctoral programme. The Master's is also, of course, an opportunity to decide that doing a doctorate is not a good idea!

There are some less good reasons for doing a Master's. The most common is that you cannot think of anything else to do. In times of substantial costs for a year of education, this probably happens less frequently. Indeed, the better, more popular Master's are oversubscribed and will not knowingly admit students with poor motivation (see below).

Types of Master's

The most obvious distinction among the many courses on offer is between a Master's 'by coursework' (or a 'taught' Master's) and one 'by research'. The former are generally one year and the latter two years in length. Master's by coursework usually include a dissertation or thesis (perhaps 15,000 words), so they too include an element of research. Terminology can be confusing: Master's by coursework may be MA, M.Sc. or M.Phil. Similarly, those by research may have the same designation. Universities vary in the terms they use. A recent development is the M.Res. degree (Masters in Research) designed specifically to teach research skills. However, many would argue that this can be better done in the context of subject-related Master's courses.

Other distinctions that can be made between types of Master's course are:

- Those that are broad and general in their coverage, and those that are more focused.
- Those that are more theoretical as opposed to the more practical or vocational.
- Full-time, part-time or those using distance learning approaches (e.g. Open University and Wye College), or even the new Virtual M.Sc. Programme in Coastal Zone Management (Ulster).
- The modular versus the non-modular Master's.

Modularity is worth emphasizing. Like most undergraduate degrees, Master's have largely 'gone modular'. The programme consists of a series of self-contained units (modules) which are assessed by a combination of coursework and exams. In contrast, some Master's retain a final exam but also have assessment by coursework and dissertation. The advantage of a modular degree is that it may offer more choice and flexibility. The disadvantage is that it may have no clear theme or development, consisting of a number of loosely related modules.

Gathering Information: What to Look Out For

Although your first degree may be in geography, you should explore courses in other associated disciplines. Geography overlaps into biological, earth and sociological sciences, plus techniques that cross disciplines (remote sensing, environmental impact assessment, GIS etc.). Also, Master's courses are increasingly recognizing that there is a demand for interdisciplinary approaches. The expanding area of the environment has spawned many Master's courses that tend to be interdisciplinary (e.g. Environment and Development at Cambridge). You should also remember that there are over 100 universities and that some of the less well-known ones have developed expertise in specialist areas – check out their websites.

Information about Master's courses is available from universities in written form and on university websites (see http://www.scit.wlv.ac.uk/ukinfo/). It is worth checking the following and making further enquiries if the information is not clear.

- How much flexibility is there to choose modules or options? Do they all run every year? Will particular modules or options be available?
- Is the course modular? How is it assessed? What is the balance between coursework, exams and dissertation?
- What is the length of the course? Some 'one-year' courses are actually 8 months; others are a full 12 months.
- How many students are admitted to the course? Are they taught in one large group; how much small-group teaching is there? Modular courses may involve students from several Master's degrees being taught together.
- Are the classes shared with undergraduates?

Funding

Funding for UK Master's students is very limited. Research Councils sponsor places on certain Master's courses. The Councils issue lists of which courses are receiving financial backing. Supported courses often advertise for applicants. Some courses have bursaries from commercial or philanthropic organizations. Some firms will support students on Master's courses but this is far more likely if you are already an employee of the firm. Because of these funding difficulties, many British students take out loans or rely on savings and parental help in order to take Master's courses. Master's courses should be demanding and intense, and students must think very carefully before committing themselves to part-time work in order to fund a full-time course.

Deadlines for Master's courses tend to be in the spring, but there is much variation related to demand for the particular course. The more popular courses have strict deadlines as a means of limiting numbers. Deadlines for application for funding are independent of course deadlines; they may be before or after. M.Sc. courses will usually hold open offers of places until news of application to funding bodies is received. Referees are a particular problem in that they may take an inordinate amount of time to submit a report. Applications for M.Sc. courses may be jeopardized by referee indolence and regular gentle reminders or polite enquiries are in order.

What Constitutes a Good Application?

Although you may be a good student, you also need to put together a good application. What constitutes a good application?

- It is legible and on time.
- It is directed at the course in question rather than being a 'catch-all' application that has clearly been designed for several courses.
- There is a CV: even if not asked for, this is a useful addition (see chapter 61 by Pauline Kneale on how to write a good CV).
- Referees' reports should arrive in good time, should be from people of good academic standing and should comment in some detail about your abilities. A reference from, for example, the local vicar, however supportive, is not likely to be very relevant. References from full-time, part-time or voluntary work employers

may be useful if the work relates to the course for which you are applying.
- Academic qualifications that are adequate and backed up with university transcripts of marks.
- Any personal statements are well written, show enthusiasm for the course and indicate why you want to do it.

It is surprising how many applications are late or incomplete. Courses that are oversubscribed (e.g. M.Sc. in Environmental Change and Management, Oxford, which has a 9:1 ratio of applicants to places) reject substantial numbers on the basis of poor or incomplete application forms.

If you are really keen to take a specific course and you believe that you are well qualified, then it is worthwhile sending a CV to the Course Director and asking to visit and discuss what the course offers. You should try to arrange to meet either the Director or tutors and, if possible, students currently taking the course. Directors are impressed by applicants with enthusiasm, relevant knowledge and ambition. Read the course documents carefully before visiting and make sure you have good questions to ask. The same advice applies for preparation for a formal interview.

Further Reading

A listing of postgraduate courses in UK geography departments can be found in L.E. Craig and J. Best (eds), *Directory of University Geography Courses 2001* (London: Royal Geographical Society, 2000). For up-to-date information on scholarships, fellowships, research grants and other awards in the arts, sciences and professions, see S. Hackwood, *The Grants Register 2001* (New York: Palgrave, St Martin's Press, 2000).

Internet Resources

- A list of all UK universities and higher education colleges can be found at: University of Wolverhampton Universities and HE Colleges, http://www.scit.wlv.ac.uk/ukinfo/

55

Postgraduate Studies in Australia

Hilary P.M. Winchester and Stephen J. Gale

Australia is a remarkable continent, with a distinctive flora and fauna, diverse geographical environments, and a tolerant and easy-going multicultural population. Much of its geomorphology and biogeography is unique. It contains the most ancient physical landscapes on Earth and its southern hemisphere location offers unrivalled opportunities for understanding past and future environmental change. It possesses physical environments ranging from subalpine mountains to coral reefs and from hot deserts to tropical rainforests. Human impact on these environments began with the arrival of the first people on the continent 60,000 or more years ago and has become increasingly dramatic in the last 200 years. Australia's population and settlement geography is complex, ranging from remote settlements in the vast and sparsely inhabited outback to highly urbanized coastal cities that contain most of the nation's 20 million inhabitants and the majority of its recent migrants. The Australian economy is currently undergoing considerable change from primary production to a fuller integration into global networks. Meanwhile, the country is increasingly shifting its orientation towards Asia. This diversity offers immense opportunity for geographical research.

Geography Departments in Australia

There are 19 principal geography departments in Australia (located at 16 universities). In addition, geographical research may be undertaken in other universities without geography departments, in schools of environmental studies, social science or applied science. Every Australian geography department has specific research and teaching strengths, with some departments stronger in particular areas than others. Table 55.1 provides an indication of major fields of research.

Geography Degrees in Australia

Geography departments in Australia offer a wide variety of internationally recognized postgraduate degrees, ranging from Graduate Certificates, Diplomas and Master's degrees by coursework, to Master's and Ph.D. degrees by research (table 55.2).

Applicants for postgraduate study at an Australian university should be of a high scholastic calibre. For example, to undertake Ph.D. research, it is essential to have a first-class or a good upper-second-class honours degree, or a Master's degree. International students will need to show evidence

University	Human											Physical/Environmental											
	Agriculture	Development geography	Economic geography/regional development	Gender and geography	Historical geography	Population geography/demography	Rural geography	Social and cultural geography	Urban geography and housing	GIS	Planning	Biogeography	Climatology	Coastal studies/management	Environmental assessment/management	Geomorphology	Hazards	Physical oceanography/marine science	Plant ecology	Quaternary/Holocene	Soils and landforms	Sustainable development	Water resources/hydrology
Adelaide	✓	✓	✓	✓		✓	✓	✓	✓	✓		✓		✓	✓	✓				✓		✓	✓
Australian Defence Force Academy	✓	✓	✓		✓			✓		✓		✓	✓	✓		✓		✓		✓			✓
ANU – Humanities		✓	✓							✓		✓			✓	✓			✓	✓	✓	✓	
ANU – RSPAS*	✓	✓	✓	✓		✓	✓	✓	✓														
Flinders		✓	✓	✓		✓		✓	✓			✓							✓				✓
James Cook						✓	✓	✓	✓	✓		✓	✓	✓	✓	✓	✓		✓	✓			
Macquarie			✓	✓	✓	✓		✓	✓	✓			✓	✓	✓	✓	✓			✓	✓		
Melbourne		✓	✓	✓	✓			✓	✓	✓	✓	✓	✓	✓	✓	✓				✓		✓	✓
Monash		✓	✓	✓			✓	✓	✓		✓	✓	✓	✓	✓	✓			✓	✓	✓	✓	✓
Newcastle		✓	✓	✓	✓			✓	✓		✓	✓		✓	✓	✓				✓		✓	✓
New England	✓	✓	✓		✓	✓	✓			✓	✓	✓		✓	✓	✓				✓			
New South Wales			✓	✓		✓	✓	✓	✓	✓				✓	✓	✓					✓		
Queensland		✓	✓			✓			✓	✓												✓	
Sydney		✓	✓	✓			✓	✓		✓	✓	✓	✓			✓	✓	✓	✓	✓	✓		✓
Tasmania	✓	✓	✓	✓			✓	✓										✓	✓				
Western Australia		✓	✓	✓	✓					✓		✓	✓			✓				✓	✓	✓	
Wollongong			✓			✓		✓		✓		✓	✓	✓	✓	✓	✓			✓	✓		

* Research School of Pacific and Asian Studies

Table 55.2 Postgraduate degrees offered in geography departments at Australian universities

University	Doctor of Philosophy	Master's (Research)	Master's (Coursework)	Graduate Diploma (GD)
Adelaide University	Doctor of Philosophy	Master of Arts	Master of Arts in Applied GIS and Remote Sensing Master of Arts in Environmental Management Master of Arts in Environmental Studies Master of Arts in Population and Human Resources	GD in Applied GIS and Remote Sensing GD in Environmental Management GD in Environmental Studies GD in Population and Human Resources
Australian Defence Force Academy	Doctor of Philosophy	Master of Arts		
Australian National University	Doctor of Philosophy	Master of Arts in Geographical Science Master of Philosophy in Geographical Science	Master of Arts in Geographical Science	GD in Geographical Science
Flinders University	Doctor of Philosophy	Master of Arts	Master of Environmental Management Master of Population and Human Resources	
James Cook University	Doctor of Philosophy	Master of Arts	Master of Applied Science	GD in Science
Macquarie University	Doctor of Philosophy	Master of Arts (Hons) Master of Science (Hons)		GD in GIS
Monash University	Doctor of Philosophy	Master of Arts Master of Environmental Science	Master of Environmental Science	GD in Corporate Environmental Management GD in Environmental Science
University of Queensland	Doctor of Philosophy	Master of Arts Master of Economics Master of Philosophy	Master of Development Planning Master of Environmental Management	GD in Development Planning GD in Environmental Management GD in GIS

University of Melbourne	Doctor of Philosophy	Master of Regional and Town Planning Master of Science	Master of GIS Master of Property Studies Master of Urban and Regional Planning	GD of Philosophy GD in Property Studies GD in Urban and Regional Planning
University of Newcastle	Doctor of Philosophy	Master of Arts Master of Science	Master of Development Studies Master of Environmental Studies	GD in Environmental Studies
University of New England	Doctor of Philosophy	Master of Science	Master of Environmental and Business Management Master of Environmental Studies	
University of New South Wales	Doctor of Philosophy	Master of Arts Master of Science Master of Social Science Master of Science (Geography)	Master of Urban and Regional Planning Master of Science and Technology (GIS) Master of Science and Technology (Remote Sensing)	GD in Urban and Regional Planning GD in Remote Sensing
University of Sydney	Doctor of Philosophy	Master of Science		GD in Science
University of Tasmania	Doctor of Philosophy	Master of Arts Master of Environmental Studies Master of Science Master of Spatial Information Science	Master of Environmental Management	GD in Environmental Studies
University of Western Australia	Doctor of Philosophy	Master of Arts Master of Science	Master of Philosophy (Urban Studies)	GD in Science (GIS) GD in Science (Geography)
University of Wollongong	Doctor of Philosophy	Honours Master of Arts Honours Master of Science	Master of Arts Master of Science	GD in Science

of an adequate command of the English language (usually an IELTS score of at least 6.5).

A wise first step for students intending to undertake a postgraduate degree in geography at an Australian university is to contact a staff member to discuss opportunities for postgraduate research. Lengthy discussions and correspondence with staff members and the international offices of universities may be necessary to secure funding and supervision and to settle on an appropriate and achievable topic. It is important that discussions with departments and possible supervisors begin as early as possible before the proposed commencement date. Contact details of geography departments in Australia are listed in the appendix at the end of this chapter.

Cost of Postgraduate Degrees

Australian government policy requires that all international students, that is, students who are not Australian citizens or permanent residents, must pay tuition fees. Although the cost of postgraduate studies for overseas postgraduate students in Australia is becoming quite expensive – with money required to cover student fees, research costs and living expenses – Australia compares favourably with many other destinations for international students, such as Canada and the United States of America.

International student fees in Australia are substantial, and can range from A$12,000 to more than A$16,000 per annum for a postgraduate geography degree. This cost makes privately funded postgraduate study almost prohibitive and means that most international students will need to secure a scholarship.

Government Scholarships for Overseas Students

Australian Development Scholarships

AusAID provides international scholarships as part of Australia's development assistance programme. Australian Development Scholarships are available only in countries participating in the scheme: that is, Bangladesh, Bhutan, Cambodia, People's Republic of China, Cook Islands, Eritrea, Ethiopia, Fiji, French Polynesia, Indonesia, Lesotho, Kenya, Kiribati, Laos, Malawi, Maldives, Marshall Islands, Mauritius, Micronesia, Mongolia, Mozambique, Namibia, Nauru, Nepal, New Caledonia, Niue, Pakistan, Philippines, Papua New Guinea, Seychelles, Solomon Islands, South Africa, Sri Lanka, Swaziland, Tanzania, Thailand, Tokelau, Tonga, Tuvalu, Uganda, Vanuatu, Vietnam, Western Samoa, Zambia and Zimbabwe.

There are two categories of scholarships: (1) *public sector* and (2) *open/equity*. In the first category, governments in partner countries nominate candidates; in the second, applicants do not need to be nominated and anyone who meets the selection criteria can apply. Details about the scheme may be obtained from the Australian diplomatic mission in the applicant's home country.

International Postgraduate Research Scholarships (IPRS)

International Postgraduate Research Scholarships enable top-quality overseas students to obtain a higher degree by research – that is, a course of study leading to a Ph.D. or a Master's degree by research – at an Australian university. The scholarships cover tuition fees and health insurance for a period of two (Master's) or three (doctoral) years. The scholarships do not include any living allowance, although students may be eligible to apply for additional funding from the university in which they will enrol. The scholarships are open to students from any country (except citizens and permanent residents of Australia and New Zealand). Approximately 300 scholarships are awarded each year. Applications for a scholarship should be made directly to a participating institution (Registrar or Scholarships Officer). Each institution is responsible for determining the selection process by which scholarships are allocated to applicants.

Australian University Scholarships for Overseas Students

Each Australian university offers a range of postgraduate scholarships to international students, although they predominantly target doctoral and Master's degrees by research. These include the following.

Postgraduate research awards

Most Australian universities offer a number of scholarships for full-time graduate research. These scholarships are determined on a competitive basis, with both Australian and overseas graduates eligible. Benefits of these scholarships vary from university to university. Many universities also offer limited numbers of scholarships to cover fees and additional living allowances. For example, some universities provide a small number of postgraduate research scholarships to IPRS recipients to assist with living costs. Competition for this funding is keen.

International scholarships

Some universities offer scholarships specifically for outstanding international students (from any country) to undertake postgraduate research in any academic discipline. These cover full fees plus an annual living allowance for the normal duration of the course. Details are available from the Registrar or Scholarships Officer at each university.

Postgraduate coursework scholarships

Some universities offer a limited number of scholarships to outstanding international students from any country to undertake a postgraduate coursework programme in any academic discipline. The benefits include full tuition fees plus an annual living allowance for the normal duration of the course. Details of the awards are available from the Registrar or Scholarships Officer at each university.

Specific details of scholarship opportunities at Australian universities can be obtained from the Scholarships homepage of the Department of Education, Training and Youth Affairs (see Internet Resources below).

Other Scholarships for Overseas Students

International guides to scholarships, such as the *Association of Commonwealth Universities Scholarships Guide for Commonwealth Postgraduate Students*, list further scholarship support through home governments, Rotary International, the World Bank, the World Health Organization, the Asian Development Bank, the United Nations, the Ford Foundation, the Rockefeller Foundation, the Fulbright Foundation and other organizations. These international guides to scholarships may be available for consultation in Australian Diplomatic Missions, Australian Education Centres, International Development Program Education Australia offices, major libraries and in various educational counselling agencies.

Applying for Admission into Australia and an Australian University

There are two quite independent requirements that must be met before commencing postgraduate studies in Australia. The first is to obtain a student visa and meet the requirements of the Department of Immigration and Multicultural Affairs. The second is to gain admission to an Australian university.

Information on admission requirements is available through the Australian Education International (AEI) website. AEI lists nine steps to gaining admission.

Contact the Australian Education Centre (AEC) or Australian Diplomatic Mission (ADM) in your country to discuss options and raise questions. Locations of Australian Education Centres and Diplomatic Missions in your country are available on the AEI Contacts page (see Internet Resources below).

Obtain application forms from the AEC or ADM for the institution(s) that are of interest to you. Alternatively, you can ring, write, fax or email the institution(s) in which you are interested.

The AEC, ADM or the institution itself will give you an application form to fill in and will inform you of the documentation you will need to complete. They will also advise whether you will be required to take an English language proficiency test.

In addition to the completed application forms, you will normally be asked to provide other information, including full details of your previous study, evidence of your proficiency in English, and the name of the person or organization who will be responsible for paying your fees.

The institution will advise you if you have been successful in your request for admission. If successful, the institution will send you an enrolment letter. Note that an offer of enrolment is usually distinct from an offer of financial support.

It is now time to begin the visa application process, which will include a medical examination. The AEC or ADM will give you advice and assistance in preparing your application.

The Australian Department of Immigration and Multicultural Affairs website also provides immigration-related information such as:

Fact Sheet 56 – Overseas Students in Australia

Fact Sheet 981i – Applying Overseas for a Student Visa

Fact Sheet 982i – Applying in Australia for a Student Visa

Fact Sheet 990i – Charges (Fees)

Fact Sheet 999i – Study Options for Visitors to Australia

At about this point (and before a visa can be issued) you will be required to pay a compulsory medical insurance fee (the Overseas Student Health Cover) to pay for any medical or hospital treatment that you may need during your stay in Australia. You may also be asked to visit an ADM for an interview as part of the visa application process.

Shortly after this you will be told whether your visa application has been successful. If so, you will receive a 'multiple-entry' visa allowing you to travel backwards and forwards to Australia during the period for which it has been granted. That period will depend on the course for which you have been accepted.

You are on your way.

Before taking any of the steps listed above, it is essential that you contact a staff member and a department at an Australian university to discuss funding opportunities, course of study or research topic, and supervision. Students intending to undertake postgraduate study in Australia should begin these discussions as early as possible before the proposed commencement date. Contact details of geography departments at Australian universities are listed in the appendix.

Internet Resources

- Scholarships homepage of the Department of Education, Training and Youth Affairs: http://www.detya.gov.au/highered/scholarships.htm#postawards
- Australian Education International (AEI) website: http://aei.detya.gov.au/default.htm

- Locations of Australian Education Centres and Diplomatic Missions are available on the AEI Contacts page: http://aei.detya.gov.au/general/contacts/internat.htm

- Australian Department of Immigration and Multicultural Affairs website: http://www.immi.gov.au/students/index.html

Appendix: Contact details of geography departments at Australian universities

DEPARTMENT	CONTACT DETAILS	
Department of Geographical and Environmental Studies Adelaide University Adelaide SA 5005 AUSTRALIA	Tel: Fax: Email: WWW:	61 8 8303 5643 (Geog) / 61 8 8303 4735 (Env) 61 8 8303 3772 (Geog) / 61 8 8303 4383 (Env) margaret.young@adelaide.edu.au http://www.arts.adelaide.edu.au/Geogenvst/
School of Geography and Oceanography University College Australian Defence Force Academy Canberra ACT 2600 AUSTRALIA	Tel: Fax: Email: WWW:	61 2 6268 8312 61 2 6268 8313 secretary@ge.adfa.edu.au http://www.ge.adfa.edu.au
School of Resources, Environment and Society The Australian National University Canberra ACT 0200 AUSTRALIA	Tel: Fax: Email: WWW:	61 2 6125 2579 61 2 6125 0746 sres@anu.edu.au http://geography.anu.edu.au
Department of Human Geography Division of Society and Environment Research School of Pacific and Asian Studies The Australian National University Canberra ACT 0200 AUSTRALIA	Tel: Fax: Email: WWW:	61 2 6125 2234 61 2 6125 4896 hgeog@coombs.anu.edu.au http://rspas.anu.edu.au/humgeog/department/
School of Geography, Population and Environmental Management The Flinders University of South Australia GPO Box 2100 Adelaide SA 5001 AUSTRALIA	Tel: Fax: Email: WWW:	61 8 8201 2107 61 8 8201 3521 geography@flinders.edu.au http://www.ssn.flinders.edu.au/geog/
School of Tropical Environment Studies and Geography James Cook University of North Queensland Townsville Qld 4811 AUSTRALIA	Tel: Fax: Email: WWW:	61 7 47 81 4325 61 7 47 81 4020 jonathan.luly@jcu.edu.au or steve.turton@jcu.edu.au http://www.tesag.jcu.edu.au/
Department of Human Geography Macquarie University Sydney NSW 2109 AUSTRALIA	Tel: Fax: Email: WWW:	61 2 9850 8382 61 2 9850 6052 humgeog@els.mq.edu.au http://www.es.mq.edu.au/humgeog/
Department of Physical Geography Macquarie University Sydney NSW 2109 AUSTRALIA	Tel: Fax: Email: WWW:	61 2 9850 8426 61 2 9850 8420 kknowles@laurel.ocs.mq.edu.au http://www.es.mq.edu.au/physgeog/

The School of Anthropology, Geography and   Tel: 61 3 8344 6339
   Environmental Studies   Fax: 61 3 8344 4972
The University of Melbourne   Email: geog-head@unimelb.edu.au
Melbourne Vic 3010   WWW: http://www.geography.unimelb.edu.au/
AUSTRALIA

School of Geography and Environmental Science   Tel: 61 3 9905 2910
PO Box 11A   Fax: 61 3 9905 2948
Monash University   Email: Bianca.Roggenbucke@arts.monash.edu.au
Melbourne Vic 3800   WWW: http://www.arts.monash.edu.au/ges/
AUSTRALIA

School of Environmental and Life Sciences   Tel: 61 2 4921 5080
The University of Newcastle   Fax: 61 2 4921 5877
University Drive   Email: margaret.lane@newcastle.edu.au
Newcastle NSW 2308   WWW: http://www.newcastle.edu.au/discipline/
AUSTRALIA   geography/index.html

School of Human and Environmental Studies   Tel: 61 2 6773 2696
The University of New England   Fax: 61 2 6773 3030
Armidale NSW 2351   Email: geoplan@metz.une.edu.au or
AUSTRALIA   urb-reg-plan@metz.une.edu.au
     WWW: http://www.une.edu.au/geoplan/

Faculty of the Built Environment   Tel: 61 2 9385 4799
The University of New South Wales   Fax: 61 2 9385 4507
Sydney NSW 2052   Email: FBE.Stu.Cen@unsw.edu.au
AUSTRALIA   WWW: http://www.fbe.unsw.edu.au/degrees/Geography

School of Biological, Earth and Environmental Sciences   Tel: 61 2 9385 2067
The University of New South Wales   Fax: 61 2 9385 1558
Sydney NSW 2052   Email: bios@unsw.edu.au
AUSTRALIA   WWW: http://www.bees.unsw.edu.au

Department of Geographical Sciences and Planning   Tel: 61 7 3365 3752
The University of Queensland   Fax: 61 7 3365 6899
Brisbane Qld 4072   Email: office@geosp.uq.edu.au
AUSTRALIA   WWW: http://www.geosp.uq.edu.au/

School of Geosciences   Tel: 61 2 9351 2805 or 61 2 9351 2886
The University of Sydney   Fax: 61 2 9351 3644
Sydney NSW 2006   Email: d.dragovich@geography.usyd.edu.au
AUSTRALIA   WWW: http://www.usyd.edu.au/su/geography/

School of Geography and Environmental Studies   Tel: 61 3 6226 2464
University of Tasmania   Fax: 61 3 6226 2989
GPO Box 252-78   Email: Admin.Officer@geog.utas.edu.au
Hobart Tas 7001   WWW: http://www.scieng.utas.edu.au/geog/
AUSTRALIA

Department of Geography   Tel: 61 8 9380 2697
The University of Western Australia   Fax: 61 8 9380 1054
Hackett Drive   Email: secretary@geog.uwa.edu.au
Perth WA 6009   WWW: http://www.geog.uwa.edu.au/
AUSTRALIA

School of Geosciences
The University of Wollongong
Northfields Avenue
Wollongong NSW 2522
AUSTRALIA

Tel: 61 2 4221 3721
Fax: 61 2 4221 4250
Email: ebryant@uow.edu.au
WWW: http://www.uow.edu.au/science/geosciences/

56
Postgraduate Studies in Canada

Christopher Keylock, Mark Lawless and Robert Schindler

After a first degree, changing continents or countries can provide a refreshing change, both academically and culturally. Canada is an obvious place to pursue Master's and doctoral studies for many English-speaking geographers due to the strength of the geography departments and the linguistic and cultural setting. If your interests lie in areas of geography unique or common to Canada, then further study there may make a lot of sense. Permafrost geomorphology and the historical geographies of first-nation cultures are two cases where Canada provides obvious opportunities for study. Furthermore, Canada has a varied physical and cultural environment that provides many opportunities for research. River systems such as the Athabasca, MacKenzie and Peace are substantial by world standards and the Yukon contains more than 20 mountains over 4,000 m. Canadian cities, although not rivalling London or New York in size, are large and diverse. Toronto, Montreal, Vancouver and Ottawa all have a population of over 1 million people, with Calgary and Edmonton close to the 1 million mark. All of these cities differ in ethnic composition and histories of development. The fact that there were some 101,000 international students in Canada in 2000 shows that Canadian institutions are keen to attract applications from overseas.

The Canadian Master's Programme

In Canada, the typical Master's degree is two years long and follows a four-year modular BA or B.Sc. undergraduate degree course. The Master's programme is integrated into the education system to a greater extent than equivalent courses in the UK. While it is possible to commence a Ph.D. course in Britain straight from an undergraduate degree, the Master's is usually a prerequisite to doctoral study in Canada. This means that in Canada, the research component of the Master's course is of great importance, with first-semester introductory courses often oriented towards research training and the second year available to carry out a research project. In contrast, a one-year Master's programme in Britain will typically consist of two semesters of courses followed by a three-month research project. Thus, if you wish to develop research skills and to test the waters for doctoral study, a Master's degree in Canada might be appealing. Furthermore, a successful Master's thesis based on two years of study may well be of publishable quality. Given that many Ph.D. students do not write their first paper until their third year, a Canadian Master's degree with one or two papers submitted or in press will be looked upon very favourably if you decide to return to your home country to pursue a doctoral degree.

Although it is difficult to generalize, a Canadian Master's course will probably consist of one year of taught courses followed by one year of research. Some of the taught courses will be compulsory for all MA or M.Sc. students (such as those that discuss research issues and training), while you will select the others in consultation with your thesis supervisor in order to facilitate your research project. These are usually, although not necessarily, taken within geography, resulting in a programme that is tailored towards individual research needs to an extent rarely possible in the UK, for example.

Canadian Ph.D.s

The Ph.D. programme in Canada, in common with all Ph.D. programmes, has as its final goal the production of an original piece of research. However, there are certain differences between typical British and Canadian Ph.D. courses that are worth noting. First, while the British Research Councils expect a Ph.D. to last three years, the typical duration in North America is at least four years. The first 18 months of the Canadian Ph.D. programme often consist of formal courses, the aims of which are to provide both generic and topic-related research skills. If you continue Ph.D. study in the same department as that in which you took your Master's, then it may well be possible to count a significant number of your Master's credits towards this aspect of the Ph.D., shortening the timeframe for the doctorate. Comprehensive examinations are an additional component to the Canadian Ph.D. system. These exams typically involve an oral defence of a series of essays that you have produced on topics related to your specific research interests. They are usually written and defended after approximately 12 months of study. Another important difference is the manner in which Ph.D. theses are examined

in Canada. In the UK the Ph.D. defence is typically behind closed doors and involves an external and internal examiner. The defence in Canada is usually public, with the possibility that anybody could turn up to ask a question. The examining panel may well consist of an external and a university examiner, in addition to your Ph.D. advisory panel and a chair. This may seem rather daunting but does provide an opportunity to present your work to a wider audience.

Funding

For all higher-degree programmes, funding is an important issue. The key thing to remember is to start thinking about things early. There are scholarships available for British students to study in Canada, and for details of the Commonwealth Scholarship programme and the Canada Memorial Foundation Scholarships browse the Association of Commonwealth Universities website (see Internet Resources). If you are unsuccessful in obtaining one of these major international awards, then there are other sources of funding that should be explored. Canadian universities are keen to attract overseas students and may have internal scholarship funds available to help you financially.

However, even if these awards are not forthcoming, it may still be possible to make ends meet through teaching assistantships. These are available to both doctoral and Master's students and involve assisting a professor with the delivery of a particular course of study. Owing to the four-year duration of undergraduate degrees in Canada, it is not unheard of to find yourself the youngest person in the class that you are instructing, a rather nerve-wracking but character-building experience! Teaching assistantships involve a formal contract between the individual and the university, where you are paid to deliver a set number of hours of teaching support per semester. Typically, this involves committing some 10 to 12 hours per week to teaching and marking, but it is important to note that the maximum number of hours per week and rate per hour will differ between institutions. The amount of income available from teaching at the departments that you are considering should be evaluated against the cost of international student tuition fees, which can vary greatly between institutions. You may find that it is possible to fund yourself exclusively from this source. It is also worth checking the reputation of the institution for dealing amicably with the teaching assistants (TAs) that it employs. One well-known Canadian university has had its reputation rather tarnished in this regard recently. Joining the campus TA union may provide benefits such as dental, optical and prescription plans, which could save you additional money.

So why study in Canada? Ideally, you will have found an area of geography that you wish to pursue further during your first degree. If the person whose work you find most interesting is at a Canadian university then you may find yourself gravitating towards further study at a specific institution in Canada for the best possible reasons – those based purely on academic criteria. However, it is rather more likely that you will be considering working with a number of different people, perhaps in more than one area. In this case, browsing on the Web will reward you with ample information about individuals' research interests and publications. In addition, the journal *Canadian Geographer*, which is available in many universities and from the Canadian Association of Geographers website (see below) contains many articles written by Canadian geographers. Volume 45, issue 1 (2001) is a special issue on the geography of Canada with review articles by a number of well-known authors. This might help you to discover the nature of contemporary research in the areas of geography that interest you.

Practicalities: Living in Canada

If you are moving to a new country for two to four years, you should select an institution that is also in an area where you would wish to live. For example, skiers might wish to concentrate their attention upon Calgary and Vancouver, while those who prioritize an active nightlife may find Toronto and Montreal more to their taste. Those who prefer smaller universities in less populated towns may find that Queen's University in Kingston, Ontario, or the University of Victoria in British Columbia are preferable to those institutions that are based in larger cities. For example, Concordia, McGill, Simon Fraser, Université de Montréal, University of British Columbia, University of Toronto and York are all located within Montreal, Toronto or Vancouver.

Climatic conditions are extremely varied across Canada, so if you wish to experience annual temperatures between $-30°C$ and $+30°C$ head for Ontario, Quebec and the prairie provinces of Manitoba and Saskatchewan. On the other hand, if more moderate temperatures are to your liking, the eastern Maritime Provinces and British Columbia might be more to your taste. Canadian cities can be reached directly from many international airports. For example, there are direct flights from London to Calgary, Edmonton, Halifax, Montreal, Ottawa, Toronto and Vancouver. The cheapest flights in the summer (the time you are likely to move) are often the Canada 3000 charter flights, while if you are flying from Europe, the Icelandair flights to Halifax are often the cheapest scheduled flights to the eastern seaboard.

Unless you are a Canadian citizen you will almost certainly have to obtain a student visa before you can study in Canada. The duration of this visa is limited to the length of your study and you will need to provide proof of acceptance at a recognized institution of tertiary education on a full-time basis. Second, you need to supply evidence that you will be able to pay the costs of living and tuition during your course of study. This often requires a letter from the institution confirming the award of grants, scholarships and/or teaching assistantship as well as a guarantee from a parent or guardian stating that they will be responsible for supporting you if the budget gets tight. The student visa affords you the right to employment upon campus, although accompanying spouses are not limited in this way. Application is completed through the Canadian High Commission and costs $125. Extending your visa can be accomplished relatively easily provided that you obtain a letter of support from the institution you are studying at. This may occur if, for example, you decide to change status to a Ph.D. from a Master's. The cost of applying for an extension is also $125, regardless of the change in duration.

A thorough overview of the immigration procedure is provided by the Immigration Canada website (see below). The processing can take between two and four months, although Immigration Canada suggests that you should allow for six months in case of problems with your initial application, or when there are particularly high application rates. Applicants should also be aware that interviews, and even medical examinations

in specific cases, may be requested, although this is rare.

Some useful sites for further information are provided below. When searching on the Internet it is worth bearing in mind that not all geographic research is done in geography departments. As such, physical geographers should take a look at earth science and geophysics departmental websites, while human geographers may find it rewarding to examine the web pages for planning and sociology departments.

Internet Resources

- A listing of Canadian universities online: http://uk.dir.yahoo.com/education/higher_education/colleges_and_universities/by_region/countries/Canada/
- Association of Commonwealth Universities: http://www.acu.ac.uk
- Canadian Association of Geographers: http://venus.uwindsor.ca/cag/cagindex.html
- Canadian Government: http://www.canada.gc.ca/main_e.html
- National Research Council of Canada: http://www.nrc.ca/
- The Official Immigration Canada website: http://cicnet.ci.gc.ca.

57

Postgraduate Studies in Hong Kong

George C.S. Lin

Hong Kong: Gateway to Asia and Vibrant Confluence of East and West

Among the several newly industrializing economies in Asia, Hong Kong is distinguished by its strategic location that bridges East and West. Until recently, Hong Kong was a major and arguably the best developed colonial outpost in Asia. It was originally chosen as a British Crown Colony in the 1840s mainly to serve the functions of an *entrepôt* for opium trade with China. Its subsequent development has, however, far exceeded the original expectation owing not only to its unique location as the gateway to China, but also to the discovery of the excellent Victoria Harbour, which is suitable for international sea transportation; the influx of capital and entrepreneurs from Shanghai and Canton, who fled the Communist take-over in mainland China in 1949; privileged access to the American and European markets; minimum tariffs and low taxation; an efficient business-oriented government; free and easy flow of capital; and, last but not least, the rule of law.

Today, Hong Kong is an international hub for business in the Asia-Pacific region and home to over 3,000 regional headquarters and offices. A recent systematic study of the 50 top-ranked cities on the basis of both global locations of headquarters of multinational corporations and their first-level subsidiaries has ranked Hong Kong as the fourth most important world city next only to New York, Tokyo and London (Godfrey and Zhou 1999: 276). Hong Kong has clearly established itself as the leading world city in Asia.

Sovereignty over Hong Kong was returned to China in 1997. Under the arrangement of 'one country, two systems', Hong Kong has been allowed to enjoy special autonomy and become a Special Administrative Region (SAR), in which the system developed under British rule can be preserved. As a result of intensified global competition in recent years, it has become increasingly evident to the government of the Hong Kong SAR that the way for Hong Kong to outperform its rivals in the regions, including Shanghai in mainland China and Singapore as well as many other Asian economies, is not only to maintain and consolidate the special Hong Kong identity within China, but also to further enhance its position as Asia's world city, building on the many advantages that have been developed over the past 100 years under British rule. The legal and judicial systems have been maintained. The common law continues to apply. Monetary and financial policies remain entirely within the control of the HKSAR. During the Asian financial crisis in 1998, the successful defence of the Hong

Kong dollar peg and the financial markets was conducted entirely from within Hong Kong. Hong Kong continues to be a full and active member of international organizations such as Asian Pacific Economic Cooperation (APEC), the World Trade Organization (WTO) and the World Customs Organization (WCO). Hong Kong investment in mainland China has continued to be treated by Beijing as 'foreign investment'. The English language remains the most important means of communication in education, legal/official documentation and business interactions. Culturally, Hong Kong is a fantastic cosmopolitan city in which to live. Vibrant, fast-paced, dynamic and contrast-filled, this sophisticated metropolis has the best of East and West, ancient and ultra-modern, and a spectrum ranging from flamboyant Chinese style to British colonial heritage. There are many new experiences in Hong Kong just waiting to be discovered.

Geography: A Key Subject of Teaching and Research in Hong Kong

Geography as a subject of teaching and research holds a unique position in Hong Kong's educational system. For historical reasons, Hong Kong has adopted the educational system originally established in the UK. Geography is one of the core subjects taught in primary and secondary schools and examined at A-level for admission into university. An estimated 8,000 high school students study geography every year, which has provided a base strong and wide enough to support substantial teaching and research in geography at university level. The physical environment of Hong Kong as an island city-state and its extensive interactions with the outside world mean that there is a great demand for knowledge about the land and the people beyond the local horizons. Historically, geography was one of the earliest subjects taught when university education was

first introduced in Hong Kong. Economic geography, for instance, was first taught in the University of Hong Kong as early as 1915. As a consequence, many government officials, business leaders and professionals holding key positions in Hong Kong today were former graduates of the University of Hong Kong majoring in geography.

For international students interested in Asian or Chinese developments, Hong Kong is an excellent choice for postgraduate studies not only because of its enormous library collection of up-to-date materials on China and East Asia in both Chinese and English, but also because of the great opportunities for fieldwork as well as interactions with a critical mass of scholars working on the region in various disciplines. Numerous seminars are offered weekly or even daily by eminent international scholars passing through Hong Kong en route to or from China. With its established tradition and widely recognized academic status, geography has been taught and researched in leading universities including the University of Hong Kong, the Chinese University of Hong Kong and Hong Kong Baptist University, all of which offer both undergraduate and postgraduate programmes.

The University of Hong Kong (HKU)

The University of Hong Kong is the first and leading tertiary educational institution established in Hong Kong. Founded in 1911, the university experienced extraordinary growth in the 1970s concurrent with Hong Kong acquiring phenomenal international stature. From its modest beginning of only two faculties, the university has grown to embrace almost all the major areas of teaching and research and is one of the best and most well-established universities in the region. The university today is a major research institution with nine faculties and more than 100 departments serving over 17,000 full-time and part-time students, of whom 6,500 are at the postgraduate level, including students from more than 48 countries (HKU website – see Internet Resources below). International students are attracted to the University of Hong Kong not only because of its high international reputation as a world-class research university, but also because of its use of the English language as the medium of instruction. There are currently over 1,800 teaching and research staff drawn internationally from such countries as the United Kingdom, the United States, Canada, Australia, New Zealand/Aotearoa and other parts of the world. They are highly visible in cutting-edge research and international publications. The university has excellent infrastructure, services and facilities to support postgraduate studies. Postgraduate studentships, scholarships, research grants and conference grants are provided to high-calibre research students. The university library is probably the best in the region, with an excellent collection of over 1.8 million books, 56,000 periodicals, 13,000 Web-based CD-ROMS, and another 15,000 rare books. The library is also equipped with great IT facilities providing online connections with numerous databases overseas. The Graduate House and graduate student hotels provide accommodation for postgraduate students as well as amenity facilities and a focal point for postgraduate activities. Postgraduate students can have access to student and medical services and sporting and leisure facilities. The university has in recent years tried to create a digital campus to encourage teachers and students to make wider use of information technology as well as to benefit from efficient electronic information exchange.

The Department of Geography, since its founding in 1954, has been offering a wide range of systematic and regional courses to enable students to develop areas of specialization within the discipline and to acquire practical and problem-solving skills highly valued by employers. Thus far, the department has produced over 2,000 BA graduates in geography and conferred 27

higher-degree awards for Ph.D., 16 for MA and 44 for M.Phil. Many of these graduates have now become distinguished academics in their own specialities, occupying prominent positions in the Hong Kong civil service and other professions. At present, the department offers research-based M.Phil. and Ph.D. degree programmes and two taught postgraduate programmes, including a one-year full-time programme for Master of Arts in China Area Studies and a two-year part-time programme for Master of Arts in Transport Policy and Planning. High-quality teaching and research are well supported by a wide range of in-house laboratory facilities, equipment and other support, including laboratories for geomorphology, soils and biogeography, GIS and GIS research, as well as an excellent map library. The geography faculty is made up of internationally recognized scholars actively involved in policy-oriented, applied and academic research, with a special focus on China (including Hong Kong, Macau and Taiwan) and the Asia-Pacific region. Three fields of specialization – urban and transport, China and the Pacific Rim, and environment and resources – dominate the department and form its research thrust. Other areas of study include geographic information systems, computer cartography, spatial analysis and quantitative methods.

Admission and financial aid

The basic qualification for admission to the M.Phil. programme is a good honours degree from a recognized university or equivalent qualification from comparable institutions. Applicants intending to apply for the doctoral programme are normally required to have a Master's degree (preferably in geography). Application for admission to study should be made at least three months before the intended date of registration to allow time for the application to be processed. Applicants should include a research proposal indicating a field of study and a clear description of the particular aspect of the field in which they are interested, explaining the particular approach that they propose to take in studying the subject matter. They are strongly advised to contact the relevant staff member(s) of the department about their proposed research before submitting their applications. Detailed information about the current research of the faculty can be found on the HKU website. Application forms and the postgraduate prospectus can be downloaded from the Graduate School, while other information relating to logistics for international students can be obtained from Office of Student Affairs (see below for addresses).

International students may apply for postgraduate studentships (PGS), which are intended to finance selected full-time postgraduate students reading for a research degree at the university during the prescribed period of study for the degree concerned: two years for an M.Phil. or three to four years for a Ph.D. Each PGS is subject to annual review. No income tax is levied on the awards. A PGS holder is required to sign an undertaking not to take up any paid employment. Violation of this requirement may lead to a discontinuation of study. The annual value of the studentships is determined by the university from time to time. There are at present four levels of PGS stipend, ranging from HK$13,500 to HK$15,600 per month. Further information and application forms for studentships are obtainable from the Academic Services Enquiry Section.

The Chinese University of Hong Kong (CUHK)

The Chinese University of Hong Kong was founded in October 1963, as a result of the federation among three post-secondary colleges: Chung Chi College (founded in 1951), New Asia College (founded in 1949) and the United College of Hong Kong (founded in 1956). In 1988 a fourth college, Shaw College, became operational. The

university adopts bilingualism and biculturalism as the basis of its teaching, giving equal emphasis to both Chinese and English languages and eastern and western cultures. The majority of the teaching staff are bilingual (Chinese and English) and have studied and/or taught in major universities all over the world. The university is international in outlook and enjoys close associations with many universities, foundations and organizations abroad.

The Chinese University of Hong Kong has devoted great attention to postgraduate studies. Postgraduate programmes were first introduced in 1965 when the School of Education was founded. This was followed by the establishment of the Graduate School one year later for the undertaking of research training and postgraduate studies. By the year 2000, 8,804 graduates had been conferred with higher degrees. The university now has ten research institutions, 49 graduate divisions and some 4,000 staff members, of whom over 1,500 are academic and research staff. It offers eight postgraduate diploma programmes, 107 Master's programmes and 50 doctoral programmes. The total enrolment in the academic year 2000–1 was 14,219, of which 9,287 were undergraduate and 4,932 postgraduate students (CUHK website – see below). The university library holds over 1.5 million books and bound volumes of periodicals in both Chinese and English. The University Service Centre of the library has perhaps the best collection of Chinese materials in the world and has always attracted China scholars worldwide to conduct library documentary research there. In addition, the library provides access to audiovisual materials, microforms, and more than 7,000 Web-based electronic journals and hundreds of electronic databases. Most of the electronic resources are networked and can be accessed on the World Wide Web.

The Department of Geography of the Chinese University of Hong Kong is one of Asia's leading geography departments. It offers a wide array of undergraduate courses, a large postgraduate school and maintains a dynamic research profile. The Department of Geography and the Geography Division of the Graduate School were among the earliest units established in the university. Geography is now one of the nine disciplines within the university's Faculty of Social Science. The undergraduate programme is organized around three areas of concentration, namely physical and environmental studies, urban and development studies, and geographical techniques. The department offers both M.Phil. and Ph.D. degree programmes.

Admission and financial aid

Applicants should state clearly the programme they wish to apply for when they write to the Graduate School for application forms. Application forms should be completed in duplicate and sent to the School, together with supporting documents as required. Each application must include the following:

- Two copies of an official transcript of the candidate's academic record (non-returnable), to be sent directly to the Graduate School by the original university/institution, not by the applicant.
- Two copies of recommendations from referees named on the application form to be sent directly by the referees to the School, not by the applicant.
- Two copies of academic/professional credentials, certificates, or diplomas, etc.
- The receipt of payment of application fee (non-refundable; HK$320 for doctoral programmes and HK$180 for Master's and diploma programmes). Applicants from overseas should pay by Hong Kong dollar bank draft/cheque, which must be crossed and drawn in favour of the Chinese University of Hong Kong.

In addition to the general qualifications required for admission to the Graduate School, candidates for the M.Phil. programme are expected to have a basic under-

graduate training in geography or other relevant subjects. The M.Phil. programme requires two academic years for completion. Each M.Phil. student must complete ten units of course work, exclusive of thesis (14 units). Candidates seeking admission to the Ph.D. programme must satisfy the university's requirements for general admission. They should also hold a Master's degree in geography or a related field from a recognized institution, and will be required to furnish proof of their research capability (in the form of strongly favourable academic references or research publications). They are required to submit a study plan together with other application materials, indicating clearly the particular study field they are interested in. They may discuss with the heads of the graduate divisions concerned matters relating to their proposed fields of research (contact person: Professor Fung Tung; email: tungfung@cuhk.edu.hk). The Ph.D. programme is strongly research-oriented. Candidates may, however, be required to take additional courses or other instructional sessions as deemed necessary. The programme must be completed within three to seven academic years (full-time) and four to eight academic years (part-time). Advancement to Ph.D. candidature is conditional upon passing a comprehensive examination, and within six months after advancement to Ph.D. candidature the student must submit a thesis proposal.

Application forms can be obtained from the Graduate School Secretariat and the Department of Geography and Resource Management (addresses below). Applications may also be made via the Internet. Applicants should quote the application number generated for their applications when they send the hard copies of their supporting documents to the Graduate School Secretariat. Applicants are encouraged to use the online application procedures to save time and postage.

International students may apply for postgraduate studentships (PGS), which is a form of financial assistance provided to full-time postgraduate students registered in research degree programmes. Holders of such studentships are not employees of the university. The studentship is available in most graduate divisions. Normally, the award of a PGS will be made known to students at the same time as they receive their admission offers, so they need not make a separate application for obtaining the award. In 2000–1, the monthly stipend for a postgraduate studentship was HK$13,615.00. Students receiving PGS awards are expected to assist in the teaching and research work of the department to which they belong. Their duties will be assigned by the Department Chairman. A PGS holder may relinquish, or the university may terminate, the studentship with one month's notice or one month's payment in lieu. The university may also suspend or curtail the award if the holder's performance is considered unsatisfactory. S/he may, however, appeal to the Dean of the Graduate School to have the case reviewed within two weeks of the notification of suspension/curtailment. While in receipt of the PGS award, a student may not undertake other engagement or employment, full-time or part-time, unless with the university's permission. PGS holders are required to observe the university's policy on research, consultancies and intellectual property. Application forms may be obtained from the Office of Student Affairs.

Hong Kong Baptist University (HKBU)

Founded as a private, Baptist denomination-sponsored institution in 1956, and becoming a public sector (government-funded) institution in 1983, Hong Kong Baptist University hosts the third largest department of geography with postgraduate programmes leading to M.Phil. and Ph.D. degrees. The university has in recent decades established its place in Hong Kong's tertiary education sector as a Christian institution of academic excellence, with a determination to provide a sound whole-person education to its students while meeting the needs and expectations of the society it serves. Pioneering and broad-based interactive programmes, academic excellence embodying a careful balance between teaching and research, an international perspective and a commitment to quality assurance have become hallmarks of the university.

The university has 6,600 undergraduates (more than 4,000 of whom are full-time), 600 course-based postgraduates and 200 research-based postgraduates, for a total of 7,400 students. An additional 3,700 students are studying in the university on programmes that are jointly offered with a number of overseas partner universities (from Australia, the UK and the United States), the undergraduate or postgraduate degree awards of which are conferred by the respective partner institutions. The bulk of the faculty (over 77 per cent) hold doctorates awarded by internationally renowned universities. The majority (78 per cent) of the faculty members are engaged in funded research (currently amounting to HK$160 million) and the annual publications to staff ratio is 5:1 (HKBU website – listed below).

The university devotes great attention to international academic exchanges and welcomes overseas postgraduate students. It has signed agreements for the exchange of students and staff as well as collaborative research with more than 100 universities on the Chinese mainland and in Taiwan, Thailand, the Philippines, Australia, continental Europe, the UK, Canada and the United States. Some of these include internships or study-abroad programmes. Currently there are 125 visiting students studying at HKBU. Full-time research postgraduate students normally receive studentships during their programmes of study in the university. Outstanding students are awarded scholarships and prizes set up by the university. Each year the university plays host to more than 100 visiting research scholars. The university library uses an integrated online library system and has a collection of 645,000 bound volumes, 4,000 journal subscriptions and more than 145,000 items of microform/audiovisual/multimedia materials. Special collections include the Contemporary China Research Collection, the European Documentation Centre and the Archives on the History of Christianity in China. In support of the university's teaching/learning and research activities, an array of electronic information resources and services through networked connections is provided.

The Department of Geography in Hong Kong Baptist University is one of the three geographic departments offering postgraduate programmes. The department has ten faculty members with diverse academic backgrounds. They hold higher degrees from renowned universities in the United States, the UK, Canada, Australia and the Netherlands. The department provides a wide range of laboratory facilities and equipment to support high-quality teaching and research activities.

The geography department currently offers both M.Phil. and Ph.D. degrees in human and physical geography as well as in GIS. Students could undertake research in various sub-fields of geography. Most of the faculty members are actively engaged in research focusing on their particular areas of interest, with special emphasis on Hong Kong and China. Their publications have appeared regularly in calibre internationally refereed journals. Facilities include a newly renovated physical geography laboratory capable of handling soils, sediments and geochemical analyses; an environmental geography laboratory equipped with microscopes and computer facilities; a GIS-cartography laboratory; a physical geography research laboratory with larger particle, sedigraphy and core logging facilities; a human geography research laboratory equipped with state-of-the-art computing facilities; and a geography resource centre comprising maps, newspaper clippings and other materials useful for research purposes.

Applicants seeking admission to the M.Phil. degree programme should possess a Bachelor's degree with honours from a recognized university or comparable institution, *or* a qualification deemed to be equivalent, and, if required, must satisfy the examiners in a qualifying examination. Normally, only first- and second-class honours graduates will be considered to have met the admissions criterion. Graduates with third-class honours degrees, with a period of relevant work experience and satisfactory results in the qualifying examination, will also be considered. Applications should be

Table 57.1 Tuition fees and studentships for international postgraduate students, 2001–2 ($HK/year)

Programme type	Tuition fees	Studentships
University of Hong Kong		
M.Phil. & Ph.D. (full-time)	HK$42,150	HK$162,000–187,200
MACHAS (full-time)	HK$42,150	
MATPP (part-time)	HK$55,000	
Chinese University of Hong Kong		
M.Phil. & Ph.D. (full-time)	HK$42,150	HK$163,380
M.Phil. (part-time)	HK$28,066	
Ph.D. (part-time)	HK$31,575	
Hong Kong Baptist University		
M.Phil. & Ph.D. (full-time)	HK$42,150	HK$162,000

£1 sterling = HK$11.5; US$1 = HK$7.8

addressed to the Research and Postgraduate Studies Section (address below).

Visa/Immigration

All international students must obtain a student visa in order to study in Hong Kong. International applicants accepted for admission should make their own enquiries about entry visa requirements from their local government, or by writing directly to the Hong Kong Immigration Department (address listed below). The Hong Kong Immigration Department issues student visas to applicants enrolling in a full-time university course. Application should be made at least two months in advance at a Chinese embassy. The applicant must nominate a sponsor who is a resident in Hong Kong, aged over 21, and who knows the applicant personally. An applicant who has difficulty in nominating a sponsor in Hong Kong may seek assistance from the Office of Student Affairs of the university to which the application has been submitted and accepted.

References

Godfrey, B.J. and Zhou, Y. 1999: Ranking world cities: multinational corporations and the global urban hierarchy, *Urban Geography*, 20, 268–81.

Internet Resources

- The Hong Kong Immigration Department, 7 Gloucester Road, Wanchai, Hong Kong; WWW: http://www.info.gov.hk/immd/

Studying at the University of Hong Kong (HKU)

- HKU website: http://geog.hku.hk
- General information on postgraduate studies at HKU can be obtained from the Graduate School, Room P403, Graduate House, The University of Hong Kong; tel: (852) 2857

3470; fax: (852) 2857 3543; email: gradsch@hkucc.hku.hk; WWW: http://www.hku.hk/gradsch

- Office of Student Affairs, Room 401 Meng Wah Complex, The University of Hong Kong; tel: (852) 2859 2305; fax: (852) 2546 0184; email: osa@www.hku.hk; WWW: http://www.hku.hk/osa
- Academic Services Enquiry Section, Room UG05, Knowles Building, The University of Hong Kong, Pokfulam Road, Hong Kong; tel: (852) 2859 2433; fax: (852) 2540 1405; WWW: http://www.hku.hk/rss/pp99
- General information about the Department of Geography: http://geog.hku.hk
- Postgraduate studies in geography at HKU: http://geog.hku.hk/postgrad
- Research-based M.Phil. and Ph.D. degree programmes: http://geog.hku.hk/postgrad/researchdegrees.htm
- Master of Arts in China Area Studies: http://geog.hku.hk/postgrad/machas
- Master of Arts in Transport Policy and Planning: http://geog.hku.hk/postgrad/matpp

Studying at the Chinese University of Hong Kong (CUHK)

- CUHK website: http://www.cuhk.edu.hk
- General information about postgraduate studies at CUHK: http://www.cuhk.edu.hk/grs
- Graduate School Secretariat, Room 303, Sui-Loong Pao Building, The Chinese University of Hong Kong, Shatin, New Territories, Hong Kong; tel: (852) 2609 8977; fax: (852) 2603 5779.
- Department of Geography and Resource Management, Room 218, Wong Foo Yuan Building, The Chinese University of Hong Kong, Shatin, New Territories, Hong Kong; tel: (852) 2609 6532; fax: (852) 2603 5006.
- Postgraduate studies in geography at CUHK: http://www.grm.cuhk.edu.hk/
- Internet applications for postgraduate study: http://grsntb.grs.cuhk.edu.hk/grs/admission.htm
- Office of Student Affairs, 1/F, Benjamin Franklin Centre, The Chinese University of Hong Kong, Shatin, New Territory, Hong Kong; tel: (852) 2609 7216; email: osa@cuhk.edu.hk; WWW: http://www.cuhk.edu.hk/osa

Studying at Hong Kong Baptist University (HKBU)

- HKBU website and general information about postgraduate opportunities at HKBU: http://www.hkbu.edu.hk
- Information and support for international students: http://www.hkbu.edu.hk/~studaff/main

- Postgraduate studies in geography at HKBU: http://geog.hkbu.edu.hk
- Research and Postgraduate Studies Section, Academic Registry, Hong Kong Baptist University, Kowloon Tong, Kowloon, Hong Kong; tel: (852) 2339 7941; fax: (852) 2339 5133; email: postgrad@hkbu.edu.hk

58

Postgraduate Studies in New Zealand

Wardlow Friesen

Why New Zealand?

New Zealand sometimes seems about as far from the rest of the world as it is possible to go to undertake tertiary studies, but it may be worth considering. New Zealand has a population of just under 4 million, and while it has a reputation as a producer of agricultural products, it also has a well-developed service sector, including a tertiary education sector with an international reputation. However, the first reason that many people have for visiting or studying in New Zealand relates to its physical attributes. Notable physical features include the Southern Alps, active volcanoes, glaciers, thermal areas with geysers and bubbling mud, temperate rainforests, beaches and rolling countryside. For physical geographers, processes and theory are often substantiated by a field trip to a nearby feature, and physical and environmental research topics are plentiful. Popular outdoor sports include swimming, sailing, kayaking, white or black water rafting, mountain climbing, tramping, skiing and many others.

New Zealand also has an interesting social environment. It is seen by many as a socially progressive country, being the first country to give women the vote, introducing state housing in the 1930s, and having an extensive welfare system. To some extent this has changed since the election of several governments from 1984 onwards that have promoted economic restructuring and trade and investment liberalizations, making New Zealand one of the least regulated economies in the world. The social outcomes of these reforms, such as reduced incomes for some, and a reduction in public health services, are often the subject for human geographical research.

Aotearoa (New Zealand) has an (unwritten) bicultural constitution based on the Treaty of Waitangi of 1840 between the indigenous Maori and the British Crown, and in recent years there has been an ongoing process of righting injustices of the past based on this treaty. There is also a growing multicultural population. The immigration of people from the Pacific Islands over several decades has resulted in Auckland becoming the largest Polynesian city in the world. Since the 1990s there has also been a large immigration of other peoples as well, especially from Asia, and the larger cities now have a multicultural environment, as evidenced by shops, restaurants, festivals and people.

New Zealand is also a potential launch pad to other places, hot and cold. From Auckland, there are direct flights to most South Pacific nations, and with some funding, a thesis research topic in the Pacific is a possibility. At the other end of the thermal spectrum is Antarctica, and Christchurch serves as the base for the American and New Zealand research stations there. A number of New Zealand geographers have undertaken research projects in Antarctica, and there may occasionally be potential for graduate involvement in these projects.

For a sense of the research being undertaken in both physical and human geography, look at recent copies of the *New Zealand Geographer*. Social, economic and cultural themes are evident in Le Heron et al. (1999) and Le Heron and Pawson (1996).

New Zealand Universities and Degrees

There are eight universities in New Zealand, some with more than one campus, including one that has been renamed as a university of technology. All but two of these universities have geography departments, the exceptions being Lincoln University (Christchurch), which has a number of programmes with geographical themes, and the Auckland University of Technology. The largest geography department is at the University of Auckland, with about 75 graduate students in any one year. In a typical year Canterbury University has about 50 graduate students in geography, while Massey, Otago, Victoria and Waikato usually have slightly smaller, but still substantial, numbers of graduate students.

There are a variety of postgraduate degrees at New Zealand universities. A one-year BA (Honours) or B.Sc. (Honours) is one alternative after a three-year undergraduate degree, and this usually involves papers and a short dissertation. Postgraduate diplomas in arts or science are also offered, but these may not receive the international recognition of other degrees. Still the most common postgraduate degrees in geography are Master of Science and Master of Arts, although in some universities variations include Master of Social Science and Master of Philosophy. Typically these degrees involve one year of papers and a one-year thesis.

Admission to a Ph.D. programme requires a Master's degree with a good mark (usually minimum B or B+) or an honours degree with a high mark. Unlike the North American system, a Ph.D. does not generally require any papers or coursework in New Zealand. A Ph.D. thesis involves a minimum of two years' full-time work, but usually takes three or four years, depending on other commitments (and degree of hard work!). Each of the six geography departments has an active doctoral programme and usually more than 10 Ph.D. students enrolled at each, with the larger departments sometimes having nearly twice this number.

Admission, Fees and Funding

Admission to graduate courses is competitive, but a Bachelor's degree from a British university with a mark in the B range or above – equivalent to an upper second – will normally allow entry. Each university has an international office, which can give details on student visas, minimum requirements, possible living arrangements and so on. A student visa requires evidence that a person is enrolled at a university and that s/he has some means of paying for living expenses, and it must be renewed each year. For degrees taking more than two years, some other requirements may be imposed. It is also necessary to be accepted by a geography department, and this is usually straightforward if you meet the general admission conditions. For a Ph.D., however, departments will also usually require a preliminary thesis proposal.

A student from a non-English-speaking background may be required to present evidence of English competence, for instance by taking an ELTS test.

In 2001 postgraduate overseas student fees ranged from NZ$13,000 to NZ$28,000 per year, depending on the specific course and university. Arts degrees are cheaper than science. Most, however, are less than NZ$20,000 (about £6,050). While this may seem high, the relatively low value of the

New Zealand dollar means that these fees are generally less than for the equivalent degrees in Australia, the United States and Canada.

At the Master's level, there is relatively little scholarship funding for international students, although some universities have a limited number of scholarships reserved for overseas students, so check the websites listed below. There are more available for Ph.D.s, but these are very competitive. Another possibility is a Commonwealth scholarship, which will fund both tuition fees and living expenses. Within most geography departments, there are a number of paid tutoring opportunities for which graduate students are given priority.

Topics of Study and Research

Different geography departments in New Zealand specialize in different themes within the discipline, so if you have a particular aspect that you wish to pursue, you should check the staff specialisms within each department. Most departments have some focus on environmental issues, and most departments have also developed a GIS capability. Physical geographers at Auckland pursue a range of topics with specialisms including coastal studies, hydrology and geomorphology, while human specialisms include economic and cultural geography. There are also strong components of environmental studies and GIS in this department. On the physical side, Canterbury is particularly strong in climatology, coastal studies and glaciology, while historical and cultural specialisms are part of a wider spectrum of human geography. In conjunction with related programmes, Massey is known for development studies, cultural geography and biogeography. Otago's focus tends towards climate and biogeography as well as development, urban and political geography. Victoria is closely aligned with development studies, and specialisms include Asian studies, urban studies, hydrology and geomorphology. Waikato is the only department which does not cover physical geography topics, although there is a separate earth sciences department there. It is well known for its focus in areas such as Maori studies, feminist geography and population studies. Despite these generalized statements about specialization, there is a great deal of diversity in each of the geography departments, and many topics and themes not mentioned above. Also, courses and research directions change from year to year, as staff are on leave, new staff appointed, or research funding is gained for particular projects. Thus, the websites of these departments should be checked to get up-to-date information on recent developments.

Living in New Zealand

All of the universities have student residences, and these are usually within walking distance of the campus. It may be a good idea to live in one of these initially to meet people and to become acquainted with a new city. Later, you may opt to go flatting with friends, and there are usually relatively low-rental areas within busing, if not always walking, distance from the campuses. The international offices of each university can help organize accommodation and other things, and their websites are listed below.

Another source of information about living in New Zealand is the New Zealand Immigration Service, whose website will tell you not only how to apply for a student visa, but also gives some details about living conditions. Further sources of local information about the cities in which universities are located are the websites of the city councils, which are also listed below. There are, of course, many other websites of New Zealand government ministries, tourist agencies and so on which may be useful in familiarizing yourself with various aspects of life in New Zealand.

Further Reading

Le Heron, R. and Pawson E. 1996: *Changing Places: New Zealand in the Nineties.* London: Longman Paul.

Le Heron, R. et al. 1999: *Explorations in Human Geography: Encountering Place.* Oxford: Oxford University Press.

Internet Resources

Geography departments

The university websites are consistently in this format: *http://www.universityname.ac.nz*

- University of Auckland: http://www.geog.auckland.ac.nz
- Canterbury University (Christchurch): http://www.geog.canterbury.ac.nz
- Massey University (Palmerston North): http://www.massey.ac.nz/~wwglobal/geog.html
- Otago University (Dunedin): http://www.otago.ac.nz/geography
- Victoria University of Wellington: http://www.vuw.ac.nz/home/undergraduate/subjects/geog.html
- Waikato University (Hamilton): http://www.waikato.ac.nz/wfass/subjects/geography

Living in New Zealand

- New Zealand Immigration Service: http://www.immigration.govt.nz

- Tourism New Zealand: http://www.purenz.com
- Auckland City Council: http://www.akcity.govt.nz
- Christchurch City Council: http://www.ccc.govt.nz
- Palmerston North City Council: http://www.pncc.govt.nz
- Dunedin City Council: http://www.cityofdunedin.com
- Wellington City Council: http://www.wcc.govt.nz
- Hamilton City Council: http://www.hcc.govt.nz

Email addresses for university international offices

- University of Auckland: international@auckland.ac.nz
- Canterbury University (Christchurch): international@regy.canterbury.ac.nz
- Massey University (Palmerston North): international.student.office@massey.ac.nz
- Otago University (Dunedin): international@otago.ac.nz/geography
- Victoria University of Wellington: international-students@vuw.ac.nz
- Waikato University (Hamilton): international@waikato.ac.nz

59

Postgraduate Studies in Singapore

Brenda S.A. Yeoh and Theresa Wong

Why Singapore?

Singapore is a city-state with a population of 4.02 million, of which about 19 per cent are foreigners (3.26 million people are citizens and permanent residents). This large percentage of non-citizens, constituted by a workforce of low-skilled foreign labour and a growing number of high-skilled foreign professionals, attests to the government's enthusiasm for foreign 'talent' that will give the country a creative and vibrant edge when competing globally. As such, over the last few years, Singapore's two largest universities have been dedicated to awarding scholarships to both local and foreign students for postgraduate study.

Besides the welcoming attitude to foreign students, however, studying in Singapore has much to offer in terms of the country's connections to both the region and beyond, which contribute to the synergies that exist and which make Singapore unique. For students keen to specialize in, or simply learn more about, Southeast Asia or the wider Asia-Pacific region, Singapore's geographic proximity makes it an ideal gateway both in an intellectual and geographical sense. Its position within the region is enhanced by excellent infrastructure. Changi airport has won numerous accolades for smoothly facilitating flows of passengers on the 63 airlines it serves. Besides being a major air hub

in the Asia-Pacific region, it is linked to 149 countries in 50 cities, with 3,200 scheduled flights a week (Singapore Infomap 2001). In 2001, the city-state was ranked the world's most 'globalized' nation by a leading American journal, the *Foreign Policy* magazine, on the basis of 'its high trade levels, heavy international telephone traffic, and steady stream of international travelers' (*The Straits Times*, 10 January 2001).

The synergies that result from the interplay of Asian and western cultures are thus tremendous in Singapore. Singapore's society is often touted as 'multicultural'. The cultures and fortunes of the early migrants to Singapore – the ethnic Indian, Malay, Chinese and European – have been entwined for a few centuries, and the result is a harmonious array of uniquely Singaporean expressions in food and speech, for example. Singapore's efforts to modernize in step with the developed countries early in its independence have also produced a nation highly plugged into the international business, communications and technological communities. The constant engagement with global–local synergies makes for a dynamic environment in which to work and live.

The university system is a reflection of society's needs and goals. The Singapore government's commitment to raising the standards of education is evidenced by the allocation of 21 per cent of the national budget to education in the year 2000 (Contact Singapore 2000: 16). There are currently three universities in Singapore – the National University of Singapore, the Nanyang Technological University and the Singapore Management University – with plans afoot for starting a fourth university over the next few years. The humanities and social science disciplines are strongest at the National University of Singapore (NUS), which offers 'broad-based' curricula combining 'the rigour of the British education system and the flexibility of the American system' (Contact Singapore 2000: 9). The NUS Library is perhaps the best in the region, boasting over 943,351 books and 39,680 periodicals, as well as online access to over 500 overseas databases covering a wide range of disciplines (NUS Library 2001).

Furthering Research Horizons

There are two departments offering geography as a university subject in the country, both of which also offer postgraduate research opportunities in the discipline. The National Institute of Education (NIE), which is part of the Nanyang Technological University (NTU), offers MA and Ph.D. opportunities in geography within its Humanities and Social Studies Unit. Although it is essentially the national centre for teacher training in Singapore, postgraduate programmes in the arts and sciences are offered. Within geography, the main research foci are: geography and environment education, urban issues, natural resources and environment, economic development, and GIS and remote sensing applications. More information is available on the NIE postgraduate website (see Internet Resources below).

At NUS, the Department of Geography is situated within the Faculty of the Arts and Social Sciences. The department has more than 20 staff members specializing in various aspects of human and physical geography, comprising locals as well as researchers

hailing from Asia, Europe, North America and Australia. About 10–20 graduate students are enrolled in the research-based Master's or Ph.D. programmes at any one time. Research in the department focuses on the following four areas:

1   Cultural and heritage landscapes and tourism.
2   Globalization and related economic and socio-cultural implications.
3   Tropical environmental change.
4   Spatial data handling.

In addition to these core areas, staff members in the department also supervise research in a number of other fields, including population studies, political geography with reference to Asia, the geography of services, the geography of gender, and urban and regional planning. The department is equipped with GIS, air photo and earth sciences laboratories and a map resource centre. All graduate students also have access to the university's computer network and associated resources. The department is highly research-active and fast developing a strong reputation, not just in the region but internationally. Staff members are regularly invited as keynote speakers to international conferences, serve on the advisory boards of prominent journals, and act as consultants with international organizations and with the governments of Singapore and other ASEAN countries. Graduate students also have the opportunity to read a number of graduate modules on broad-based topics ranging from globalization and tourism to environmental management issues.

In addition to pursuing graduate studies in geography, there are several multi-disciplinary departments and programmes which may be of interest to geography students wanting to specialize in a certain region in Asia or in interregional comparative studies. Most of these offer MA and Ph.D. programmes by research. Geography students from the UK may be interested in the following programmes at NUS: Southeast

Asian Studies, South Asian Studies, Japanese Studies, Chinese Studies and Malay Studies. Up-to-date details on the aims and objectives of the programmes, research foci, courses and contact information may be accessed from their respective websites listed at the end of this chapter. The faculty is also host to research initiatives with particular cross-disciplinary themes, for example, Asia-Pacific tourism. Research scholarships (see below) are available to encourage research in these fields, while students from overseas universities working in these areas may also apply for short-term fieldwork support. Other research opportunities at NUS for those with a Ph.D. may be found in the form of postdoctoral fellowships offered by the various departments of the Faculty of Arts and Social Sciences.

Master's Coursework Opportunities

MA coursework opportunities are also available at NUS, although they are currently not available in geography. However, students trained in geography who are keen to pursue more specified paths may be interested in one of the following offered by the Faculty of the Arts and Social Sciences:

- Master of Arts (Southeast Asian Studies)
- Master of Social Sciences (International Studies)
- Master in Public Policy
- Master in Public Management (in collaboration with Harvard University)

Admission requirements to the above coursework-based degree programmes are generally a good degree (equivalent to an upper-second honours degree) and/or a minimum of two years of work experience in the relevant fields. For more information on these programmes, applicants may refer to the Graduate Studies website of the Faculty of the Arts and Social Sciences.

The School of Design and the Environment (SDE) offers two coursework programmes which aim for a broad-based, multi-disciplinary approach to environmental management and urban design. The Master of Science (Environmental Management) covers related topics in environmental law, planning, economics, technology, business, assessment and management. This programme likewise requires a good degree and relevant work experience. The Master of Arts (Urban Design) is a full-time intensive course aimed at studying the city in 'more complex and inclusive terms' (School of Design and the Environment 2001).

Admission and Funding Opportunities

Admission to research and coursework programmes is competitive and generally requires a good Bachelor's degree from a recognized university. Besides the basic degree, applicants for research programmes whose native tongue or medium of undergraduate instruction is not English must submit TOEFL/IELTS scores as evidence of their proficiency in the English language, the medium of instruction at NUS and NTU. Coursework programmes normally require a good degree, and are enhanced by relevant work experience. The international student fees for the above-mentioned courses are summarized in table 59.1.

Scholarships are available for research-based postgraduate programmes. At NUS they range from SG$1,200–1,400 a month, and at NTU they range from SG$1,300–1,500. At NUS, research scholars may supplement their stipend with occasional teaching and tutoring jobs within their department of study. The Research Scholarship Augmentation Scheme, available to students from the G8 nations (including the UK), provides a supplementary top-up of up to SG$500 a month for Master's students, and SG$900 for Ph.D. students, over and above the usual research scholarship

Table 59.1   International student fees for the different course types in NUS and NTU*

Programme type	Fees per year (in Singapore dollars)	Fees per year (in £UK)[a]
NUS Faculty of the Arts and Social Sciences[b]		
Coursework (full-time)	SG$4,350	£1,667
Coursework (part-time)	SG$1,980	£759
Research (full- and part-time)	SG$3,150	£1,207
NUS School of Design and the Environment – MA (Environmental Management)		
Full-time	SG$4,350	£1,667
Part-time	SG$2,180	£835
NUS School of Design and the Environment – MA (Urban Design)		
Research and tuition	SG$3,960	£1,517
NTU National Institute of Education – MA (Geography)		
Both full- and part-time	SG$3,100	£1,188

* Fees are subject to a 10% increase each year for the next few years as well as other possible changes.
[a] Based on an exchange rate of £1 = SGD2.61 as of 8 May 2001.
[b] *Source*: FASS brochure.

stipend. Scholarships tend to be more competitive than the normal admission to degree programmes. The Graduate Studies websites of the respective universities are useful sources of such information.

Living in Singapore

An excellent starting point for finding out more about living in Singapore, and for obtaining information about the relevant entry procedures, is through Contact Singapore, an organization set up by the government of Singapore to help both Singaporeans maintain contact with the country while living abroad, as well as for eigners coming to live and work in Singapore. In the case of the latter, Contact Singapore provides a range of services to facilitate a smooth transition to living in Singapore. This includes an information pack containing general and practical advice, a website with up-to-date information including estimated cost of living, an advisory service, online job-matching services, and work attachment programmes. Contact Singapore has an office in London (address below).

Other very useful sources of information relating to accommodation, immigration, health and financial issues and insights into living in Singapore may be found on the websites of the NUS International Student Services, the Office of Student Affairs (at NUS) and Singapore Immigration and Registration (SIR). For example, graduate students may apply for housing through the Office of Student Affairs, at a cost of approximately SG$260–300 per month on a twin-sharing basis or SG$400 for a single room. Accommodation on the open rental market provides another alternative but will cost more, in the range of SG$800–1,000 for a small apartment. For more general topics such as Singapore's resources, the economy, the government and the people, the Singapore Infomap website offers a useful introduction.

References

Contact Singapore 2000: *Discover a Future Beyond the Horizon.* Singapore: Contact Singapore.

Ministry of Information and the Arts 2000: *Country Profile Singapore 2000: Transport and Communications* [online], http://www.sg/flavour/profile/Transport/aviation.htm#Airport/

Ministry of Information and the Arts 2001: Singapore Infomap: The Official National Website and Internet Gateway to Singapore [online], http://www.sg

NUS Library 2001: About Us: http://www.lib.nus.edu.sg/about/about.html

School of Design and the Environment 2001: Master of Arts (Urban Design): http://www.sde.nus.edu.sg/DRGS/higherdegree/bycoursework/MA(UD)-appln%20info%202001-2002.htm

Internet Resources

Studying at NUS

- General information on postgraduate opportunities at NUS: http://www.nus.edu/nushome/faculties_index.html
- Postgraduate opportunities at the Faculty of Arts and Social Sciences (FASS), NUS: http://www.fas.nus.edu.sg/graduate.htm
- Postdoctoral fellowships at FASS: http://www.fas.nus.edu.sg/info.PDF
- Department of Geography: http://www.fas.nus.edu.sg/geog

- Southeast Asian Studies: http://www.fas.nus.edu.sg/sea/Frame(Postgrad).htm
- South Asian Studies: http://www.fas.nus.edu.sg/sas/program.htm#postgrad
- Japanese Studies: http://www.fas.nus.edu.sg/jap/MA_PhD_Programmes.htm
- Malay Studies: http://www.fas.nus.edu.sg/malay/wpostgrd.htm
- Master of Science (Environmental Management): http://www.sde.nus.edu.sg/MEM/
- Master of Arts (Urban Design): http://www.sde.nus.edu.sg/DRGS/

Studying at NIE

- Postgraduate opportunities at NIE: http://www.nie.edu.sg/nieweb/Main/Graduat99/default.html
- Geography at NIE: http://www.soa.ntu.edu.sg/geo/index.htm

Living in Singapore

- Contact Singapore: http://www.contactsingapore.org.sg
- Contact Singapore London Office, Charles House, Lower Ground Floor, 5 Regent Street, London SW1Y 4LR, UK; tel: + 44 (0)20 7321 5600; fax: + 44 (0)20 7321 5601; email: london@cs.org.sg
- NUS International Student Services: http://www.nus.edu.sg/NUSinfo/iguide/
- NUS Office of Student Affairs: http://www.nus.edu.sg/NUSinfo/osa
- Singapore Infomap: http://www.sg

60

Postgraduate Studies in the United States

Michael C. Slattery

Coming to America

In the lighthearted comedy *Coming to America*, comedian Eddie Murphy is cast as pampered African prince Akeem, who rebels against an arranged marriage and heads to America to find a new bride. Murphy's regal father (James Earl Jones) agrees to allow the prince 40 days to roam the United States, sending the prince's faithful retainer Semmi (Arsenio Hall) along to make sure nothing untoward happens. Well, you now find yourself in Akeem's shoes: you too are thinking of coming to America, albeit to pursue a graduate degree in geography rather than in search of a spouse. This may seem a rather silly analogy, but I use it here to make a point: the chances of landing in the United States and stumbling across the perfect partner are about as remote as stumbling into the perfect graduate programme (of course, in the film Akeem *does* find the perfect bride, but only in Hollywood, right?). The decision to attend graduate school in the United States and the choice of an institution and degree programme require serious consideration. The time, money and energy you will expend during graduate work are significant, and you will want to analyse your options carefully. Before you begin filing applications, you should evaluate your interests and goals and know precisely what programmes are available. This short chapter will, I hope, give you some usable information to get you started.

First off, you need to ask yourself why you want to pursue a graduate degree at all. My guess is that you are probably thinking along one of two lines: either you want to get a Master's degree on your way to the Ph.D. or a Master's in a specialized field that necessitates advanced education, such as spatial data management or GIS; or you simply want to expand your knowledge and develop your critical thinking skills. In either case, you are on the right track! But why specifically the United States for graduate work? The fact is, many people decide to earn their undergraduate degree at one institution and then select a different university or a somewhat different programme of study for graduate work. My experience in both applying for jobs and chairing search committees is that those candidates who have acquired a broad background, and who have been exposed to different educational systems, very often progress farther down the interview road. It may just be perception, but we do find candidates who have been to different institutions more attractive, especially if they have been to a first-rate, international university. This is particularly important if you intend to teach in the United States. The make-up of a university degree is very different in the United States to that found in many other countries, and you will certainly acclimatize more quickly if you have been through a Master's or Ph.D. programme yourself. But more importantly, the structure of these programmes invariably forces you to take courses outside of your own narrow interest or specialism. This helps build breadth of knowledge and, I believe, gives one a head start in the field of class preparation.

My first piece of advice is to research the programmes very carefully. There are many sources of information you can and should make use of in choosing a programme. The best way to begin is to consult the Association of American Geographers' *2001 Guide to Programs in Geography in the United States and Canada* (ISBN 0-89291-253-7, $25 for AAG members and $50 for nonmembers; order from the AAG website). This volume will tell you what programmes exist in the field or fields you are interested in and, for each one, will give information on the degree, research facilities, the faculty, financial aid resources, tuition and other costs, application requirements, and so on. Certainly, talk with your university adviser and professors about your areas of interest and ask for their advice about the best programme to research. Besides being well informed themselves, these faculty members may very well have colleagues at institutions you are investigating, and they can give you inside information about individual programmes and the kind of background they seek in candidates for admission. The bottom line is that being aware of who the top people are and where they are will pay off in a number of ways. A graduate department's reputation rests heavily on the reputation of its faculty, and in many ways it is more important to study under someone of note than it is to study at a college or university with a prestigious name! Moreover, graduate funds are often tied to a particular research project and, as a result, to the faculty member directing that project. As a graduate candidate, you will have to pick an adviser and one or more (normally three) other faculty members who form a committee that directs and examines your work. Many times this choice must be made during the first semester, so it really is important to learn as much as you can about the faculty before you begin your studies. Other important questions you need to ask that are of a more general nature are:

- What kinds of students enrol in the programme?
- What are their academic abilities, achievements and skills?
- How many complete the programme and what is the average completion time?
- What are the programme resources?
- What kind of financial support does it have? Dig a little deeper here – i.e., are there small pots of money available to support student research, conference travel, etc.?
- What are the library facilities like?
- What laboratory and computing facilities are available?
- What is the nature of the interaction between students and faculty – i.e., is it friendly or very formal; are there field trips; does the department have social get-togethers like an annual banquet; is there an active students' geography club?

This last point may seem inconsequential, but I have seen many students leave a programme because of the internal politics of a department or because the place is, quite simply, dull. Graduate school is more than coursework and a thesis. Student life, as we refer to it, really counts.

Now on to the application process. My advice here is start early, at least one year before the anticipated date of enrolment. Each programme sets its own deadline for applications, many using 31 December as the cut-off for the following autumn, but there are external constraints on deadlines, such as national scholarships or graduate admission tests. Remember you have to:

- Research the programme and the institution.
- Investigate funding.

- Obtain letters of reference.
- Take the required graduate admission tests.
- Write an application essay, if required, and so on.

It is always a good idea to make contact by writing to the department chair briefly describing your training, experience and research interests. He or she will most likely put you in direct contact with one or more of the faculty best suited to your particular area(s). And never, ever, send a preprinted letter with someone's name filled in by hand, or a letter addressed 'To whom it may concern'. It normally takes all of 30 seconds to file those!

Many programmes will require you to take the so-called GRE (Graduate Record of Examinations). You will most likely have to take the general test (and possibly the subject test, depending on the programme of study) consisting of three sections: verbal (V), quantitative (Q) and analytical (A). The good news is that all of this is available online; the bad news is that it costs $130 to take the test. Log on to the website and go to the section 'GRE At A Glance' for more information, including free test preparation software.

Admission committees also require official transcripts in order to evaluate your academic preparation for graduate study. The rigour of your course load, the courses taken and the reputation of your undergraduate institution are all scrutinized. However, the most important part of your application package will be your letters of recommendation. Most departments require three letters of reference and these are always given considerable weight in the evaluation process. My advice here is not to overlook non-academic references. It is much better to have a letter from someone who has known you for ten years rather than three months. But if you have taken several courses from a professor or have been advised in an undergraduate project or thesis, then by all means use that person. At the very least, you should

have someone who can judge you academically and who can rank you among your peers or other students that s/he has taught over the years. Ideally, the person writing the letter should hold you in high opinion and have seen you in more than one area of your life. It is always advisable to ask for a confidential letter of reference that will be mailed directly to the graduate admissions office of the institution. Most important of all, ask early and give faculty time to write a thoughtful recommendation. You know how things tend to get buried on our desks!

Many universities require an application essay or composition. Writing such a personal statement is often the most difficult part of the application process. While there is no set formula here, it should be a succinct statement showing you have a definite sense of what you want to do and your enthusiasm for the field of study you have chosen. It must reflect *your* ideas and goals. I always look for that so-called 'spark', a sense that the student is really turned on by geography. Ask yourself: what was the moment that really made me think about graduate school? A particular experience on a field trip? The personality of a faculty member? Be original, creative and personal. The essay should also reflect your writing abilities, clarity, focus and depth of your thinking.

As an international student, you follow the same application procedures as other graduate school applicants. However, you will have to meet additional requirements. First, if English is not your native language, you will be required to take the TOEFL (Test Of English as a Foreign Language) or similar test. In addition, your application must be accompanied by a certified English translation of your academic transcripts. You may also be required to submit records of insurance and certain health certificates. You should contact the university health centre to learn if there is a student health plan.

You will almost certainly have to provide documented evidence of financial support at the time of application. Universities are getting increasingly concerned that students are leaving programmes because they run out of funds during their course of study. For example, at Texas Christian University we require a document signed by the sponsor (e.g. parent, guardian, spouse) affirming that they will be responsible for all tuition, fees and living expenses during the academic year. In addition, we require, from their financial institution on bank letterhead, a statement to the effect that the sponsor has the financial resources to support the student, along with the current balance and mean monthly balance of the account for the past 12 months. We also require a letter from the sponsor's employer attesting to their current salary. It sounds insurmountable, but is well worth doing first to ensure a smooth application process. Since you will probably be applying for financial assistance from the graduate school as well as other sources, you may get conditional acceptance prior to submitting these financial statements, but the norm is to submit them with your application. In many schools, ours included, your formal application package also serves as your scholarship application, and you will be considered for financial aid and various grants and scholarships ranging from very little to full tuition. Make sure this procedure applies to your particular graduate school. In all honesty, you cannot over-research the financial aspect of coming to the United States for graduate work. As an international student, you will be paying out-of-state fees, which can be very high and, in many cases, can approach those of a private school. You may also be offered a TA (teaching assistantship) or RA (research assistantship, if you are working on a particular grant-funded project), and you can also work on a student visa, but these rules are campus-specific. Many universities will allow you to work no more than 20 hours per week on campus. After being here one full year, you can apply for off-campus work, which is usually granted. But again, make sure of the rules at each institution.

There is a variety of funding sources for foreign nationals studying in the United States beyond the particular institution. The problem is that international students and their US educational advisers often find it difficult to identify sources of financial aid. Most grants directories offer little or no information on the availability of awards to citizens of other nations. One exception is *Funding for United States Study: A Guide for Citizens of other Nations* (ISBN 087206-219-8, $39.95), published by the Institute of International Education in New York (see its website for more information on this excellent resource).

Once your application has been received, or you have been formally admitted, the school should send you INS Form I-20, which you must then present at your nearest US embassy or consulate, along with a valid passport, in order to obtain your student visa. Again, you will have to prove to the consulate that you have the financial resources required for your education and stay in the United States. It may well be worth checking the INS website for more information regarding the visa application process.

If you have no particular feel for where you want to study, but know that you want to end up at a nationally recognized school, then one resource at your disposal is 'the list' ranking graduate programmes in geography. Actually, there are a number of rankings, all controversial, but the one most widely referred to is the National Research Council's *Research-Doctorate Programs in the United States*. This is probably the most comprehensive, examining 3,634 programmes in 41 fields at 274 institutions. The details, including all the statistics, can be found at http://www.research.sunysub.edu/research/nrcdata.html or http://www.nap.edu/readingroom/books/researchdoc/.

I have included the rankings for the top 36 departments at the end of this chapter, along with the websites for the top five. It makes for interesting reading but, again, don't be blinded just by the numbers. There are many excellent programmes in smaller schools and, importantly, many top-notch geographers in other departments. It is worth remembering that many graduate programmes are interdependent and interdisciplinary in the United States. You will find geographers in geology, geosciences, earth sciences, environmental sciences, and so on, so it would be remiss to only look in geography programmes.

Finally, let me point you in the direction of the Princeton Review, an excellent website that has a wad of information on studying in the United States. Your first stop should be the link under 'Better Schools' that says *Grad School*. This covers everything from hunting for scholarships to ensuring great letters of recommendation.

Good luck. Happy hunting. I strongly encourage you to look into coming here. You never know – you may even pick up an accent!

Internet Resources

- Association of American Geographers' website: http://www.aag.org
- GRE (Graduate Record of Examinations): http://www.gre.org
- TOEFL (Test Of English as a Foreign Language): http://www.toefl.org
- Institute of International Education, New York: http://www.iie.org and http://www.iiebooks.org/granandfel.html
- INS website: http://www.ins.U.S.doj.gov
- Ranking of graduate programmes in geography: http://www.research.sunysub.edu/research/nrcdata.html http://www.nap.edu/readingroom/books/researchdoc/
- The Princeton Review: http://www.review.com

Quality Assessments in Geography

Pennsylvania State University (1) http://www.geog.psu.edu
University of Wisconsin Madison (2) http://www.geography.wisc.edu
University of Minnesota (3) http://www.geog.umn.edu
University of California Santa Barbara (4) http://www.geog.ucsb.edu
Ohio State University (5) http://thoth.sbs.ohio-state.edu
University of California Berkeley (6.5)
Syracuse University (6.5)
University of California Los Angeles (8)
Clark University (9)
University of Washington (10)
State University of New York Buffalo (11)
University of Colorado (12)
Rutgers State University New Brunswick (13)
University of Texas Austin (14)
Arizona State University (15)
University of Illinois Urbana-Champaign (16)
University of Iowa (17)
Louisiana State University (18)
University of Arizona (19)
University of Kentucky (20)
University of Georgia (21)
University of North Carolina Chapel Hill (22)
Johns Hopkins University (23)
University of Florida (24)
Indiana University (25)
University of Kansas (26)
Boston University (27)
University of Oregon (28)
University of Maryland College Park (29)
University of Hawaii Manoa (30)
University of Wisconsin Milwaukee (31)
University of Nebraska Lincoln (32)
Oregon State University (33)
University of Utah (34)
Kent State University (35)
University of Cincinnati (36)

61

Creating a Good CV

Pauline E. Kneale

All geography graduates have skills that employers want and are employed in a vast range of areas (see box 61.1). The real problem is that most employers do not know that geographers have bucketloads of skills, and quite possibly neither do you – yet. The trick with a CV (curriculum vitae) is to impress on both fronts in two smartly presented sides of A4. Here are some advice and suggestions, starting with: get a pen and paper and make notes as you go. Set aside 15 minutes a day to work on content ideas and research options, and then create or update your CV.

First, list your skills, interests and activi-

ties. Use this to create your CV and as an *aide mémoire* before interviews. Its big benefit is in raising your awareness and confidence in your personal skills. You *are* skilled in most of the following:

- From your geography degree: IT skills, organizational skills, presentation skills of all sorts and specialist subject knowledge. What else can you include: laboratory experience, fieldwork, expeditions…?
- From part-time and vacation employment most students acquire communi-

---

Box 61.1 Careers for geography graduates

Accountancy	Courier Services	Law	Retailing
Advertising	Diplomatic Service	Logistics	Sales
Air Traffic Control	Environmental	Market Research	Social Services
Aircrew	Consultancy	Marketing	Social Work
Archaeology	Environmental	Meteorology	Software Development
Architecture	Monitoring	Mining	Sports Management
Archivist	Farming	Museum Work	Stockbroking
Army, Navy, Air Force	Forestry	Ordnance Survey	Systems Analysis
Banking	Geology	Photography	Teaching
Cartography	GIS	Planning	TEFL
Catering	Hotel Management	Police, Fire Service	Theatre
Charity Management	Human Resources	Pollution Control	Tour Management
Civil Service	Hydrology	Post office	Transport
Complementary Medicine	Information Science	Public Relations	Travel Agent
Computing	Journalism	Publishing	TV, Films
Conservation Services	Laboratory Analysis	Recycling	Voluntary Work
Countryside Officer	Languages, Translating	Research	

cation, team-working, time management, self-management, motivation and people skills. People skills include working with people from diverse backgrounds. What else: driving, language, financial…?

- Just by attending university you acquire skills and competence in financial (debt) management, decision making, stress management, self-management and motivation skills. And…?

Make some notes about the circumstances where your experiences, holiday and work activities exemplify each skill on the list. These are data for the CV and when an interviewer asks what you learned at university, you can confidently reply: 'I learned to work in teams of five and above completing projects to tight deadlines'; 'I discovered I was really good at organizing when I took the Snowboarding Society to Switzerland for a week'; 'I found that I can persuade people to work with me on a project that involved…'. It is vital to have a real example for each activity.

Knowing what you want to do when you graduate is difficult and there is an enormous choice of careers. Your first advice point is your University Careers Centre. In the absence of a Careers Centre, check the AGCAS (2001) site. Ask yourself: 'How does what I know about myself, my personal skills and attitudes influence what I want to do after I graduate?' What sort of organization do you want to work with, and what are your aims and ambitions? Why wait until you graduate before thinking about careers? Get involved with your Careers Centre from year 1. They have information about summer placements and internships as well as career options. Be paid to work in environmental consultancy, accountancy, sales or banking over a vacation and find out if it is really for you.

Whether in your Careers Centre offices or surfing electronic career networks, use these points to research potential careers:

1. Pick a career/occupation you think you might like.
2. Make a quick list of the things you think this job will involve.
3. Find literature/websites for three potential companies – what skills do they say the job will need?
4. Make a quick list of the skills and the evidence you can quote to show you have these skills.
5. How did your original ideas about this occupation match your first impressions?
6. Is this the right type of company for you?
7. Now research another career. (Back to 1)

Making Applications

A CV is a persuasive marketing statement that presents the personal information that an employer needs to know. It should always be accompanied by a covering letter. A CV tells an employer about your past and current qualifications and experience. The covering letter links to your CV. You may use it to say more about the skills and experience that will make you the ideal person for a job. Use a CV and covering letter when applying in writing for a specific post or when approaching a company where you would like to work. Where an application uses a standard or employer application form, do not add your CV unless requested. The main sections of a CV are listed in box 61.2.

Always remember to tailor your CV to each job. For postgraduate and research posts focus on your geographical skills, but for generic jobs focus on your skills and the work and student experiences that are particularly relevant. Most importantly, make sure your CV is:

- Two sides of A4 only.
- Professionally presented, using your word-processing skills.
- Punchy and precise – no waffle allowed.
- Consistent in explaining what you have done – no time gaps.

Box 61.2 The main sections of a CV

Personal Details	Full name, address, telephone number. (Under Equal Opportunities, date of birth, nationality and marital status are optional.)
Education	Degree results (key details: relevant projects/modules undertaken). A-levels and grades. GCSEs in brief: 9 GCSEs, 4 As, 5 Bs, or 9 GCSEs, including English and Maths.
Work Experience	Include work placements, summer jobs, voluntary work and any permanent or part-time work. Place in reverse order, with the most recent first. For the most recent and relevant jobs, use two or three sentences to state the content or skills you used and developed. Be concise.
Computer Skills	Most geographers have enough experience to highlight this. Remember the packages used in practicals and projects, statistics, mapping, GIS, programming, Excel, ACCESS; and also word-processing, Powerpoint, Internet searching, Electronic Library skills etc. You may not feel very expert, but many applicants have fewer skills.
Other Skills	This can be helpful, especially if you feel you have limited work experience skills. Include driving licence, first aid and language skills. State your transferable skills. Ideally, use examples rather than lists: *'Working with children helped me to be organized and patient.' 'I realize that I need to be aware of safety issues and mindful of colleagues' activities.'* *'Negotiating for funds with the Student Union showed that I could be persuasive and present my case well.'*
Interests and Activities	No boring lists, please (and don't make them up). *'I played violin (Grade 8) with the university and county orchestra.'* *'I was a member of the ski club for 3 years, the secretary in year 2, and organized a trip to Switzerland for 35 people.'* *'I played football for my Hall team and helped organize the transport. At present I prefer to concentrate on aerobic training for skiing but intend to play more football in future.'* *'I raised £3,000 to fund my place on Operation Raleigh/Tall Ships race...'* *'I really enjoyed the team work involved in ... and in having to rely on one's own resources in awkward situations.'*
Referees	It is usual to cite two referees, one academic and one from your work experience, or two academic. Always ask first. Giving referees a copy of your application will help them write a more focused letter.

- Consistently ordered within sections. Put recent experiences first.

Your covering letter should also be short, positive and pointed. One side is ideal. Do not repeat information from your CV, although you may expand on really relevant matters: 'As you see from the CV, I have considerable experience of...'. Use the letter to state which job you are applying for, why you are interested, and the skills and experience that you can offer. Show that you are a career-minded person by stating how you see your career developing. Be positive about the times you are available for interview.

Make the most of your skills, personal, transferable and academic. Be upfront and fident about them. Tell employers that have given 20 presentations to groups of 5–50 in the skills section, or that you have created OHTs, slides and Powerpoint display materials. Practical skills are easy to list on a CV, but often seem so obvious that they are left out. Check you are presenting yourself fully. You are familiar with different word-processing packages, spreadsheet and statistical packages, and perhaps GIS and other software. The business may not use the packages you mention, but by listing your familiarity with software an employer becomes aware that you have more computer skills than many graduates.

Building and updating a CV throughout your degree course will save time in your last year. There are plenty of texts on CV design. Do a key word library search, or check out the Careers Service. There may be an online CV designer, while Vadas (2001) has links to examples of CVs.

Type letters and CVs. Always use formal English rather than a casual style. As more applications are written and sent electronically, it is easy to drop into a casual, email writing style which will not impress human resources managers. This is especially important when applying for positions online. Be very careful to treble-check every entry before hitting the 'Send' icon.

The Interview

There is no excuse for not researching a company before an interview. Check out their literature in your Careers Centre and on the Web. Read carefully anything they send you. Before an interview, think through (with a pen) some answers to the following questions, which are often asked at interview.

1 Why are you choosing to work as a sales executive/solicitor/accountant/personnel manager/marketing manager etc.?
2 What prompted you to apply to this company?
3 What do you know about the company?

4 What interests you most about the job you have applied for?
5 Why does working in this area/organization/country attract you?
6 Why do you want to work in a large/middling/small organization?
7 Why should we offer you a job with us?
8 What were the three most significant/tough/traumatic things that happened in your year abroad?
9 Variations on: What are your strengths and weaknesses? What can you offer the company?

Check out Careers Centre sites that can help with interviews. Topgrads (2001) has example answers to difficult questions asked at interview and other advice.

Exploring careers options and preparing a CV is time-consuming, but it will pay off later. Ignoring career research while at university may lead to your returning home to carry on with vacation work (on vacation pay scales) rather than kick-starting the rest of your life with new opportunities. Think of yourself as a very marketable product. Geography graduates possess many skills that employers seek, and it is a matter of articulating them clearly to maximize your assets. Get yourself and your CV well prepared for the job-seeking process.

References

AGCAS 2001: Careers Services: http://www.prospects.csu.man.ac.uk/student/cidd/carserv/index.htm
Topgrads 2001: Does your future start here? [online], http://www.topgrads.co.uk/
Vadas, A. 2001: Advice on crafting your CV [online], http://ukjobsearch.about.com/cs/curriculumvitae/

Internet Resources

Websites from around the globe with good advice for geographers on CVs and seeking jobs include the following.

- University of Washington Department of Geography, Career Resources: http://depts.washington.edu/geogjobs/
- Geography Career Prospects: http://www.geog.canterbury.ac.nz/geog/dept/careers.html
- GeoJobs: http://www.ssn.flinders.edu.au/geog/geojobs.htm
- Careers in GIS: http://www.gis.com/resources/careers/
- Developing career skills, an action plan for students: http://w2.vu.edu.au/careers/DevelopingCareerSkills.html

Index